THE GREAT MATH WAR

THE GREAT MATH WAR

How Three Brilliant Minds Fought for
the Foundations of Mathematics

JASON SOCRATES BARDI

BASIC BOOKS
New York

Copyright © 2025 by Jason Socrates Bardi

Cover design by Alex Camlin
Cover images: Photograph of L. E. J. Brouwer Courtesy of Beeldbank Blaricum; History and Art Collection / Alamy Stock Photo; © desertsands / Adobe Stock
Cover copyright © 2025 by Hachette Book Group, Inc.

Hachette Book Group supports the right to free expression and the value of copyright. The purpose of copyright is to encourage writers and artists to produce the creative works that enrich our culture.

The scanning, uploading, and distribution of this book without permission is a theft of the author's intellectual property. If you would like permission to use material from the book (other than for review purposes), please contact permissions@hbgusa.com. Thank you for your support of the author's rights.

Basic Books
Hachette Book Group
1290 Avenue of the Americas, New York, NY 10104
www.basicbooks.com

Printed in the United States of America

First Edition: November 2025

Published by Basic Books, an imprint of Hachette Book Group, Inc. The Basic Books name and logo is a trademark of the Hachette Book Group.

The Hachette Speakers Bureau provides a wide range of authors for speaking events. To find out more, go to www.hachettespeakersbureau.com or email HachetteSpeakers@hbgusa.com.

Basic books may be purchased in bulk for business, educational, or promotional use. For more information, please contact your local bookseller or the Hachette Book Group Special Markets Department at special.markets@hbgusa.com.

The publisher is not responsible for websites (or their content) that are not owned by the publisher.

Library of Congress Cataloging-in-Publication Data has been applied for.

ISBNs: 9781541605008 (hardcover), 9781541605015 (ebook)

LSC-C

Printing 1, 2025

Dedicated to
{JB}
∧
{Lucy Lee & Albert D}
∧
{Lenny}

CONTENTS

Dramatis Personae ix
Preface: Everything My Parents Taught Me About Logic xiii

CHAPTER ONE
1900: The Judgment of Paris 1

CHAPTER TWO
Adventures of a Wooden Puppet: 1883–1884 19

CHAPTER THREE
Infinity Loves Company: 1884–1899 45

CHAPTER FOUR
Guns, Gold, and the Gruesome Math of War: 1899–1902 63

CHAPTER FIVE
All Cretans Are Liars: 1903–1908 87

CHAPTER SIX
The Day of Judgment: 1909–1911 111

CHAPTER SEVEN
Airplanes and Sunshine: 1912–1913 129

CHAPTER EIGHT
Boston Loves Bertie: 1913–1914 147

CHAPTER NINE
1914: The Sins of August 159

CHAPTER TEN
Relativity + Turnips: 1915–1916 173

CHAPTER ELEVEN
Collective Influenza: 1917–1918 187

CHAPTER TWELVE
All Eyes on Albert Einstein: 1919–1920 203

CONTENTS

CHAPTER THIRTEEN
The Pied Piper of Paradise: 1921–1924 225

CHAPTER FOURTEEN
Nobody Loves a Revolution: 1924–1928 245

CHAPTER FIFTEEN
1928: The Battle of the Frogs and Mice 263

CHAPTER SIXTEEN
Goodbye to All Math: 1929–1932 281

CHAPTER SEVENTEEN
The Great Migration: 1933–1935 293

CHAPTER EIGHTEEN
$\forall_{\text{THINGS}} \exists_{\text{ND}}$: 1935–1938 307

Acknowledgments 319
Annotated Bibliography 323
Index 379

DRAMATIS PERSONAE

THE GAUNTLET THROWERS

L. E. J. Brouwer (1881–1966) — Dutch mathematician ("Intuitionism")
David Hilbert (1862–1943) — German mathematician ("Formalism")
Bertrand Russell (1872–1970) — English philosopher and writer ("Logicism")

THE MAJOR PLAYERS

Georg Cantor Jr. (1845–1918) — Invents set theory and redefines infinity
Richard Courant (1888–1972) — Directs Göttingen's math institute in the 1920s
Albert Einstein (1879–1955) — The face of physics and father of relativity
Gottlob Frege (1848–1925) — Analytical philosopher who reinvents logic
Kurt Gödel (1906–1978) — Viennese logician known for "incompleteness"
Sofya Kovalevskaya (1850–1891) — First woman in Europe to get a PhD in math
Ottoline Morrell (1873–1938) — London socialite and World War I peace activist
Emmy Noether (1882–1935) — A leading mathematician after World War I
Queen Math (b. ~20,000 BC) — A "cold and unresponsive love," says Russell
Hermann Weyl (1885–1955) — Coins the phrase "foundational crisis"
Alfred North Whitehead (1861–1947) — Bertrand Russell's principal collaborator
Ludwig Wittgenstein (1889–1951) — Philosopher and Russell's would-be protégé

THE MINOR PLAYERS

Ludwig Bieberbach (1886–1982) — German mathematician and Brouwer ally
Otto Blumenthal (1876–1944) — Math editor and chief champion of Hilbert
Emil du Bois-Reymond (1818–1896) — Famous for his "triumphant pessimism"
Constantin Carathéodory (1873–1950) — A famous modern Greek mathematician
Helen Dudley (1886–1932) — American poet and Russell's girlfriend
Epimenides of Crete (~seventh c. BC) — The most famous liar in human history
Euclid (~third c. BC) — Writes the most famous book on math ever
James Franck (1882–1964) — Nobel laureate who protests Nazi takeover
Paul Gordon (1837–1912) — Noether's mentor, the "king of invariants"

DRAMATIS PERSONAE

Lize de Holl (1870–1959)	Dutch pharmacist and Brouwer's wife
Felix Klein (1849–1925)	German mathematician and Hilbert colleague
Gyula "Julius" König (1849–1913)	Hungarian mathematician from Budapest
Diederik Korteweg (1848–1941)	Dutch mathematician and Brouwer's mentor
Leopold Kronecker (1823–1891)	German mathematician known as "the doubter"
Hermann Minkowski (1864–1909)	Friend of Hilbert and teacher of Einstein
Gösta Mittag-Leffler (1846–1927)	Swedish mathematician and journal editor
Giuseppe Peano (1858–1932)	Italian who is Russell's source of inspiration
Émile Picard (1856–1941)	Isolationist French mathematician
Henri Poincaré (1854–1912)	Most famous mathematician of his day
Alys Russell (1867–1951)	Feminist writer, activist, and Russell's wife
Karl Weierstrass (1815–1897)	Mathematician and Kovalevskaya's mentor
Evelyn Whitehead (1865–1961)	Good friend of the Russells in 1900
Zeno of Elea (~fifth c. BC)	Greek philosopher of turtle v. Achilles fame
Ernst Zermelo (1871–1953)	Mathematician who puts the "Z" in ZFC

POLITICIANS, ARTISTS, ANARCHISTS, SOLDIERS, AND LEADERS

H. H. Asquith (1852–1928)	British prime minister in 1914
Winston Churchill (1874–1965)	Famed British politician and PM
T. S. Eliot (1888–1965)	Famed poet and Bertrand Russell's student
Franz Ferdinand (1863–1914)	Archduke whose murder sparks a war
Robert Graves (1895–1985)	English writer who fights in the trenches
Sophie von Hohenberg (1868–1914)	Duchess whose murder starts World War I
D. H. Lawrence (1885–1930)	English author who falls out with Russell
Frieda Lawrence (1879–1956)	D. H.'s wife and Ottoline's nemesis
Horatio Herbert Kitchener (1850–1916)	British Boer War general and military chief
Paul Krüger (1825–1904)	Last president of the Transvaal Republic
David Lloyd-George (1863–1945)	English war secretary and prime minister
Philip Morrell (1870–1943)	Ottoline's husband and liberal politician
Logan Pearsall Smith (1865–1946)	Russell's brother-in-law and later enemy
Gavrilo Princip (1894–1920)	Serbian nationalist who starts a war
Lady Frances Russell (1815–1898)	Bertrand Russell's grandmother
Ferdinand Springer (1879–1965)	Famed German publisher of mathematics
Alexandrina Victoria (1819–1901)	England's queen during the nineteenth century

DRAMATIS PERSONAE

AXIOMS, FALLACIES, HYPOTHESES, PARADOXES & MYTHS

Axiom of Choice (Chapter 5) — "For any family of non-empty sets there exists a correspondence that associates to each of these sets one of its elements," according to Zermelo.

Burali-Forti Paradox (Chapters 3, 5) — If you create a set of all the possible "ordinal numbers" reflecting the sizes of different sets, then it will contain an ordinal larger than itself. But that's impossible.

Cantor's Paradox (Chapters 3, 5) — Says that if the set of all sets includes all sets, it must also include its own superset. This is a paradox because it means one subset of the set of all sets would be larger than the set of all sets itself.

Continuum hypothesis (Chapters 3, 5) — Cantor's statement that no infinite set has a cardinality or size in between that of the infinite wholes and the infinite reals, written in nongeneralized form as $2^{\aleph_0} = \aleph_1$.

Desperate deliberateness (Chapter 7) — A narrow-minded illusion where a person thinks their actions and those actions alone account for their successes.

Exuberant solutionism (Chapters 1, 12, 13, 16) — David Hilbert's gospel that any problem can be solved given enough time and energy.

Fallacy of pure process data (Chapters 4, 5) — A psychological distortion of reality that informs a pathological approach to strategic planning where decisions are foolishly based purely on process data at the expense of outcome or cost considerations.

Fallacy of rich data (Chapter 12) — The illusion that the more data we happen to use in an analysis, the more valid our results appear to be.

Fallacy of seeming (Chapter 18) — Says that the more you invest in an idea, the more true it seems. And the more true an idea seems, the more meaningful it becomes, the more you cling to it, and the harder it is to let go once that idea turns out to be false.

Fallacy of unimportant results (Chapter 4) — Holds that the more trivial a problem seems, the less urgency there will be to find an answer or solution to it.

DRAMATIS PERSONAE

Isaac Newton complex (Chapter 3)	A loathsome and pathological hatred for publishing informed by the paranoid certainty that your contemporaries are out to backstab you.
Law of Noncontradiction (Chapter 5)	Holds that A & ¬A (not A) are never both true.
Liar's Paradox (Chapter 5)	Says, essentially, *This statement is false.*
Russell's Paradox (Chapters 5, 14)	Consider the set of all sets that don't belong to themselves—does that master set belong to itself? *(Answer: It does if it doesn't & it doesn't if it does.)*
Triumphant pessimism (Chapters 1, 13)	The philosophy that embraces "unknowable unknowns"—questions we can't ask, problems we won't solve, and answers we'll never know. This is captured in the statement *Ignoramus et ignorabimus* (We don't know, and we shall never know). David Hilbert hates it.
Well-ordering theorem (Chapter 5)	Ernst Zermelo's so-called *Wohlordnungssatz*, the underlying basis of his Axiom of Choice. Well-ordering says that sets, even infinite sets, can be ordered based on the smallest elements of their subsets.
Zeno's Paradox (Chapter 2)	How is it that Achilles fails to catch the turtle in a footrace? Why does the arrow never reach its target?
Ozymandias (Chapters 1, 18)	The king of kings, whose crumbling heel stands as testimony to the faded glory of *once-upon-a-time* greatness.
Icarus (Chapters 3, 5, 6, 15, 16)	The waxy-winged Cretan prison escapee from mythology who flew too meltingly close to the sun and tumbled, humbled, back to Earth.

PREFACE

EVERYTHING MY PARENTS TAUGHT ME ABOUT LOGIC

> *I want to go on.*
>   *The hour is late.*
> *I want to go on.*
>   *The night is dark.*
> *I want to go on.*
>   *The road is dangerous.*
> *I want to go on...*
>
> —Carlo Collodi,
> The Adventures of Pinocchio

My father passed away while I was working on this book, and while clearing out the attic of his house in Gettysburg, Pennsylvania, I discovered a thick paperback titled *The Basic Writings of Bertrand Russell*. I was already deep into research on Russell, one of the main subjects of *The Great Math War*, so I kept the book, set it aside, and promptly forgot it. I did not find it again until more than a year later, when it resurfaced by chance while I was writing about Russell's theory of types (described in Chapter 5, "All Cretans Are Liars").

Finally flipping the book open, I was floored to find a personal note from my mom addressed to my dad, "JB," on the inside cover. She had apparently given him the book when they were both in college, two years before I was born. He was a musician, a math major, an aspiring philosopher, and a college football star. She was a journalism major and student activist. Her inscription inside the book was curious: *To my intellectual better! / Love, Lucy / 1968.*

What does that mean, and what did she mean by it? There were lots of ways to read her words, none of them completely satisfying. Was it a straightforward compliment? An inside joke between young lovers? Was she being ironic? Sarcastic? Were there missing air quotes? My mother was, and is, whip-smart and fiercely proud. She is a strong feminist. She's not given to casual bow-to-the-male-ego prostration. Of course I asked her about the book, some fifty-five years after she had purchased it as a gift, but she couldn't recall ever buying it or giving it to him—much less what she meant by the inscription.

I was born a little more than a year after my mom wrote the inscription, in early 1970, and 125 hours later, Russell died. He was ninety-eight years old when he passed, and I was five days. My neonatal condition was strangely parallel to Russell's own physical circumstances at birth a century before. He was 8 lb., 11 oz., and 21 inches long. "Not 1 child in 30 is as big & fat," his doctor declared in 1872 upon seeing the newborn. I was cut to almost exactly the same proportion: 8 lb., 5 oz., and 20.75 inches long. So fat, my mom says, that my facial features were scrunched and my toes looked like "little bunches of carrots."

When I started this project some fifty years later, I discussed Russell at length with my dad a few months before he died. He was shocked by what I told him. He knew the Bertrand Russell of the 1960s. Russell was an old man then, a living legend, an old-school philosopher who made headlines protesting nuclear arms and the Vietnam War. He accused American leaders of war crimes (once claiming, in fact, that President John F. Kennedy was worse than Hitler). He advocated for free love long before it became a popular trend in the Age of Aquarius of the 1960s. Russell had called for temporary, childless, "trial" marriages since the 1920s. Premarital sex would improve university life both intellectually and morally. The "musty Moloch" of traditions, laws, and moral beliefs that stood in the way should be dismissed. "Nothing is gained by continuing to pay lip service" to such beliefs, he said.

PREFACE

Russell faced a lot of blowback for his cultural views. In 1940, after he became a professor at City College in New York, an Episcopalian bishop in the city accused him of conducting a propaganda war against religion. One New York State Supreme Court justice condemned Russell, ruling him unfit to teach. The New York County registrar called for him to be tarred and feathered during World War II. But he was undeterred and would soon win the Nobel Prize despite his controversial, unconservative views. My father, always appreciative of the brash, radical, and outspoken, loved Russell for his outlandishness.

"I wanted to name you Bertrand Russell," my father told me on Father's Day, a few months into the COVID-19 pandemic in 2020. I was working on my book proposal at the time, and this was one of the last conversations we ever had. He had been serious, he said, discussing this possibility with my mom several times while she was pregnant with me in 1969.

My father had just started graduate school in Ohio then. Vietnam protests were in full swing, and like so many in their generation, my parents were a part of the antiwar movement. My father had stirred controversy on his undergrad campus some months before by writing an antiwar editorial as editor-in-chief of his college paper. A number of students on campus objected to his essay and organized a protest calling for him to step down. My dad defied the call and attended his own protest. He stood in the back of the room. People noticed him. Whispering. Buzzing. Pointing. Jeering. One speaker at the podium addressed him directly. "Will you step down?" he asked. "No," my father shouted back across the cafeteria, "I won't."

Russell was a rock star in my dad's eyes for his own trailblazing righteous obstinacy. He made antiwar crusading cool during World War I, long before war protests were popular—in fact, when they were grossly unpopular. Plus, he was a mathematical philosopher given to espousing free love and ethics, both squarely in my dad's corner. But my father's love of Russell went deeper than admiration from afar. In Bertrand, he saw JB—himself. His adult life was, like Russell's, governed by an endless search for knowledge, a love of all wisdom, and a long, endless search for human connection.

PREFACE

My father, like Russell, raged with "unbearable pity for the suffering of mankind," to borrow Russell's words. It's no wonder my dad wanted to make me his hero's namesake.

But my father was floored after I told him what Russell had been like more than a century earlier. My book was covering the young Russell of 1900, not the old-man activist Russell my dad loved in the 1960s. The Russell he knew was pro–free love and antiwar—the so-called Pablo Picasso of modern philosophy. The twenty-something Russell I knew was a different breed, groomed in aristocracy, weaned on a diet of rah-rah imperialism, just starting out as a green, unknown junior professor, and subscribing to an outlook of *every-part-of-the-world-should-be-governed-by-some-European-power* and *great-empires-do-more-good-than-small-ones*. People change, and nothing reveals that more than politics.

Not that it would have mattered, since my mom put her foot down. Bertrand Russell Bardi. *Bertie* Bardi? "No, no, no!" she said in 1970 after giving birth to me. That wouldn't do. "We'll give him a regular name," she insisted, "a *normal* name."

My father tried several times to push the Bertie brand, but he finally gave it up. He hatched a new plan instead, pushing to name me Plato Aristotle Bardi. "No, no, no," my mom objected again. Finally, they struck a compromise. I would get the middle name Socrates upon my birth—given the name of the father of philosophy to appease a father who was a philosopher. (*Oh, the Places You'll Know!*)

Someone asked me once what this book was about, and I said it was about history, math, love, warfare, and infinity. *Wow*, they said. *That's quite a mix. What ties all those ideas together?* And my answer was, simply: *Lies*. This is a book about lies—not really the lies we tell other people so much as the ones we tell ourselves. My subject is the foundations of things. Buildings and houses have foundations. By sheer coincidence, as I was drafting this book, the house next door was sold and demolished. I watched as they put planks in place, laid rebar inside, and poured concrete over it to lay new foundations.

PREFACE

This was the right way to lay a foundation, presumably. There are lots of wrong ways as well—though maybe not so much for modern buildings, which must adhere to strict codes. Many other things have foundations, however—love, war, peace, friendships, the future of humanity, the security of democracy, and innovative intellectual ideas. All complex relationships and conceptual constructs have their foundations, and it's a fallacy to think those foundations are always solid. Some relationships are built atop reinforced concrete, solid and truthful. Others stand on loose sand. Some intellectual ideas rest on bedrock. Others on dirt.

What we talk about when we talk about the foundations of anything is danger. Loose sand. Lies. Faultiness. You never think about the foundations of anything when they seem rock solid. You don't concern yourself at all with a building's foundation unless you see cracks. Hear creaking. Feel shakes or vibrations. Fear the ground moving beneath you. And then, when you sense imminent collapse, that's when you think hard about how to fix the problem. That's basically what happened in mathematics at the dawn of the twentieth century. Cracks appeared in the foundations of math, the whole subject shook, and everyone grew worried.

Mathematics is special among human endeavors in that it gives the ability to transform shifting-sand supposition into rock-solid proof. Thus, when those cracks in the foundations of math appeared, they were seen as fundamentally threatening to mathematics and its position as the Queen of the Sciences. So at the dawn of the twentieth century, when several mathematicians set out to fix math's foundational flaws, there was a great deal of hope initially. These efforts to solve the foundational flaws evolved into three camps: logicism, formalism, and intuitionism. Predictably, the coexistence of those competing worldviews led to a nasty fight, which I call the Great Math War.

How did these sides differ?

In the years leading up to World War I, Bertrand Russell and his collaborator Alfred North Whitehead set to work trying to fill the yawning hole in the foundations of math with logic—advancing the approach known as logicism. They succeeded somewhat and won renown for their incredibly rich and complex three-volume book *Principia Mathematica*, which is considered a triumph in mathematical philosophy to this day. But it failed to fix the flaws in the foundations of math, and everybody knew it. Their work ended prior to World War I, and during the war years of 1914–1918, much of the research in the European mathematical community ground to a halt. Then, in the years after the war, two other competing worldviews came into view.

Formalism was pioneered by the German mathematician David Hilbert, who developed a grand scheme to turn math into a sort of game where mathematical objects are the playing pieces and axioms are the formal rules of play that determine outcomes. His vision was large and sweeping, basically a plan to fix the foundations of math by gamifying it.

Prior to Hilbert, when people thought about math, they thought about the so-called privileged objects of math—things like numbers, lines, and geometric shapes—the stuff of external reality. They had obvious, unambiguous, and objective meaning—not just as part of math but also synonymous with our understanding of the subject. Hilbert was not trying to remove that privilege. Instead, he wished to ignore it. His vision was to divorce mathematics from reality and treat mathematical objects as "meaningless marks on the page." That would allow mathematicians to focus on the rules for deducing mathematical truth and solve problems abstractly. It was powerful. Penetrating. And with it there was nothing that could not be solved, he thought. That was formalism.

Intuitionism was developed as an alternative worldview by the Dutch mathematician L. E. J. Brouwer. Forget meaningless marks on the page, Brouwer said. His idea was to develop a "constructivist" approach to math in which a mathematical object is like a flowchart construction of the human mind rather than an abstraction penned on paper. He saw a mathematical object more as a becoming than as a being, to put it in purely mystical terms. It was a new and original way of approaching math, one that seemed to Brouwer and a few others to be rich and righteous. He derided Hilbert's work as "empty" formalism, and he absolutely refused to admit that anything in mathematics was true or even existed unless it could be explicitly constructed mentally. Those differences put Brouwer in conflict with Hilbert. Once friends, they grew to bitterly detest the other's work—and eventually each other.

The Great Math War unfolds before, during, and after World War I, covering a fifty-five-year span from 1883 to 1938. There's one striking parallel between the fight over the foundations of math and the orders-of-magnitude-more-awful horrors of World War I: Both were motivated by perceived fundamental flaws, believed foundational cracks, supposed existential threats, competing worldviews, and all the lies we tell ourselves. The fight for supremacy in World War I was thought to be a necessary evil to settle things geopolitically, and the same sorts of motives are behind the Great Math War and its attempt to resolve foundational flaws mathematically.

This book is also about faultiness in general: all the imperfection, incapacity, inefficiency, ineptitude, insufficiency, paucity, shortcoming, weakness, and ineffectiveness we perceive in the world—and all the kind and destructive things we try to do to fix those faults. This is a story of cruel war, bright hope, forgotten pasts, and exaggerated futures. Of success and failure. Of trial by fire. Of bitter-cold comforts. Of ambition. Bravery. True love. Endless heartbreak. Loathsome gossip. Enduring friendship. Peace. Pettiness. Horror. And human justice.

How often in the last few months, while finishing this book, did I wish I could have shared it with my father. That hippie. Philosopher. Musician. Radical. Lover of wisdom. He would have loved it. I was happy to share a draft with my mom. And just before I did, I discovered the source of her inscription in that paperback copy of *The Basic Writings of Bertrand Russell*. It's an apparent reference to something Russell said to his most famous student, Ludwig Wittgenstein, in 1912. He was scolding Wittgenstein for his presumptuousness—though exactly what my mother meant in repeating the line to my father, I will never know.

CHAPTER ONE

1900: THE JUDGMENT OF PARIS

> *My name is Ozymandias, King of Kings;*
> *Look on my Works, ye Mighty, and despair!*
> *Nothing beside remains. Round the decay*
> *Of that colossal Wreck, boundless and bare*
> *The lone and level sands stretch far away.*
>
> —Percy Bysshe Shelley

The Judgment of Paris, an engraving by M. A. Raimondi after a sixteenth-century Raphael. Cropped from a public domain image via Wellcome Collection.

It's a hot day in Paris that late-summer morning in 1900 when they arrive. The city is already full. Restaurants have lines out the door. Hotels are booked. The streets are packed. Something's happening, and the crowds know it. They feel it. They yearn to gnaw and taste it.

The air is humid, filled with awful scents. It's bad air. Hot, banging air. Those malarial humors of superstition and legend. A forgotten stink. Some oppressive stank. Sweaty crowds in August. That off-putting old-man smell of Paris before the war: baked baguette and BO. It burns the nose, swells the lungs, phlegms its way back up, and hacks its way out.

There's a sweet perfume to the city that summer as well—an imagined scent, perhaps, but unmistakable—a thin patchouli whiff of hope. People know this smell. They've been breathing it in a lot. Some aromas are hard to recall but impossible to mistake, like fresh-cut grass. That, my friend, is what the future smells like: the fresh-cut grass of perfumed hope! That's why most of the people in the city are here. They've come from far and wide to breathe it in. They've given up country cottages, emptied outlying villages, and abandoned their summer gardens to the weeds. Everybody's in Paris on this hot, glorious, crowded, stink-of-a-city day. They love it. And why not? What a day it is.

Here is a breezy hotel courtyard packed with small tables and tall waiters. Fancy Paris—the Ritz. Cut fruit. Jellied birds. Shucked oysters. Auguste Escoffier himself, the so-called king of modern French chefs, is newly installed as chef de cuisine. He fled London's Savoy under a cloud of scandal—but no need to mention that. Nearby is another café, beloved but harder to find. Deep in the seedy underbelly of old Paris. Dark and lovely, that place. Near the dance halls and hookers, it is a dim spot where night turns to day turns to night and where Henri de Toulouse-Lautrec sits drawing furiously. The present is the past.

Also nearby is the half-finished Byzantine basilica—Sacré Coeur. It is haunted in 1900 by a teenage Pablo Picasso. He is about to hit his blue phase. What a time it is for art! Back in Picasso's Barcelona, the visionary architect Antoni Gaudí is throwing up curved masterpieces. In the American Midwest, Frank Lloyd Wright is building bioinspired

1900: THE JUDGMENT OF PARIS

architecture—beacons of light on the dark prairie. Ethel Barrymore stars on the New York stage. Sarah Bernhardt is in London. A teenage James Joyce is getting ready to leave Dublin, and Jack London is settling back in California after the Yukon gold rush. Anton Chekhov, Upton Sinclair, Margaret Sanger, Mark Twain, Leo Tolstoy, Henrik Ibsen, Joseph Conrad, Ida Tarbell, H. G. Wells, W. E. B. Du Bois, Rudyard Kipling, and Robert Frost are all busy writing and publishing.

But people are here in Paris on this hot August day in 1900 to catch a glimpse of the future, not the past.

THE 1900 PARIS EXPO

More gorgeousness than the eyes can bear—that's what the crowds have been sold. That's what everyone has waited eight years to see. They've come for the promised glory of the Paris Expo—a once-in-a-lifetime electromechanical, technological, gee-whiz wonder-of-science event. And now it's spread before them: the future, like a warm blanket. It's international. It's universal. And it's here. The moment has arrived. L'Exposition universelle internationale de 1900. And it's everything they said it would be. Delivery is the fondest form of deliverance.

Nearly forty million people trudge through the streets of Paris this summer. They flock, pack, cram, push, nudge, queue, jam, sandwich, crush in, squeeze by, pull through, and shove on. They come to see it all. The monster machines. The mechanical marvels. The electric dreams. Every which way is lit by brand-new buzzy incandescent lights. One palace is aglow with five thousand bulbs, a new record, they say. Marvelous! Fabulous! So much to look at. So impressive and blinding. The future in 1900 is like some modern steampunk vision of the past—horrific yet alluring, like an outlaw-biker-gang patch on the back of a leather vest. You can't not stare.

Whole new buildings welcome the masses. In tech pavilions, consumer toys are on display. Eastman Kodak's first Brownie box camera. The first plastic wrap, that tiny kitchen miracle. Over here are the world's first wireless radios. Over there are some of the first X-ray films. People can't wait to stand for a scan. Zap! Moving walkways carry them about. Zap! Tractors, cars, taxis, trucks. Zap! Gas, electric, leather, and crank. Zap! The world's first escalators. Zap! Rolling and rolling. Up, down, in, out. Zap!

And the artwork—oh, the artwork!

Galleries. Gardens. Paintings. Sculptures. This is *the* event of the year, one art magazine covering the expo claims. The importance of L'Exposition universelle internationale de 1900 cannot be overstated. Nothing like this has ever been seen, and no installation this grand will ever be unveiled again. Dozens of artists. Hundreds of galleries. Thousands of paintings. A smorgasbord of marble. The world's most celebrated painters and sculptors. Édouard Manet. Claude Monet. Henri Matisse. Paul Cézanne. Edvard Munch. Auguste Rodin.

That same art magazine in 1900 asks, was the exhibition a success? And then it answers its own question gleefully: "We venture to say that no intelligent person who studies these pages can answer the question otherwise than by an emphatic, and even an enthusiastic, affirmative."

"Art-loving people throughout the world found in the Exhibition a collection of works such as has rarely, if ever, been brought together in the history of the human race," the magazine reports.

All summer long, Paris is the place to be. *Zap!*

THE BIRTH OF FUTURISM

Inside a stuffy university lecture hall nearby, the future is also on display this hot August morning. The heat is oppressive, but there is buzzing in the room. He's here! The speaker is here! A small crowd settles, excited. The speaker makes his way to the podium. He is one of *their* Matisses, *their* Manets, *their* Monets, *their* Cézannes, *their* Escoffiers, *their* Bernhardts, and *their* Barrymores. All that greatness rolled into one.

Another event is taking place in Paris this summer besides the Expo, this one in a stuffy room: the Second International Congress of Mathematicians. It's a splendid smorgasbord of scrumptious scholarship—"all to the mustard," as they loved to say in 1900.

Conference organizers think, at least initially, that the Paris Expo will be a draw. They sing its praises. A delightful diversion. The future of tech. Fabulous art. The organizers initially expected some 1,000 mathematicians and an additional 680 family members to show. But far fewer came. So now, once the meeting is underway, they take turns slamming L'Exposition universelle internationale de 1900. It's too loud. The city is distasteful when crowded and packed. Too many tourists. Too few hotels. The "great heat" of summer. Distractions, distractions. One magazine covering the math congress, the *Bulletin of the American Mathematical Society*, will explicitly blame the Paris Expo for the math meeting's low attendance. Many have stayed away, their reporter says, "who would otherwise have been present." *A shame!*

Still—it's their loss for not attending. The hundreds who have come to the meeting are treated to something special, something unique: a vision of the future of mathematics that mirrors the steampunk futurism of the Paris Expo tech and electric light displays.

Futurism—can you even call it that yet? It's taking its first baby breath in 1900. The word "futurism" won't even be coined for another nine years, toward the end of the decade—just a few years before the horrors of World War I begin—when Italian poet Filippo Tommaso Marinetti will write a manifesto to all things fast, tech, and violent.

But even if people don't know what to call their light bulb dreams of the future in 1900, they nevertheless cheer its bright-burning glow. The future is electric. It's exciting. It hangs in the city air. People consider it. They love it. It's glory. It's genius, that future. *Zap! Zap! Zap!*

———◆———

Futurism in 1900, as today, is dominated by vagueness: that faint perfume of fresh-cut grass and hope. Then, as now, the future we are sold is a pulled-wool vision of technology. It's monstrous. It's shiny. It's metallic and full of promise—or is it fantasy? The future we imagine and the future we are sold is one where the inexplicable transforms the impossible. One that promises to make our lives unrecognizably better and infinitely more convenient. And why not? The future is good. The future is bright. The future is healthy. It's just. It's tech. It's speed. Better medicines. Cool machines. Lights to guide our way. And machines to move us.

Planes are coming in 1900. Cars are coming too. The first automobile race just took place in 1900. The very first auto show is soon to open in

New York City. Paris this summer is full of hand-built, cloud-belching, choke-belly cars and trucks. They fill the streets. Their noise fills the air. Everybody marvels. People hear the engines roar. They feel them kick. They smell the fumes. Ignition. Pistons. Smog. Exhaust. The crowds go crazy. Roar and roar and roar go the machines. Onlookers ooh and aah. They love it. Most of them have never even seen a car before. *Only monsters make sounds like that!*

THE BEAUTY OF MATH

The mathematicians meet for a formal reception at Café Voltaire the first day of the congress. The second day, they convene at the Palais des Congrès on the Expo grounds for the opening ceremonies. Famed French mathematician Jules Henri Poincaré is there. He carries wisdom the way some rappers wear bling. It's dripping from him. Mathematics, he says, should be cultivated for its own sake. Math is not just some dull tool for digging but a precision instrument—a power tool, we would say today. Mathematics is no rusty shovel. Poincaré feels that it's an intellectual technology of incomparable importance. A thing of beauty. *An art!*

On the third day of the meeting, the attendees settle into their seats for a plenary lecture in a stuffy Sorbonne University lecture hall. Here they are, and here he is. They have come to hear this thirty-eight-year-old German mathematician talk about something—geometry, maybe?

The room groans with dull relief as he takes the stage. They can't wait to get started—the sooner the better. And why not? It's hot. And in these dawning days of scientific conferences in 1900, talks are typically bad and worse. Difficult topics. Monotone voices. Opaque ideas. Experts, experts. Awkward deliveries. Hesitant. Exhausting. And talks are almost always, according to British mathematician Charlotte Angas Scott, who is there covering the meeting for the American Mathematical Society, "tedious and incomprehensible." Brilliance is always harried, never hurried.

But not today.

No one in the audience has any idea what's about to happen. How can they? As German mathematician David Hilbert stands at the podium, he is about to reinvent the professional lecture. This will be *the* lecture that will

1900: THE JUDGMENT OF PARIS

redefine what we talk about when we talk about science. This lecture marks the moment when we move beyond what *has been* done in science and focus more on what *will be* done or what it should *become*. It's the talk that will make scientific lectures famous—the original TED Talk taking place nearly a century before such things even existed.

But how can the audience know that? Oh, sure—Hilbert's a rising star. Everybody sees that. Some claim he's the greatest mathematician alive. He's just published a famous book, the *Grundlagen der Geometrie* (*The Foundations of Geometry*), based on a few bawdy lectures he had given to shell-shocked grad students at the University of Göttingen, where he is a professor. It's a dreamboat of a success as math books go. He gives some lectures, hastily transcribed. His talks are formatted into a book. It's rushed into print. And what appears is odd and unexpected—a little book, but huge. And successful—*very* successful. The first copies are handed out at a public event at his university in Göttingen, Germany. Within months, a French edition appears. And everyone wants a copy. By the time of the Paris Expo, the book is all the buzz in math circles. New and interesting, they say. *A revolution!*

What has he done that is so impressive? In his book, Hilbert revisits the simple secondary school subject of geometry and rebuilds it fresh, from the ground up, with new foundations. He does not content himself with treating the subject the same way all the schoolmasters, monks, Islamic scholars, ancient Egyptians, and every forgotten Greek mathematician since the hoary old days of Euclid have always done. For thousands of years, geometry was tamed. Unsullied. Unblemished. Unaltered. But it is not Hilbert's intention to leave it untouched.

---◆---

The ancient Alexandrian mathematician Euclid pioneered the axiomatic approach to geometry, originally defining his five famous "postulates" or axioms at the beginning of his book *Elements*—perhaps one of the five most famous books of all time. In 1900, Euclid is still the reigning heavyweight champion of math authors. But in 1899, Hilbert imposed a newly expanded set of postulates on the subject. He groups them according to how they pertain to things like geometric continuity, connectivity, congruence, and order. So as not to get bogged down in the detail, suffice it to say that with outstanding creativity and impressive rigor, he impressively reworks the more than two-thousand-year-old subject of geometry.

People *are* impressed. According to Hilbert's biographer Constance Reid, one American review of his 1899 book says it will "do much for the logical treatment of all science and for clear thinking and writing in general." Overnight the book becomes a mathematical bestseller. People heap praise on *Grundlagen*. They call Hilbert's book a jewel. A gem. An intellectual masterpiece. It's "radically innovative," reflects modern British philosopher Peter Simons. *Pure art!*

And now, on this hot Parisian summer day in 1900, Hilbert is here. At long last. The air is hot. The attendees drift in. They find their seats. They settle. Some dab their foreheads with kerchiefs. Others mill about the edges of the room. Everyone is expecting a deep dive down that same geometric hole. Blah-blah-blah axioms and blah-blah-blah Euclid. Maybe he will trace the genesis of *Grundlagen*. Maybe he will get into geometric shapes. Who knows? The title of his talk is simply "Mathematische problemes" (Mathematical problems).

Only one person in the world actually knows what Hilbert's going to say—his good friend Hermann Minkowski, another mathematical legend.

EINSTEIN'S PHYSICS TEACHER

As a teenager, Minkowski had won a highly prestigious mathematics prize from the French Academy of Sciences for work on complicated functions known as quadratic polynomials. The prize was controversial at the time because the French academicians awarded it jointly to him and an older and much more established Irish mathematician named Henry John Stephen Smith, who had been working on these quadratic polynomials for years, even before Minkowski was born. So sharing the prize seemed a shame, if not a slight. British-French relations being what they were in the 1800s, a formal complaint was filed with the French academy from London and afterward duly ignored.

1900: THE JUDGMENT OF PARIS

Minkowski will become more famous in a few years for something even more random: having the singular distinction of being Albert Einstein's physics teacher in college. But Minkowski is a genius in his own right. Anyone who has ever studied Einstein's special relativity theory will know his name, for one thing. The mathematics involved with describing four-dimensional spacetime is sometimes called "Minkowski spacetime" because of his contributions. His partial differential treatment of space as a 4D "manifold" that includes time as well as three-dimensional space coordinates is an achievement that will forever associate Minkowski with all the strange and wonderful results of his student's special relativity theory. Length contraction and time dilation, where an imaginary space traveler could fly at 99.9 percent of the speed of light to a distant star system and return a few years later to find their twin decades older than when they left, are beautiful outcomes of that partial differential equation treatment.

In early 1900, before the meeting, Minkowski was thrilled to learn what his good friend Hilbert was planning to say on this hot day in Paris. The structure of his talk is simple. Here are several compelling problems, Hilbert says. Now, solve them! Hearing this, Minkowski is floored. He loves Hilbert's plan. This will make you famous, he tells Hilbert. It will become the lecture people will realize they always wanted but never knew they needed. "You could have people talking about your lecture [for] decades," Minkowski tells Hilbert before the meeting.

The problems Hilbert chooses are not things anyone has recently solved. The talk will be about not what *has* been done but what *should* be done—it's about not just problems but *huge* problems. Quandaries that have bedeviled mathematicians for years. Knots that have tied up the field for decades, even centuries. Fickle math pickles in desperate need of a slice. He plans to talk about hard problems—the harder the better. "A mathematical problem should be difficult," Hilbert says from the podium.

Hilbert presents problems that "invite attack," the journal *Nature* will report after the meeting. Solving these problems is "most likely to prove profitable," driving the field forward, *Nature* reports.

Minkowski was right to predict the talk's success. It's something special. It's the mathematical talk of the year. The decade. The century—nineteenth, twentieth, or both. It's perhaps the best talk of the millennium.

THE GREAT MATH WAR

Some will later say it is the single most important lecture by a mathematician in human history. He dazzles and charms the dozens of people in the hall, and millions more who read about it after the fact will also marvel. Anyone lucky enough to hear it will remember having been there the rest of their life. It's a mold-breaking, mind-bending, future-shaping, *say-hello-to-my-little-friend* triumph.

THE TWENTY-THREE PROBLEMS

Many today still consider the 1900 Paris lecture to be one of Hilbert's greatest achievements—a scandalous claim, considering how many substantive original ideas he will contribute to mathematics in the decades to come. His name still adorns dozens of concepts in use today: Hilbert spaces, Hilbert inequalities, Hilbert transforms, Hilbert invariant integrals, the Hilbert irreducibility theorem, Hilbert axioms, Hilbert class fields, and many more. And that's the least of his legacy.

Some forty people will receive PhDs under Hilbert in the next thirty years. Many will remember him as the single greatest mathematician they ever met. German physicist Max von Laue, for instance, will come to hold Hilbert's brilliance above even his own—despite the fact that von Laue will win the Nobel Prize in 1914 for discovering X-ray diffraction in crystals, launching the fields of structural biology and structural chemistry and helping set the stage for modern structure-based approaches to drug discovery. Notwithstanding his own brilliance and success, von Laue believes that Hilbert is the true wizard. Hilbert "lives in my memory as perhaps the greatest genius I have ever laid eyes on," von Laue will say.

The sun never sets on David Hilbert, according to his friends and students, and that all begins on this hot August morning in Paris.

◆

The influence of Hilbert's speech is hard to exaggerate. In the early 1970s, the American Mathematical Society holds an entire symposium at Northern Illinois University in DeKalb to explore the consequences of the problems (ten listed in his original talk but expanded to a set of twenty-three

1900: THE JUDGMENT OF PARIS

by the time he publishes the companion paper a few months later). The DeKalb conference concludes that a "wealth" of new methods and results have sprung from the problems, which was exactly Hilbert's goal.

Today, in fact, most of his problems have been solved—or otherwise resolved by showing that they have no solution. One of Hilbert's own students provides the first solution to one of the twenty-three problems a few months after his lecture—a complicated conjecture involving how to cut a tetrahedron into a cube. Over the coming decades, more and more are solved. Some of the stories about their solutions are fascinating. Here are a few.

Hilbert's thirteenth problem falls to a group of Soviet mathematicians just after World War II, including Vladimir Igorevich Arnold, a nineteen-year-old undergraduate at Moscow State University. The thirteenth seeks something akin to the quadratic equation as a solution for all polynomials, not just those with two variables. Solving this gives the Soviet Union a propaganda win and elevates Arnold's status—at least initially. He will run afoul of communist authorities in the 1960s when he signs a letter criticizing the treatment of dissidents. (And as a modern footnote, the problem has been revisited in recent years by a topologist at the University of Chicago who claims that it's still unsolved.)

Hilbert's tenth problem is tackled by a group that includes Hilary Whitehall Putnam, a philosopher who in his later career will pioneer the now familiar idea that communication is influenced by context. As an undergraduate math major, he is part of a team that attempts to solve the tenth problem, which asks whether there's a way to find general solutions to a certain multivariable algebraic equation named after the ancient mathematician Diophantus of Alexandria. The team will fall short. In the 1960s, Putnam will become a hippie/scholar who lives on a commune, organizes anti–Vietnam War protests, and flirts with communism. After Russian mathematician Yuri Matiyasevich actually solves Hilbert's tenth problem in 1970, Putnam will write the foreword to his book.

In 1997, University of Michigan mathematician Thomas Callister Hales develops a proof of the eighteenth problem, known as Kepler's conjecture, which hypothesizes that the most efficient way to stack oranges, cannonballs, or any other spherical objects to maximize their average density is "face-centered cubic," with each globe sitting on top of a pyramid base made by three other globes. Hales solves it via brute force, using supercomputers and geometrical analysis to carry out exhaustive calculations and compare myriad possible stacking configurations. It takes him ten years and a massive amount of computer time, but he finally does it.

An odd footnote to his proof is that it's so complicated, at 250 pages, that nobody has any way of knowing whether it's right. Who will proofread the proof? people wonder. *Annals of Mathematics*, where he sends his paper for publication, sets up a team of twenty top experts to review it. The task is so daunting that they hold a special symposium just to lay out a strategy. Then, one by one, the referees drop out. In 2004, the journal finally pulls the plug and punts, claiming it is 99 percent sure that the proof is correct, despite having discontinued its peer review.

Finally, it's worth noting as inspiration for any budding mathematician readers that Hilbert's eighth problem, known as the Riemann hypothesis, has *still* never been solved. This fact is not often missed by the media today, who have taken in recent years to labeling it *the most important unsolved problem in mathematics* with not the least hint of irony. Solving the Riemann hypothesis is also a Millennium Prize Problem—a set of challenges inspired by Hilbert's twenty-three problems, sponsored by the nonprofit Clay Mathematics Institute, and backed by a $1 million bounty. So solve it!

That's easier said than done, of course, and Hilbert himself guesses that the Riemann hypothesis won't fall so easily. Someone asks him years after his famous lecture what he would do if he could travel five hundred years into the future. Whom would he see? What would he say? The first thing he would do, Hilbert says, is find a mathematician and ask, "Has anyone proved the Riemann hypothesis yet?"

The answer today, 125 years after his talk and more than one-fifth of the way into his five-hundred-year thought experiment, remains the same: *No!*

EXUBERANT SOLUTIONISM

The year 1900 is a new dawn and a new day for Hilbert's sweeping vision and never-ending mathematical ambition, which flower in his famous

1900: THE JUDGMENT OF PARIS

talk. His lecture is religion. Revelation. This is a church revival, and he is a charismatic pastor preaching a simple, compelling sermon. He's selling a vision of the future where any mathematical problem can be solved.

I call this the gospel of "exuberant solutionism." If you can logically state a problem, Hilbert claims, you will eventually solve it, given enough hard work, time, energy, and pure reason. His message to the Paris audience is simple: Find a problem, solve it, and move on. It's *the perpetual call*, he says.

Not everyone agrees with him. Many of Hilbert's contemporaries belong to a competing, constrictive camp popular at the end of the nineteenth century that subscribes to the concept of unknowable unknowns. Things we cannot answer. Problems we could never solve. Questions we may not even be able to ask. And answers we could never hope to know. This competing philosophy is called "triumphant pessimism," a hopelessly negative but nevertheless popular intellectual worldview due to the influence of one man: the French philosopher and physiologist Emil du Bois-Reymond, who had pioneered the idea a generation before.

By day, du Bois-Reymond was a basic physiology researcher of no small repute. He was the father of electrophysiology, in fact, famous for having discovered action potentials, those measurable electrical "firings" of nerve cells that form the physiological basis of neuronal transmission and all things cognition, from neuropathic pain to memory recall to reasoning to taste. But by night, du Bois-Reymond spun philosophy and spoke of unknowable unknowns. He gave a famous talk in 1872 in which he stated his basic philosophy: *Ignoramus et ignorabimus* (We don't know, and we shall never know). The talk resonated for decades. By the time Hilbert takes the stage twenty-eight years later, he has to swim upstream against all those *ignoramuses*.

Hilbert's whole lecture and, some have claimed, his entire approach to mathematics for his entire life are one big reaction to du Bois-Reymond. He sees mathematics as a single cloth made of separate fibers—geometry, algebra, number theory, logic, mathematical physics—all forming one congruous, intertwined fabric that covers everything and can uncover anything. Du Bois-Reymond is no mathematician, he doesn't see what Hilbert sees. In fits and starts, mathematics has pushed its way into every corner of science to the end of the nineteenth century. After thousands of years of success and triumph, especially in the last several hundred years post–scientific revolution, mathematics now fills the creases and seals the cracks of human knowledge.

People have taken to glorifying mathematics by calling her the Queen of the Sciences. Hilbert's whole twenty-three problems lecture, in fact, is one big, throaty defense of his queen. She *can* do anything. She *is* everything. If his talk is the greatest call to arms science has ever known, then it's a triumph for old Queen Math. "In mathematics," Hilbert says, "there is no *ignorabimus*."

And yet, like all fanatics, he hides a secret doubt. Eternal optimism sometimes conceals quiet desperation. Queen Math is in trouble in 1900, and Hilbert knows it.

All reigns end eventually. Rulers are defeated. Or they are deposed. Or they're term-limited out. Or they retire. Or they die in office—sometimes of natural causes, sometimes not. Inanimate objects fail as well. Things break. Items lose relevance. They become neglected. Buildings fall. Cities crumble. Nations cease to exist. Eras come to an end. Empires die.

Even intellectual ideas cannot escape such deaths. Songs go unsung. Stories go untold. Memories lapse. Moments end. Knowledge is forgotten. Names are lost. Greatness fades. Paradigms are overthrown. Time destroys every absolute monarch. Even the greatest Ozymandias of his day crumbles into dust. Hilbert fears that sort of fate for old Queen Math. Her golden days could be at an end. Mathematics in 1900 is feeling a bit like a crumbling heel. *Look on my Works, ye mighty, and despair!*

That's why Hilbert is here in Paris. That's why he's talking about the future. That's the point of all his optimism. The future should be bright, not dark. Hilbert is worried, and he won't leave the queen's fate to chance. He wants to set the agenda for the next century, and with this lecture, he aims to do it.

His queen is dying, and single-handedly, he means to save her.

"NO EYE COULD BEAR MORE GORGEOUSNESS"

Across the English Channel, another queen sits on a lonely throne. Her world is also in crisis. Alexandrina Victoria is at the tail end of a long, glorious reign. She's been queen since 1838, making her the second-longest-ruling sovereign in British history (until Elizabeth II will surpass her in the 2020s). She is a ruler like no other, though the bar wasn't exactly high when she first took the stage. When she ascended to the throne at age eighteen on June 20, 1837, her three predecessors were George III, George IV, and William IV—considered a motley of fools by some experts. Modern American journalist/historian Robert K. Massie calls them "an imbecile, a profligate, and a buffoon."

Victoria was determined not to channel the Tumbledown Dick ineptitude of her predecessors. "I am very young," she declared upon taking the throne, "but I am sure that few have more goodwill and more real desire to do what is fit and right."

The empire she inherited was elastic. Ecstatic. Great Britain responded to its queen and expanded under her reign—the best six decades the empire ever knew. She inherited a colonial stretch that was a scrabble bag of random, far-flung colonies. Gambia. Calcutta. Hudson Bay. Newfoundland. The South African Cape. Australia. But by 1900, nearing the end of her reign, her empire has grown much larger. It now encompasses a thousand lakes, two thousand rivers, ten thousand islands, and more deserts, mountains, coastlines, and marshlands than you could survey in a lifetime. During the last twenty-five years of the nineteenth century alone, Britain added 4.75 million square miles of land and brought ninety million people under its imperial umbrella. (For context, the total world population in 1900 is 1.9 billion.)

By 1900, Great Britain has far-flung colonies all over the world. It now owns more than one-fifth of all the land on Earth. Britain is everywhere. Queen Victoria's mighty empire rules the high seas. Union Jacks fly over ports of call from Weihaiwei to Wellington, from Halifax to Gibraltar, from Cape Town to the Falklands, and from *Where did you go?* to *Hell if I know!*

British laws govern cities. British ideas penetrate culture. British influence changes men's attire. The dark wool suit, so Italian to the modern mind, was actually the invention of London's nineteenth-century fashion houses. It has taken hold by 1900. American envoys. Hungarian waiters. Russian musicians. Japanese bankers. Brazilian undertakers. German mathematicians. Mexican presidents. All the ego-prancing, self-important K Street/Wall Street preeners of our modern day. They all have nineteenth-century golden-age-of-the-British-empire fashionistas to thank for their de rigueur business formal wear.

One in every four people alive in 1900 falls under British rule. They are treated to an English tongue. A common law. A vast international banking system. Free trade. Supply chains up the wazoo. The odd cricket pitch or two. And that thin vanilla whiff of liberty.

If the sun never sets on Great Britain, it *really* never sets on Queen Victoria. People call her "stability incarnate." Most have never known another ruler. When Victoria's diamond jubilee, the sixtieth anniversary of her accession to the throne, was celebrated in 1897, the city of London seethed with people. Red carpets rolled. Monarchs and politicians flocked. It was the greatest display of military vainglory the world had ever seen. Some forty-seven thousand troops from around the empire paraded in proud pomp and procession. Pristine uniforms. Gold-and-red plumage. Dragoons and dirks. Zaptiehs and Dayaks. Guayaberas and fezzes. Feathered scabbards and camel corps. Hussars and high-step marches. Oh, what splendor! "No eye could bear more gorgeousness," one witness said.

"The mightiest and most beneficial empire ever known in the annals of mankind," the London *Times* declared. "Practically unassailable," the Berlin-based German newspaper *Kreuzzeitung* wrote—of the empire if not its queen. *Practically unassailable!*

But three years later, in 1900, just a few months before Hilbert will take the stage in Paris, the ceremonial hats are gone. The celebrations have faded. The British mood is black, and Queen Victoria's is somber. In early 1900, she is photographed in a horse-drawn carriage, wearing a coal-colored suit and a white-feathered hat. She is sitting for shots on a photo junket in Dublin, Ireland. Britain is at war, and she is recruiting. Things are going very, very badly.

It was not supposed to be this way. Britain has almost never known defeat in the queen's day. Her empire has fought some seventy-two "little" wars during Victoria's reign and has always come out on top. Her army is modern. Well-equipped. Formidable. It has seasoned soldiers. Professionals. They are well trained. Her infantry is made up of Tommy Atkins heroes. Her officers are the cream of the imperial aristocracy. They are ever ready to drum-beat their way to glory's peak, one bloody step at a time. *Glorious!*

But in 1900, Britain is ensconced in something a lot more than a little war. In South Africa, it faces an enemy like it's never seen before—one that doesn't drill. Never marches. Has no formal military training. Wears dust-rag clothes. An undisciplined militia—less an army than a rabble. Sack lunches and glue-factory ponies. A tattered band of "weather-beaten old men and half-grown boys" fighting a farm-to-battlefield war. That is what the British public has been sold.

That narrative never really fit. It seems like a lie in 1900. Actions taken based on false foundations are often doomed. The war has quickly become Britain's longest, most expensive, and deadliest conflict in memory. *God, Queen, and England* is beginning to sound wheezy, like a heaving cry of desperation. The queen wears black in Dublin. Agony is her fur coat. Silence is her shivering shawl. Her entire empire is mourning.

◆

Somewhere in South Africa, British soldiers are disembarking. Down the gangplanks they go. Across the docks. They are boys, really, not men. Laughing. They board trains. They *ketchunk-ketchunk* hundreds of miles up the tracks. These are special trains. Military trains. Troop trains with armored cars. Machine-gun turrets at the ready. Rifle slots cut on each side. Hospital cars in tow. Up and down the dusty veldt of South Africa they go. War is out there somewhere. The truth is out there too—the gruesome math of war. Bodies. Death. Glory?

War is often about the lies we tell others, the ones we don't believe. But sometimes it's about the lies we tell ourselves—the ones we do believe. Salvation. Pristine khakis. Stiff collars. New uniforms. Crisp and clean at first. Then rumpled. Then dusty. Then dirty. Then red. Then black.

The queen is dying in 1900. And there is no glory in a dark sunset. Fezzes have faded. Dragoons have been dirked. Guayaberas are gone.

THE GREAT MATH WAR

Dublin in 1900 is the beginning of the last year of Victoria's life. The rest of the European world senses the future that summer—all the marvels of art and tech at the Paris Expo. But if 1900 seems like a birth of wonder, the coming years will show that it's really a dawn of horror. Europe will soon be shaken to its foundations when the Great War arrives. Nothing will ever be the same again—not in politics and certainly not in mathematics. Everything sits on a soft foundation.

It is the dawn of violence and hope!
But the future's not looking so dope!
It soon ushers in decades of horror!
One only marked by war after war!
This will not be humanity's delight!
Following the twin queens' twilight!
     That summer...
       that day...
         that crowd...
           that expo...
             that speech...
               1900 in Paris

CHAPTER TWO

ADVENTURES OF A WOODEN PUPPET

1883–1884

> "
> The poet must see what others do not... the mathematician must do the same.
> "
>
> —Sofya Kovalevskaya

Sofya Kovalevskaya in 1885 (second row, center, seated in white) with a theater troupe in Stockholm wearing Viking dress. Cropped from a public domain image via Nordiska Museet.

Four things happen in 1883 that set the stage for what I call the Great Math War. Four things amounting to four crossroads reached by four people—well, three people and a wooden puppet.

Pinocchio is carved into our consciousness when his eponymous *Adventures* by Carlo Collodi first appears in print in Italy in 1883. The book doesn't exactly shake the earth, but it will half a century later, once Walt Disney silver-screen immortalizes that nose-grows, donkey-eared, lovable hunk-of-wood liar. He has awful judgment, poor taste in friends, and even worse luck. But why mention this motley fool? What's a marionette Ulysses got to do with the history of math? Not much on the surface! But this is a book about foundations and lies—the lies we tell other people and, more importantly, the lies we tell ourselves. Pinocchio isn't just an apt metaphor for this story. He's the living wooden incarnation of it. *A real boy!*

Around the same time that Pinocchio gets his first taste of life, eleven-year-old Bertrand Arthur William Russell gets his first taste of mathematics in his boyhood England when he becomes suddenly obsessed with it in late summer 1883. His brother, Frank, is home on holiday, wasting time before heading off to Oxford. One August morning, he decides to show his younger brother a thing or two from the most famous math book ever written—Euclid's *Elements*. It's the same classic, if terse and tedious, two-thousand-year-old standard text on geometry that David Hilbert will reimagine sixteen years later in his own book *Foundations of Geometry*.

"I gave Bertie his first lesson in Euclid this afternoon," Frank recalls. "He did very well indeed, and we got half through the definitions."

This is one of the greatest moments of Russell's life—and one of the most disappointing. "As dazzling as first love," he claims. But as soon as he hits that first blush of lust for math, he's let down by it. Euclid's axioms, the so-called five postulates that form the foundations for the rest of the book, fall short in his mind.

ADVENTURES OF A WOODEN PUPPET

How do we know if these postulates are true? young Bertie asks his brother.

There is no proof, Frank says. You just have to accept them.

That, Russell will later recall, is crushing. "My hopes crumbled," he says.

Nevertheless, this little bro moment sends him down a long and tortuous path toward philosophy, math, and the search for the ultimate logical truth when his initial reaction of disappointment turns to desperate determination. "Why should I admit these things if they can't be proved?" he will later recall thinking. Pretty soon, proof must come.

◆

Russell wastes no time once he goes off to university. He throws himself into math at Trinity College, Cambridge, and becomes obsessed with logic. It seems a strange path for someone with his background, an aristocrat hailing from the preening, pressed-suits-privilege set. He comes from a great family, according to his biographer Ray Monk, an old, important, and well-heeled family whose fortunes date back to the English Reformation. Centuries before, when King Henry VIII broke with the Catholic church and claimed vast amounts of land, money, and property in the process, many of the proceeds went to his loyalist supporters, including Bertrand's ancestors. The Russells became fabulously wealthy overnight. The head of their clan was named the Earl of Bedford and given large estates in Devon, Cornwall, and Dorset.

Centuries later, "Bertie," as his friends and family would call him, is born the youngest son of Viscount John Russell, or Lord Amberley, and Viscountess Katharine Louisa Stanley Russell, Lady Amberley. His grandfather Lord John is a famous politician, a progressive member of parliament, and the last member of the liberal Whig Party to serve as prime minister. Lord John was famous for a few key legislative achievements, reforming Britain's death-penalty laws, pushing for workers' rights, jamming through public health policies, and extending voting rights to many in Great Britain (though not to women). Charles Dickens was so enamored

THE GREAT MATH WAR

with Lord John that he dedicated his book *A Tale of Two Cities* to him "in remembrance of many public services and private kindnesses."

Lord John's wife is Bertrand's grandmother Lady Frances Anna Maria, the Countess Russell. She continues to have great expectations for the boy in 1883, and she has poured all her hopes for the future into him from the time he was a toddler. She holds fast to the notion that he will achieve greatness and leave behind towering accomplishments as his legacy. And why not? Obviously, he'll be prime minister one day. Countess Russell knows it, and she is known to say it aloud whenever the mood strikes without the slightest hint of irony. Her influence upon Bertie is strange, profound, lifelong, and puissant—a potent mix of high-minded noblesse oblige shaken with melancholy self-denial, given a twist of religious zeal, and served cold, on the rocks.

"The sense of public duty," Ronald Clark, one of Russell's late-twentieth-century biographers, will write, "permeated the household like the smell of hops in a brewery."

KATE & JOHN AND DOUG & KATE

Russell's childhood is marked by loneliness, and not just because of the isolating intensity of impossibly high grandmotherly expectations—or Bertie's own sad, solitary disposition. There is something much deeper and more disturbing going on. Bertie's early life is marred by tragic loss. His parents, Kate and John, are passionate, warm, and outspoken liberals. In the decade before he was born, they were on the political cutting edge of issues like women's rights. Bertie's father, John, was a young member of parliament who ruined his career almost as soon as it began, before Bertie was born, after he advocated legalizing birth control, a highly controversial topic in nineteenth-century Victorian England. He was skewered in the press, and the ensuing scandal brought his short parliamentary career to an ignoble end after a single sad term.

After that, things got strange. By the time Bertie was born, the Amberleys had acquired a live-in houseguest named Douglas Spalding, an amateur biologist who makes a minor splash in experimental psychology by imprinting newborn chickens on himself, posing as the mother hen after they hatch. They follow him around, a gaggle of chicks, clucking as though he can listen.

The whole Russell household is awash in Doug's little experiments. Bertie has strange memories of them years later, almost like fever dreams. All kinds of psycho weirdness. Live salmon, cut open in one cruel experiment. Spalding chopped the heads off wasps. He kept little birds in tiny boxes from the moment they were born, never once allowing them to flap their wings—all so he could see if they were still able to fly as adults. His experiments seem sad and cruel by today's ethical standards, and they would probably never withstand the scrutiny of a modern institutional review board. It's just as well that much of his work has been forgotten.

Spalding's boldest experiment in those early days of Bertie's life involves Kate and John. Kate is Doug's assistant, and they are sleeping together—openly and apparently with the consent of her husband. Doug has active tuberculosis, that withering bacterial lung infection known as consumption in those forgotten days. It has ravaged Europe and elevated that gaunt, pale, flushed appearance into something of the heroin chic of its day. Charlotte Brontë once called it "a flattering malady," and women often dressed in hip-hugging corsets, white-powdered faces, bright rouged cheeks, and other fussy consumptive flair. But the Russells are far more fashionable. If the fad of looking like you have active tuberculosis is quite the rage, the hot-ticket buzz of having an actual live-in TB lover is even more en vogue.

Colorful and unconventional is the ménage à trois that his parents enjoy. But it ends abruptly following an unrelated tragedy. After the family returns home from a trip abroad, Bertie's older brother, Frank, develops a bad sore throat. Doctors diagnose it as diphtheria, a highly contagious bacterial infection that's now somewhat rare in western Europe because it's preventable by vaccination. Frank is isolated, and he seems to recover. He's given a clean bill of health and promptly sent home. But it's too soon. He's still infectious, and he spreads the disease to his mother and sister. Frank makes a full recovery, but Kate and his sister, Rachel, die from the disease. John is crushed—worse, destroyed. The two most important women in his life are gone. Russell's father suffers a dark depression, describing his fate as "purest happiness turned into the bitterest pain." He loses his will to live, and several months later, he dies.

◆

The two surviving children, Bertie and Frank, are handed over to the care of Lady Russell, their grandmother. Doug is dismissed, Frank is sent away to boarding school, and Russell spends the remainder of his childhood subject to his grandmother's whims—strict, regimented, cold, and puritanical. She loathes comfort, according to one account. Cares nothing for cuisine. Hates wine. Begins each day with an ice-water bath and a mournful morning routine of piano practice and solemn prayer. She is aloof—incapable of love, Bertie later claims. Her favorite saying, he will recall, is the constrictive, closed-book circular construction:

> *What is mind?*
> *No matter.*
> *What is matter?*
> *Never mind.*

Lady Russell also has an existential fear of formal education. She decides Bertie needs to be protected from school so as to be preserved—"pure, religious, and affectionate," she says. It's too late for Frank, she feels, so he is sent off to boarding school. But Bertie is special, she thinks. He should be homeschooled. "He must be fitted to take his grandfather's place as prime minister and continue the sacred work of reform."

By the time Frank gives his younger brother those first few lessons on Euclid and geometry in 1883, Bertie has been homeschooled for years in an emotionally bereft *Grey Gardens* gloom. It's as sad as it is transformative. "I try to make some sort of human contact," he says at one point in his childhood, "but it is impossible."

"I know myself doomed forever to lonely impotence," he adds.

Learning Euclidean geometry in 1883 is a lone light in that derelict darkness. "I had not imagined that there was anything so delicious in the world," he later recalls. It's a taste he will relish in the decades to come. A taste for life shared by others as well.

THE SHORT LIFE OF SOFYA KOVALEVSKAYA

Across the North Sea, sweet mathematical success is tasted in 1883 when Russian mathematician Sofya Vasilyevna Kovalevskaya is hired by the University of Stockholm, becoming the first female professor in Europe—at least in the sciences. (A few women had earlier won professorships teaching humanities in Italy.)

In the typical selfish sexist fashion of the times, Kovalevskaya faces an uphill battle. She is forced to accept the job conditionally and without pay for a one-year probationary period. And even then, her appointment is controversial. Misogynist views of women are common at European universities. Most institutions won't even allow women to set foot in the classroom—never as a student and definitely not as a teacher.

"A female professor is a pernicious and unpleasant phenomenon," one prominent Swedish writer says in a local newspaper at the time, penning an op-ed in opposition to Kovalevskaya's appointment. "Even, one might say, a monstrosity."

Many men agree with this writer. Some call Kovalevskaya the queen of mathematics—a name not meant as a compliment. German chemist Robert Bunsen calls her "dangerous" because of her intelligence. (That, I suppose, is what you call a Bunsen burn.) What an idiot!

None of this resistance in 1883 is new for Kovalevskaya. She has stoked sexist fires her whole adult life. A decade earlier, she overcame almost insurmountable barriers when she became the first woman in Europe to obtain a PhD in mathematics. It was a remarkable but odd feat—she had to do her thesis remotely and obtain her degree in absentia because even though the university was willing to grant it, they would not allow her to attend classes.

She will also become the first woman to work as an editor for a major academic journal and the first woman ever admitted into the staunchly elitist Russian Imperial Academy of Sciences. Its bylaws actually have to be rewritten after her election because under its old rules, no woman is even allowed to set foot inside the building. She literally has the door slammed in her face the first time she tries to enter.

◆

Kovalevskaya is born the middle child of a rich, important Russian army general named Vasily Vasilyevich Korvin-Krukovsky who owns a huge private estate with thousands of acres of land near the Lithuanian border in Russia. It has a sheep ranch, a dairy farm, a vodka distillery, a large forest filled with game, and lakes stocked with fish ready to be caught and sold. When Sofya is a little girl, the general retires, and the whole family relocates from St. Petersburg to the country estate.

A random misadventure in interior design determines the course young Sofya's life takes. The family accidentally miscalculates how much wallpaper they will need to refurbish their country estate, and they wind up one roll short of a complete decor. They briefly consider the possibility of sending someone hundreds of miles back to St. Petersburg to fetch another roll, but they decide that would be too extravagant. So they resolve to simply decorate Sofya's room differently—using newspaper instead of wallpaper. Then someone discovers a bunch of old lecture notes dating back decades from math courses General Vasily Vasilyevich took during his training to be an artillery officer. Even better! They paper her walls with the math notes instead. Muzzle velocities. Firing angles. Arcs and ranges. Falling objects. The effect of gravity on the horizontal path of a projectile. Years later, she will claim this was how she learned math—spending hours and hours staring endlessly at her father's old fusilier notes pasted on all the walls around her room.

"I began scrutinizing the walls very attentively," she later says. "I could make no sense of any of it at all then, and yet something seemed to lure me on."

When she expresses her interest in math, however, her father is horrified. He's a traditional Russian man of the old school. He doesn't have any refined taste for well-educated women. Even his daughters—especially his daughters! He forbids her to study mathematics. She does it anyway, borrowing an old algebra book from one of her tutors and secretly reading it at night while everyone else is asleep.

But her opportunity comes when a physics professor who lives nearby visits the family. He leaves behind a copy of a textbook he's just written. It's a gift, and the old professor is hoping the old general will take an interest in his work and be impressed. But it's Sofya who picks up the book instead. She completely lacks the necessary mathematical background in trigonometry, however, so she soon finds herself completely lost. Always resourceful, she finds a workaround. She develops a way to approximate trigonometric functions numerically. When she plugs those approximations in, they work. The next time the professor comes to visit, Sofya is excited to discuss his book with him. The old professor isn't sure what to make of her. He is amused at first. Then politely dismissive. Next slowly put off by her persistence. And finally completely annoyed. *She read his book? Is she joking? Come on! She couldn't possibly have understood it.*

She didn't understand much of it, Sofya admits. She tells him she had to substitute her own original numeric approximations for the trigonometric functions, and he rolls his eyes. "Now you're bragging, aren't you?" he says. But when Sofya shows him her notes, he can't believe it. She wasn't kidding. She managed, completely on her own, to develop the frameworks of trigonometry in a way remarkably similar to how certain other mathematicians had done it centuries before. He's blown away. And more. The girl is a genius! The old professor now begs her father the general to let her study math.

Eventually the old general relents. These are changing times in Russia. He consents to allow his teenage daughter to attend high school in St. Petersburg, where she is finally allowed to study math—at least for a time. Though she is a brilliant student, she cannot enroll in college. The Russian university system is completely closed to women except for one awkward path to attendance: Some progressive professors allow women to informally audit classes. They can't take degrees, but they can sit in on lectures. And so she does, sometimes having to sneak into the back of the lecture halls so as not to call attention to herself.

But then a bigger barrier emerges.

THE GREAT MATH WAR

◆

The 1860s and 1870s, when Sofya comes of age, are a time of revolutionary fervor in Russia. The country's czarist leadership is convinced that there's a connection between educated women and revolutionary frenzy. So the government moves to ban women from the classroom entirely. This creates a small exodus of young women (who are well educated and of independent means) who make their way to Western countries in search of greener pastures and more liberal opportunities. That includes nineteen-year-old Sofya and her older sister. But after moving to Berlin, she finds only another door slammed in her face. "The capital of Prussia proved to be backward," Kovalevskaya later says. "Despite all my pleadings and efforts, I had no success in obtaining permission to attend [the University of Berlin]."

She meets her mentor there, however—the German mathematician Karl Theodor Wilhelm Weierstrass. He understands underdogs and loves a long shot, having been one of both himself. Years earlier, he rose to become one of the greatest mathematicians of the nineteenth century after toiling away for the early part of his career as a lowly high school math teacher. "Too poor to afford postage for scientific correspondence," according to the mathematician T. A. A. Broadbent.

Weierstrass, who is also from an older generation, brushes her off at first. But she persists in asking him questions, so he gives her a problem set, probably thinking that will get rid of her. The problems are the same graduate-level challenges he gives to all his advanced students, and he doesn't expect her to complete them. But she does. And when he sees her work, he is more than a little impressed. He immediately begins pleading with the University of Berlin to enroll her. But the school is unwilling. So Weierstrass takes it upon himself to give her private lessons, and eventually he helps her obtain her PhD in absentia from the same university where Hilbert will later work. Sofya becomes the first woman in European history to receive this degree.

While she is wrapping things up, the political situation in Russia worsens. Growing paranoid that the furtive seeds of future revolutions are germinating in the fertile soils of Western Europe, the heavy-handed Russian government completely loses its taste for women being educated in the West and demands they all return, within six months. Some refuse. Others rush to finish their degrees. One woman drives herself to death in desperation to finish. Many women abandon their studies and dutifully return, only to be met with contempt as they come home.

Kovalevskaya is fortunate to be well advanced in her studies. She finishes her degree, returns to Russia, but then languishes there for five years. The unofficial government policy is that none of the women returning from the West shall be allowed to take jobs at universities, hospitals, or government agencies. The only jobs open to Sofya back in Russia are low-level positions teaching rudimentary math at all-girls high schools (think multiplication and takeaways). But she has neither the taste nor the aptitude for this, since she has spent all her time on higher-level math and never actually mastered the basic stuff. So she becomes a writer instead—anything to fill the time.

She does theater reviews, writes a successful autobiography, produces a play, pens a novel, and becomes a sort of tech reporter for the local newspaper, covering new discoveries like the telephone. Kovalevskaya is unusually gifted and amazingly successful as a writer. Her autobiography, *A Russian Childhood*, is considered a major literary achievement when it comes out. And as her fame grows on the heels of that success, people start calling her the girl genius. "People would often stop in the street to stare," one of her biographers later says.

"IN THE HIGHEST DEGREE"

The hand of fate intervenes again in 1880, when an organizer of a scientific conference in St. Petersburg begs her to give a talk. She agrees, dusts off some old unpublished work from half a decade before, and presents it to an audience of men stunned by her brilliance.

Swedish mathematician Magnus Gösta Mittag-Leffler is in the audience, and he has a secret mission. He is another former student of her mentor Weierstrass in Berlin, and before Mittag-Leffler left for Russia, Weierstrass begged him to seek Kovalevskaya out. Talk to her, Weierstrass asked. Convince her to come back to the West. After Mittag-Leffler hears Kovalevskaya's talk, he vows to do more than that. He promises her that he will pave the way for her return. He will find her a faculty position in Sweden, and failing that, he will create a new one. He lives up to his word, bringing her there in 1883.

She gets out of Russia just in time. Bad business decisions and financial speculation drove Kovalevskaya and her husband into financial despair. In the late 1870s, her husband invested all their money in a speculative business, building family homes with attached commercial bath houses. But the bottom fell out of that market, she will later recall, "and brought us to utter ruin." Two weeks after Kovalevskaya impresses the audience with her talk in St. Petersburg, all their possessions are sold at auction. This has a devastating effect on her husband and torpedoes their marriage. They divorce, and in 1883, just before Sofya leaves for Sweden, her now ex-husband, Vladimir, becomes embroiled in another financial scandal and commits suicide by chloroform. Sofya is devastated, and in the months before moving to Sweden, she tries to starve herself to death.

Kovalevskaya suffers awful bouts of depression for years after that, made worse by the overwhelming grief she feels when her beloved sister dies after a horrible illness. This weighs heavily on her for the next few years, and then she also dies a lonely death. At Christmastime in 1890, she travels to Italy, and on her way back to Sweden, she has a stopover in Denmark. But she arrives in the middle of the night and has none of the local Danish currency she needs to hire a porter, so she is forced to walk with her bags through the cold January rain. Back in Sweden, she develops pneumonia and dies a week later at age forty-one. Her brain is removed, weighed in autopsy, inspected, and judged to be "advanced" according to the unscientific methods of the time.

Newspapers in Stockholm run stories declaring that her brain is developed "in the highest degree" and "rich in convolutions, as might have been predicted, judging by her high intelligence." Thus ironically, in death, Kovalevskaya finally receives the recognition for her brain that she deserves.

———◆———

The greatest legacy one can claim for Kovalevskaya's short life is her trailblazing success in throwing open the doors of universities to women. Had she lived, she may have left an equally great legacy through her work. She is said to have had extraordinary talent and great potential. Her university in Stockholm makes that clear when, within six months of her conditional appointment, it gives her a full five-year contract. And when that expires, she receives a lifetime appointment.

More direct evidence of her brilliance shows when she enters a competition sponsored by the French Academy of Sciences for a prize called the Prix Bordin, which awards cash for the best solutions to challenging problems—though the problems are often so hard that they go unclaimed. In fact, in fifty years of offering the Prix Bordin, only ten have been awarded by 1888, when the academy announces a new challenge that Kovalevskaya enters.

The challenge in 1888 is to solve something called the "mathematical water nymph" or mermaid problem, a name reflecting its almost mythically elusive nature as well as its beauty. It's surprisingly simple to picture but amazingly hard to solve: What are the equations describing the rotation of a heavy, rigid body around a fixed point—such as a pendulum or a gyroscope? For decades, some of the top mathematicians in Europe tried and failed to find solutions.

Some months after announcing the prize, the academy judges fifteen solutions, but only one is accepted—Kovalevskaya's. Her solution is spectacular, offering not simply a suitable formula but a general framework for considering rotating rigid body problems. The work is so impressive that upon seeing it, the academy scales up the prize money from 3,000 to 5,000 francs. And then it unmasks the winner and award Kovalevskaya her prize. "The result was beyond what I had hoped," she says.

Her legacy remains one of awkward taste and promise, however. She is noteworthy, for sure, but not transcendent. Had she lived, would we have seen more of the same brilliance she showed with the mathematical water nymph? Sadly, we will never know.

GOD'S GIFT TO MATHEMATICS

The final thing that happens in 1883 that puts the taste of the coming Great Math War on the tongue involves a German mathematician named

Georg Ferdinand Ludwig Philipp Cantor. In 1883, he finishes his monograph *Grundlagen einer allgemeinen Mannigfaltigkeitslehre* (*Foundations of a General Theory of Sets*). It's a major milestone for him, marking the fact that his life's work is now almost complete.

Finishing *Grundlagen* in 1883, Cantor hopes, isn't just the launch of a book but the end of a chapter. The book describes in detail what he calls set theory—a basic framework he formulates for mathematics based on the concept of sets. Modern experts consider it one of the most successful advances in the history of science. "It introduced an entirely new field of inquiry to mathematics," Cantor's biographer, the American historian Joseph Warren Dauben writes. But Cantor's success is a Pyrrhic victory: "One which promised new gains, while carrying with it the hidden seeds of trouble yet to come," Dauben says.

That's hardly surprising when you think of what set theory represents. Sets are collections of objects with associated rules that define which objects belong to a particular set. Think of all the workers on a job site. All the hammers in Home Depot. Or all the nails in a coffin. "Many, which can be thought of as one," Cantor once says, describing sets. "A totality of definite elements that can be combined into a whole by a law."

That all seems reasonable, but the trouble comes when you consider larger and larger sets. In 1883, Cantor's set theory, especially his work in infinite sets, is controversial. For the five years leading up to 1883, Cantor's ideas have been subject to criticism, partly because they are so original. Set theory allows mathematicians to revisit, revise, and reinforce the fundamental concept of numbers in a way that has been called the most radical extension of arithmetic since its beginning. But that's not all. Cantor dares to defy convention by dealing with the concept of infinity, treating infinite sets as just another type of mathematical object. That is radical. And anytime you lick infinity, you are bound to taste both bitter notes as well as sweet ones.

———◆———

For thousands of years leading up to 1883, infinity belonged more to the realm of philosophy than to math. With set theory, Cantor takes on

ADVENTURES OF A WOODEN PUPPET

infinity and forever redefines it. And he makes no secret of the source from which he drinks his knowledge. He compares himself to Euclid and Archimedes and claims, with unwavering conviction, that he is divinely inspired. "I am only an instrument of a higher power, which will continue to work long after me," he writes to Mittag-Leffler in 1883, just as the latter is welcoming Sofya Kovalevskaya to Stockholm.

There's almost no way to view Cantor's life as anything but tragic—or rather as a long series of tragedies. Like most great tragedies, it begins in his childhood. Cantor's mother came from a family of musicians, including one famed uncle who founded and directed a major music conservatory in Vienna. As a boy, Cantor was an outstanding violinist. *Sonata. Glissando. Staccato. Spiccato*—he gets it. He could have been a musician if only his father had let him. He was a promising artist as well, able to render gorgeous line drawings of objects and animals (although only one piece survives: an impressive pencil sketch of a dog curled up and resting against a tree that has a slightly Salvador Dalí feel to it).

Cantor's father, Georg Sr., was not having any of it. He was a hard-driving businessman. His multinational wholesaling firm, Cantor & Co., was a smashing success in St. Petersburg, with satellite offices in Hamburg, Copenhagen, and London—and dealings with Brazil and the United States. But the company collapsed before Cantor was born. The family struggled, and when Georg Jr. was eleven, his father developed tuberculosis. That forced the family to leave Russia for the relatively cleaner air of small-town Germany. There Georg Sr. built a successful new venture in the few years he had left.

Like all businessmen, Cantor's father had big plans, and naturally he wanted to bring his son and heir into the fold. Historians say he took a special interest in his son's future, but they are somewhat divided on the significance of this. Some say Georg Sr. was an enthusiastic but caring father who loved his children deeply. They credit him with instilling a thirst and an almost inhuman drive and doggedness in his son that sustained the boy for his entire adult life, carrying him to great heights. Others see the old man's posture as problematic, if not pathological—a tiger dad whose drive to succeed at any cost pushed his son over the edge. According to this narrative, Cantor holds himself to unreasonably high expectations and sets himself up for a lifetime of unfulfilled promise, personal disappointment, and future failure.

Is that true? It's hard to say. We have more speculation than facts. What we know is that even though young Cantor was a gifted musician, his father pushed him to pursue business or engineering, and Cantor did as his father wished. He wasted two miserable years in the meaningless pursuit of an empty degree at a trade school before finally gathering the courage to tell his father that he wanted to switch gears and enroll in a mathematics program. Georg Sr. reluctantly consented and then died of tuberculosis a few months later. As his father lay dying, Georg Jr. wrote him a letter explaining his passion for his new subject. "My soul, my entire being lives in my calling," he assured his father. "I am happy when I see that it will no longer distress you if I follow my own feelings in this decision."

"A SO-CALLED RUINED GENIUS"

Georg Sr. wrote his son a letter as well—a *neither-a-borrower-nor-a-lender-be* missive full of blistering warnings about the unknown challenges that surely awaited him. The letter advised Georg Jr. how to withstand these threats. Beware the "jealousy and slander of open or secret enemies," it warned. "How often the most promising individuals are defeated after a tenuous, weak resistance in their first serious struggle following their entry into practical affairs," Georg Sr. said. "Their courage broken, they atrophy completely thereafter, and even in the best case they will still be nothing more than a so-called ruined genius!"

There are two ways of reading this letter, depending on how tiger-dad forward you take Georg Sr. to be. The letter could be saying *don't be* a ruined genius or... you *will be* a ruined genius. It's not clear how Georg Jr. interpreted it, but he kept the letter on his person at all times for the rest of his life and often referred to it.

Distant fathers, whether aloof, dead, or divorced, are both attractive and mysterious. They're somewhat like fixed stars on a January night—navigational guides. Dark, faint inspiration. They could be huge gas giants or mighty quasars up close—blisteringly hot, bright, and massive with time-bending and space-warping gravitational pulls. Fixed stars are so bright that they can be seen from many, many light years away. But they're too far to convey any real feeling. They're remote. Distant. Dim. Fiery gas giants seen a thousand light years away are witheringly dull. Speckish points of light in a cold, cold sky.

After his father died, Cantor attended graduate school at the University of Berlin, the same school that would reject Kovalevskaya's application a few years later. Upon finishing his PhD, he began teaching at the University of Halle—a fine achievement, though at a less prestigious university than he would have liked. There, in the 1870s, he writes his first paper on infinite sets. And in the next eight years, culminating with his book in 1883, he drives himself mercilessly, perfecting his theories, redefining the concept of infinity, and pushing himself to exhaustion in the process.

―――◆―――

Infinity has been around for a long time, and mathematicians in 1883 are certainly no strangers to it. But the concept of infinity until that time has been empty. You can blame Aristotle for that. His definition of infinity, which rules Western science until Cantor arrives, draws a distinction between *potential* infinity, which we can imagine, and *actual* infinity, which we cannot.

Philosophically his argument is sound. You could argue, as Aristotle more or less did, that our finite brains are not properly equipped for dealing with infinity. Humans are finite people, and we live limited lives. Our brains are specifically tuned to a temporal existence, and we are hardwired to focus on the transient. The limited. The incremental growth. The sudden decline. The ebbs and flows. The highs and lows. And all those selfish little peaks and valleys of life. We think of discrete data. We connect points into lines and numbers into series. We analyze data to see trends. We think of aspects like rates of change. And we predict future outcomes. Evolution may have seen to our ability to do that.

Okay, but why not infinity as well? In some ways, we humans are perfectly well equipped to deal with the concept. For thousands of years, long before we were ever challenged with dealing with infinity mathematically, we could nevertheless easily imagine what Aristotle deemed a potential infinity. It's trivial to conceive of an infinite list of discrete counting numbers added to a series ad infinitum in a never-ending, uncountable list. Infinity is intellectual glory. Even small children who are completely ignorant of higher math get this. They can envision the basic concept of an infinite amount of something and then take the next cognitive leap and invoke *infinity plus one*.

That's the genius of Aristotle's conceptual innovation: He acknowledged that a *potential* infinity was within human reach—just not an *actual* one. Mathematically, his conception was influential. For thousands of years after Aristotle, the most anyone ever said about infinity was that it was infinite. We could imagine infinity but never actually achieve it. Post-Aristotle, infinity was always idealized, never realized—a philosophical construct at best. As something that could never be reached, infinity could never be treated as a proper mathematical object, most believed. Through the millennia, some of the top philosophical and mathematical minds of their day turned their attention to the concept, pondered it at length, and inevitably gave it up.

"The infinite, whether the infinitely large or the infinitely small," is troublesome, University of North Carolina professor J. W. Lasley Jr. will later say in 1942. More than that, he adds, it "seems to carry disaster in its wake."

Cantor will come to understand that disaster better than anyone. But he is undeterred. He wrestles with Aristotle's conception and fully tackles it. He even has a name for the great philosopher's distinction: He calls the idea of an idealized, abstract infinity an "improper infinity." What Cantor is interested in are *proper* infinities—completed infinities—infinity not just as a concept but as an actual object that can be treated mathematically. That's the revolution he launches with set theory: infinity as a bird in the hand, not a bogey in the bush.

But as I said, Cantor tips toward the Pyrrhic. His work is outstanding, but some of his ideas are way too taboo for many mathematicians. Completing an infinity implies somehow mastering an impossibly long iterative process, like counting every whole number all the way to infinity, including all the really, really large numbers we can't imagine and don't even have names for. Count to a googolplex of googols multiplied by many more googols. It can't possibly be done, many mathematicians think, so best to leave infinity hiding in the bush.

"I realize that in this undertaking I place myself in a certain opposition to views widely held concerning the mathematical infinite," Cantor says. But that doesn't bother him in the least because of one other thing: He firmly believes he's on a mission from God. For him, the bird lies not just in the bush but in a burning bush.

ACHILLES AND THE TURTLE

Blame the ancient Greek philosopher Zeno of Elea for first exposing the hazards of infinite thinking before even Aristotle. His famously fickle paradoxes are still taught today in college philosophy courses, the bane of students everywhere struggling to make sense of stupid turtles winning races and deadly arrows hung in midair.

Zeno's so-called Dichotomy Paradox claimed motion cannot exist. Why? Because for an object to be moved over a given distance, it must first arrive at an infinite series of midpoints. Hitting a target was impossible, then, because in order to reach it, the arrow had to traverse half the distance—half the distance to the midpoint, half the distance to the quartile, halfway to the eight, and half again and half again until the end of time. Given an infinite number of such midpoints, you theoretically never actually reach your destiny—or so said fickle Zeno.

Zeno also picked on poor Achilles, that hero of the Trojan War and mightiest warrior in the ancient world. Fearless. Proud. Unrivaled. Unbeaten. Unbowed. But Zeno turned Achilles the hero into Achilles the heel—casting him as the loser of a footrace against a slow, dawdling turtle. At the start of the race, Zeno said, the old reptile has a sporting head start. Achilles is fast, there's no doubt. But to catch the turtle, he first has to close half the distance, then half again, and half again, and half again—until either he gives up in disgust or the turtle dies of boredom. At best, Achilles never finishes. At worst, he loses the race.

The real winner, of course, is Zeno. His spiteful paradox was really just a clever attack on the Pythagorean school of thought, which embraced a primitive concept of actual, completed infinities. As Zeno was strongly opposed to completed infinities, he rained piss on Pythagoras's head from a very great (though not infinite) height. In doing so, he warned hundreds of generations of mathematicians and philosophers that the very concept of infinity was fraught. Beware! Approach at your own peril! And his warning worked.

After Zeno, Aristotle salvaged infinity to some extent by setting up his distinction between actual and completed infinities and imaginary and potential infinities. The cleverness of Aristotle's treatment was that it resolved Zeno's paradox by avoiding it. He essentially punted, kicking infinity downfield into the *things-that-are-interesting-but-you-never-have-to-deal-with* realm. All infinities are potential, he claimed, existing only in the mind. And since we can only conceive—but never actually complete—a process that has no end, like counting to infinity, why worry about it?

That more or less settled things. For two thousand years after Aristotle, it was absolute infinity, sashay away! Potential infinity, shan't you stay.

Although actual infinities came up again and again through the centuries, people generally accepted Aristotle's potential-and-only-potential framing. Christian scholars in particular adored Aristotle because his framing implied that actual infinities, by which one naturally means a divine and absolute God, lay far outside the mortal realm. St. Thomas Aquinas agreed that actual, unchanging, incomprehensible, completed, godlike infinities cannot be grasped by mere humans. Doing so would challenge the absolute, infinite nature of the divine. Only God knows the mind of God and the scope of the absolute, Aquinas said.

Galileo Galilei considered all this and came to more or less the same conclusion. He also opposed completed infinities. He laid out a strong criticism on mathematical grounds. If completed infinities were allowed, he said, then two infinite sets would be equal. That means the collection of just even integers would be the same size as the collection of all whole numbers—odd and even integers combined—even though common sense says the latter set should be twice as large. (Galileo didn't use the word "set" because it was 250 years too soon for that.)

That's ridiculous, Galileo said. How could all the evens and odds combined be the same size infinite set as only the evens? So it seemed to him like QED—*quod erat demonstrandum* (what was to be shown)—all over again. And he was not alone. Other philosophers like John Locke, René Descartes, and Baruch Spinoza all came to the same conclusion as Galileo and Aristotle, holding fast to the ancient creed *infinitum actu non datur* (the actual infinite doesn't exist).

Thus, infinity had acquired an awful reputation by the time Cantor took it up. And for some, he only seems to make it worse.

To understand infinity as Cantor sees it, do this thought experiment: Imagine turning out your pockets and emptying everything in them into a gigantic bucket. While you're at it, throw your clothes in there as well. And your shoes—everything you've got on your person. Now imagine throwing into this bucket every object you touched today, from your coffee mug to your bedside lamp. But don't stop there. Throw in everything you even looked at today. Add every single thing you thought about, whether real or imagined, from cool dragons to hot doughnuts.

How large would your bucket be? What sort of space would it fill? A stadium? A city? An ocean? The entire planet? What if you filled a bucket like this every single day for your entire life? How large would that collection of thirty thousand or so buckets filled every day of a long average life be? What would all those buckets look like together? If you stacked them one on top of the other, would they reach past the Oort cloud? Even if they did, what would your selfish little bucket stack be compared to the entirety of the universe? A drop in the ocean? A heartbeat in a hurricane?

This is where Cantor's genius shines through. Instead of simply throwing up his hands in despair, he goes about reconciling infinity with his abstract concept of a set. The collection of all those things you touched in a day is but one example of a set—a strictly defined group of sometimes thinly related things.

A set, that bucket, is defined by what's inside—its elements—and sometimes some of those things are themselves collections. Think of the set that contains the days of your life—all your thirty thousand or so daily buckets. Or think of an army. An army as set would contain separate groups of privates, corporals, sergeants, lieutenants, captains, majors, colonels, and top brass. There could be groups like civilian contractors and civil servants seconded from other branches of government. And there may be some individuals in your army as well—a four-star general or the commander in chief.

What if you have an infinite army, however—an infinite number of soldiers falling into an infinite number of ranks? It's not easy to consider. No such thing could ever exist in the world. Our minds and our outlooks are grounded in the finite. We do not live in the infinite. Nor do we ever touch it, which was sort of Aristotle's point. But that's exactly what set theory allows you to do: contemplate the scale of things beyond all human scales. And then, using mathematics, to actually touch them.

Cantor began to appreciate this in the 1870s, and after 1883, many others began to appreciate it as well. Cantor's set theory flips the script in a sense. He shows that infinite sets, including the collection of all whole numbers, are completed infinities. They are actual infinities. You can touch them, in other words, even if they're not countable in the traditional sense. And you can be touched by them.

GRUNDLAGEN EINER ALLGEMEINEN

Because Cantor puts actual infinity within reach, this has strange consequences and starts huge controversies. Why? For one thing, he rigorously proves that there's more than one type of infinity, some larger than others. And he works out a way to compare them.

The smaller infinite sets, according to his theory, are those equivalent to the collection of all the whole numbers—$\{1, 2, 3, 4, 5, 6, 7, 8, \ldots\}$. Cantor proves exactly what Galileo fears: You can have a subset of whole numbers that is equal in its infinite size to its parent set. So the set of all even numbers is exactly the same size as the set of evens and odds combined. What once made Galileo groan now makes Cantor grin.

But this astoundingly nonintuitive result seems unreasonable to many, and for several years leading up to 1883, Cantor's ideas are routinely criticized. Why? Think of it from Galileo's point of view: Imagine taking every possible whole number and placing it in a bucket—a single, completed act. Now imagine doing the same thing with a second bucket. Put just some of the numbers from the first bucket into that second bucket, say all the primes (numbers larger than one divisible only by one and themselves: {2, 3, 5, 7, 11, 13, 17, 19, 23,...}. Or take all odd numbers: {1, 3, 5, 7, 9, 11...}. Or all evens: {2, 4, 6, 8, 10, 12...}. Or consider just Fibonacci numbers—the set of numbers invented to describe how rabbits reproduce where each number in the series is the sum of the previous two numbers: {1, 1, 2, 3, 5, 8, 13, 21,...}. Put all the sets aside in separate buckets and compare them. Which bucket is heaviest? Common sense would say the first. It has *all* the numbers, so it *must* be heavier. Galileo desperately wants that to be true.

This makes sense in an everyday, finite way. If you have a bucket filled with, say, all the hammers in Home Depot, and then you compare that to a subset like all the hammers in Home Depot *that are pink* (a small subset indeed), common sense would say the whole bucket far outweighs the pink one. And it does. But that's a mistake based on limited, real-world, finite thinking and is actually incorrect when applied to infinite sets.

Infinite sets are not like finite sets at all. They're infinite. And in infinity, Achilles catches the turtle. Why? When you compare two completed infinite collections of numbers, one being a subset of the other, they are both equally infinite. In fact, the very definition of a completed, countably infinite or "denumerable" set, as Cantor calls it, is specifically that its members can be matched in correspondence one-to-one with the whole numbers. Cantor not only argues that one infinite subset is exactly equivalent to its infinite parent set—he actually proves it. WTF, Cantor?

This is where things get truly interesting, because Cantor doesn't stop there. He also shows that not all infinities *are* equal. There is a larger type of infinity: the set of all decimals or "real numbers." He says the set of real numbers is uncountable, or "nondenumerable."

Think of the number line. There is an infinite number of whole numbers on the line, but there is also an infinite number of decimal expansions you can do between any two given numbers: {1, 1.000001, 1.00001, 1.0001, 1.001, 1.01. 1.1...}. Did I say this is where things get interesting? Really it's where they get truly bizarre. Cantor has already challenged what seems like common sense with infinite sets of whole numbers. Next he will violate one of the most basic tenets of geometry held since the dawn of mathematics—and basically challenge reality itself.

"JE LE VOIS, MAIS JE NE LE CROIS PAS!"

By the time of the ancient Egyptians, mathematics was naturally defined by geometry, concerning itself primarily with shapes and space. People always assumed a basic, intuitive, and what some call "naive" if plainly and completely obvious concept of 3D space. Then came the scientific revolution and the innovations of Pierre de Fermat and especially Descartes, whose development of the Cartesian coordinates enabled the algebraic representation of points in geometric space. Then the early nineteenth century saw the explosion of the sometimes-regarded-as-beautiful results of "number theory." Then along comes Cantor. With set theory, he espouses a whole new way of thinking about both space and numbers.

The naive view of 3D space prior to Cantor assumes that just as there are an infinite number of points on a line, as you expand into more dimensions, the same is also true. A plane is made up of an infinite number of lines, and 3D space is filled with an infinite number of such planes. So which has more points in it: a line, a plane, or 3D space? The answer seems intuitively obvious. All the infinite points on a single line are nothing compared to the vastness of an entire plane made up of an infinite number of lines. And any single plane falls flat when compared to the infinite stack of planes that would constitute 3D space. It's almost too stupid to have to even ask the question *Which has more points?* The line is greater than the point. The plane is greater than the line. And 3D space is greater than the single, sorry, petty plane. That's right, isn't it?

Not with Cantor. And not with set theory.

By the late 1870s, Cantor had reached the strange result that the size of an infinite set of real numbers is *independent of dimension*. This meant that the infinite collection of points in 3D space is exactly equivalent to the infinity of points in a single 2D plane, which is the same again as all the points captured by a single line in that plane.

Again, this makes no sense to the finite mind. How in the world could one line contain as many points as an entire plane? But they do, Cantor says—and he shows it. The line and the plane are the same in that they are equally uncountable. Real number sets are all equally big infinities regardless of whether they're arrayed in a line, scattered on a plane, or spewed across the vastness of 3D space. And Cantor demonstrates that the set of whole numbers cannot be aligned in one-to-one correspondence with the real numbers. Big infinity is not countable. It's nondenumerable. And it's vastly different than denumerable, countable small infinity.

When Cantor first discovers this, he declares, "*Je le vois, mais je ne le crois pas!*" (I see it, but I don't believe it). His reviewers don't believe it either—nor are they kind in their contempt. People object to his work because it freaks them out. Some think Cantor has gone too far. But he doesn't stop there.

———◆———

People have always held that the whole is greater than the part—so much so that we often even say the whole is greater than even the sum of its parts—like somehow the act of becoming whole adds information. But set theory refutes much of what we find self-evident. It says the whole is not greater than the part. Sometimes the whole and the part are identical in size. Other times the part is larger than the whole.

His haters believe Cantor should be ashamed of what he's done. But he's not. If anything, he's cocksure. He shouts his discoveries from the rooftops. Cantor is a dreamer of big dreams, and he's plagued by troublesome things—*infinity!* He knows his work will transform the world.

The world, however, has different plans.

Things are never easy for Cantor after 1883. His methods are strange. His results are weird. He strikes some of his colleagues as odd and presumptuous. And his mental state is erratic. Even as his ideas explode on the European math scene, they are awkwardly received. As science historian Loren Graham and mathematician Jean-Michel Kantor will later say, "They were not welcomed." Many mathematicians reject set theory. By the time Cantor finishes his book in 1883, he has to spend as much time defending his ideas as presenting them. And that leaves a strange taste in his mouth. He soon begins a long, strange, Pinocchio-like journey of madness and misfortune.

Maybe he is Pinocchio. Or maybe he's more of a Prometheus. He steals the fire of the gods for humanity, and like the hero of Aeschylus's 2,500-year-old play, he pays the same sad price that any would-be savior or warm, selfless giver in a cold, selfish world eventually does.

Prometheus's lesson is that there's an evil, eagle-pecking price to be paid for even the greatest charity of genius. Hello, wisdom—goodbye, liver. Tragedy is a lesson Cantor is forced to learn well: *No good deed goes unpunished!*

CHAPTER THREE
INFINITY LOVES COMPANY

1884–1899

"
Nothing capable of proof ought to be
accepted without proof.
"

—Richard Dedekind

Prometheus bound to a rock, his liver eaten by an eagle and his torch dropped from his hand. Cropped from a public domain image of unidentified author etching via Wellcome Collection.

People love rejecting set theory. Many are confused, if not outraged, by Cantor's methods. Others question his results. Some just love to hate. Or they loathe his theories altogether—not least among them the great French mathematician Henri Poincaré, who calls set theory "a disease," one he hopes his field will someday be cured of.

Another famous French mathematician named Charles Hermite lobbies against even translating Cantor's work into French. Better it remain unknown in some foreign tongue than poison the eyes and ears of any respectable French mathematician. Instead of Cantor's *I-see-it-but-don't-believe-it* leap of faith, Hermite preaches pure resistance. His motto is more like *"Je ne le vois pas, et je ne le crois pas"* (I don't see it, and I don't believe it).

Here, there are probably geopolitical tensions at play as well. In the late nineteenth century, even for a far-from-the-toxic-bloom-of-politics field like mathematics, the French are French, the Germans are German, and rarely do the two kumbaya. The true birth of scientific internationalism doesn't start until the end of the nineteenth century, thanks in no small part to Cantor's own international efforts. But even then, withering suspicions on both sides of that long, awkward Franco-Germanic border persist well into the mid-twentieth century—especially after World War I.

Ironically, it's not the French but one of his countrymen who gives Cantor the most trouble—German mathematician Leopold Kronecker, whom Cantor knows from his student days in Berlin. The parallels between the two men are remarkable. They both study in Berlin (though Kronecker a generation before Cantor). They both finish their PhDs when they are twenty-two. They both marry in their late twenties. They have six children each and both pursue their mathematical research while raising their kids. Those similarities could have been enough for Kronecker to be a sort of father figure for Cantor, but they are too much at odds intellectually for that to ever come true. Kronecker could never have ever been a father figure except in the worst possible sense: if you remove any blush of familial love and think solely of a mean father in his most tough-love tiger aspect. Unfulfilled expectations. Quiet disappointment. Angry disapproval.

INFINITY LOVES COMPANY

Kronecker's fatherly terroir has a strong nose of sour disposition, a bouquet of dour daddy doubt, and astringent notes of toxic hate. Kronecker won't look on Cantor with friendly respect, let alone a mentor's pride. Their worldviews are in direct opposition. Kronecker is the foremost "finitist" of his day, a man who basically hates infinity. His preference is for a throwback approach to the subject. He thinks all of mathematics should be based on whole numbers—the only numbers worth talking about. Counting. Basic arithmetic. Simple algebra. In his opinion, infinity is a subject worthy of rejection, if not outright derision and scorn. He admits no infinity in step, size, or collection, and he is sometimes called "the doubter" because of this.

Kronecker the doubter rejects any concepts in math whose existence can't be demonstrated by straightforward construction. That puts him at odds with many in his generation and causes him to stake out strange positions. He rejects irrational numbers, for instance—those nonrepeating, nonterminating decimal numbers like π and $\sqrt{2}$ that the Pythagoreans discovered thousands of years ago. Unlike so-called rational numbers, which can be formed by taking a ratio of two whole numbers, irrational numbers have infinite decimal expansions and cannot be reduced to a simple ratio. Kronecker doesn't like that idea at all. He irrationally rejects irrational numbers and declares that only natural numbers should be the objects of analysis, And if he feels that way about irrational numbers, you can be doubly sure that he hates Cantor's set theory in all its ugly glory.

"God made the integers," Kronecker famously says. "All else is the work of man."

GOD MADE THE INTEGERS

In developing set theory, Cantor invents the concept of transfinite numbers to quantify the size of infinite sets and the elements they contain. He calls them "transfinite" numbers because they are unusual, capturing infinity and larger than a finite number yet still written into algebraic equations and treated as numbers.

He defines the transfinite cardinal $\aleph_0$ (aleph-naught), for instance, to denote the size of the "denumerable," countable little infinity of whole numbers. And he defines the transfinite cardinal $\aleph_1$ (aleph-one) to denote the size of the "nondenumerable," uncountable big infinity of real numbers. These two transfinite numbers are related in equation form through an equivalence Cantor develops as part of his continuum hypothesis, reflected in this simple equation:

$$\aleph_1 = 2^{\aleph_0}$$

The continuum hypothesis captures Cantor's attention and becomes his chief conviction from the late 1870s until the end of his life during World War I. All infinite sets, he is convinced, come in two and only two sizes: aleph-naught and aleph-one. Little infinity and big infinity. Denumerable and nondenumerable. Infinity of the size of whole numbers or the real numbers. The continuum hypothesis says all infinite sets fall into one bucket or the other and never in between the two.

Proving the continuum hypothesis will become a major hobgoblin to the minds of mathematicians after Cantor. Hilbert, in fact, in delivering his famous twenty-three problems of the future lecture in Paris in 1900, will consider proving the continuum hypothesis to be so important that he makes it the first problem on his list.

But long before Hilbert takes the stage in Paris, Kronecker is not buying it. He roundly rejects infinite sets as well as transfinite cardinals. Numbers that are not numbers are no numbers, he says, along with his favorite *God-made-integers-all-else-is-the-work-of-man* mockery. (Another of his favorite derisive maxims is "Interesting, but not mathematics.") This presents a problem for Cantor because none of these insults are lobbed in a vacuum.

Kronecker's opinion matters a lot because he is an influential mathematician and the editor of *Journal für die reine und angewandte Mathematik* (*Journal for Pure and Applied Mathematics*)—sometimes just called *Crelle's Journal* after its founder, August Leopold Crelle. It's the oldest math publication in Europe, and in the late 1800s, when Cantor is first publishing, it is also the most prestigious. In his powerful editor's role, Kronecker manages to hold up publication of one of Cantor's key papers in the 1870s. This work demonstrates the dimensional invariance of real numbers—the idea that the infinite number of points on a line are "equinumerous," or the same size set, and equivalently infinite with all the points on an infinite plane.

Cantor's results are shocking and astounding, to be sure, but it's a bridge too far for Kronecker. He simply can't accept it. Nor will he allow his precious journal to accept it. So he does everything in his power to kill the paper. It languishes unpublished for months. No explanation. Cantor starts blaming Kronecker and begins forming the opinion that his older colleague is deliberately thwarting his efforts. Insulting him behind his back. Wielding undue influence over his career. Kronecker shows him a finger in the eye if not a knife in the chest. And Cantor is not wrong. Kronecker calls him a renegade, a charlatan, and a "corrupter of youth." To anyone who will listen. He unabashedly tells Cantor to his face that he has been badmouthing him, telling him how he has dismissed set theory while huddling with the mathematician Hermite, one of Cantor's French critics. "Humbug" is what Kronecker calls set theory. *Humbug!* It may seem a mild if not half-charming Dickensism today, but I get how it must have been a serious, *them's-fightin'-words* insult in the 1800s.

Ultimately Cantor withdrew the paper on the dimensional invariance of real numbers in 1878, published it elsewhere, and vowed never to send anything to *Crelle's Journal* again. To hell with them, and to hell with Kronecker! He also broke ties with Kronecker, professionally and personally, at least for a time. His father's warning, that letter, had finally come true, at least in part: Beware the jealous lies "of open or secret enemies." *Beware!*

Still, Cantor is overly sure of himself—on a mission from God and all that. Regardless of Kronecker's insults and machinations, set theory is secure. Cantor knows this. It will stand up to any criticism, whether mathematical or philosophical, Kronecker or no Kronecker. There's only one desperate problem remaining: proving the continuum hypothesis.

THE GREAT MATH WAR

In the fall of 1883, Cantor gives a lecture describing his mathematical journey as one that "strives to be carried out in the imagination or in dreams." And at this point in his career, what he dreams about most is solving the continuum hypothesis, the last little bit of unresolved dust to settle.

Expressed in a gorgeous-looking if somewhat impenetrable equation, the continuum hypothesis basically insists on the *one-or-the-otherism* of infinite sets. It says infinities come in one of two sizes: small infinity, $\aleph_0$, the countable, denumerable, infinite set of whole numbers and its equivalents (evens, odds, primes, Fibonacci numbers, etc.)—and large infinity, $\aleph_1$, the uncountably infinite, nondenumerable set of real numbers. The aleph-one real numbers have the power of the continuum in Cantor's philosophy.

The *one-or-the-otherism* of the continuum hypothesis means that all infinite sets fall into one bucket or the other: $\aleph_0$ or $\aleph_1$. No infinite set exists that has more elements than the whole numbers but fewer than the reals. Cantor first articulates this concept in 1878 but hits a wall trying to prove it. Over the next few years, he takes on an air of desperation and tries repeatedly to prove the hypothesis, failing every time. Again and again, he thinks he has the answer, and again and again, he discovers he's wrong. And so it goes. He tries. He fails. Another attempt. Another failure. One day, he writes a letter announcing a proof. The next day, he retracts it—on and on in an endless, painful cycle. Wash. Rinse. Repeat. Announce. Retract. Lament. *Torture!*

Kronecker, meanwhile, quickly becomes the reigning queen-bee troublemaker in Cantor's life. At the end of 1883, he writes to Mittag-Leffler, their mutual friend (all three were at the University of Berlin at the same time). In Sweden, Mittag-Leffler has just started the world's first international math journal, the *Acta Mathematica*—the same one that enlists Sofya Kovalevskaya to be the first female editor in the history of scientific publishing. The inspiration for this journal, according to American mathematician Einar Hille in 1953, is healing the rift between French and German mathematicians, still so strained after the Franco-Prussian War.

What Kronecker offers Mittag-Leffler for his journal in 1883 is something like a juicy op-ed. Editorial content explaining his perspective on set theory. Would Mittag-Leffler be interested? Kronecker promises a paper that will show how Cantor's whole body of work is of "no real significance."

INFINITY LOVES COMPANY

The thing is, Mittag-Leffler is good friends with Cantor. They shared the same adviser in graduate school. Of course he tells Cantor. Of course Kronecker knew he would tell Cantor. And of course Cantor rages, recoiling in anger and paranoia. First Kronecker freezes him out of *Crelle's Journal*, and now this! Kronecker's paper never materializes, but the whole exchange stokes Cantor's desperate "anger-noia." And his response? He redoubles his efforts to solve the continuum hypothesis once and for all.

WASH, RINSE, REPEAT

Early in 1884, Cantor thinks he's found proof. Again, it turns out he's wrong. He begins to think he's made no progress at all, and that's when his troubles start. He has a mental breakdown of some seriousness in the spring of 1884, causing him to be institutionalized for a bit. The episode lasts a month, wrecks his nerves, concerns his wife, and terrifies his kids.

Historians disagree about the exact cause of his breakdown. It may have been a severe episode of clinical depression—the mechanisms of which are poorly understood even today. Some attribute it to his father, their relationship having left him psychologically damaged. Others put it down to his troubles with Kronecker. Others say it could just be the simple stress. A hard professorial career. Busy home life. The frustration of failing to prove the continuum hypothesis.

Some experts cynically state that the resistance to set theory in his lifetime and his subsequent mental health problems (assuming the former was indeed the cause of the latter) were an ultimate good for humanity. Cantor's troubles forced him to work hard for many years and produce an amazing body of work—this argument claims—even if they ruined his mental health in the process. One of his biographers, Joseph Warren Dauben, will go so far as to argue that his psychological woes pushed his productive efforts, fueling his advances during manic phases. "Far from playing an entirely negative role," Dauben writes one hundred years later, "they may have driven the energy and singlemindedness with which he promoted his theory."

For me, that's a cruel, coldly utilitarian outlook—akin to making excuses for the mean bully by saying his awful taunts make the person he

THE GREAT MATH WAR

bullies stronger. Would we have set theory without Kronecker? Yes. Would we have set theory without Cantor's mental health problems? Yes, I think we would. The corrupt narrowness of utilitarianism often destroys lives and makes the world worse, despite the hopeful lip service of its *ends-justify-the-means* nonsense. God save us from good intentions, as they say—those cobblestones on the road to hell. Even the best intentions should not excuse the many awful things in life.

In addition, there's something much darker than utilitarianism at play. Even as he wildly succeeds in mathematics, Cantor is deeply haunted by a tragic sense of failure. He comes very far very fast but along the way is forced to defend his work against many rabid critics. Worse, his massive personal and professional traumas cause him so much mental anguish that he has to be committed to professional care in institutions time and time again.

That all has a pernicious effect on Cantor's loving family, which is the main reason why all that *for-the-good-of-mathematics* whitewashing of his mental problems is so troubling. History often treats people who live with and around its central figure as fringe characters. Window dressing to be ignored except inasmuch as they reveal something about the subject. But all people throughout history are real, with private lives and personal feelings. This is no different in Cantor's case. His family loves him, and that informs a key part of his tragedy.

Whenever someone suffers a mental breakdown, it can be deeply scarring for their family and friends, who must watch in horror as the person they know is transformed, sometimes into a complete stranger. A neurologist I once interviewed in San Francisco told me mental illnesses often strike as multiple separate diseases—a primary disease that physically afflicts the sufferer and a secondary disease that psychologically impacts their loved ones.

------♦------

Cantor is far from done fighting, however. After he is institutionalized the first time for one month in the spring of 1884, he is determined to set things right with Kronecker. He decides Kronecker's main objections are not personal but rather philosophical—concerned solely with some minor aspect of his work. So Cantor writes Kronecker.

"I am of the opinion that the greatest part of what I have done scientifically in the last few years," he says, "is not so very much opposed to the demands which you place upon 'concrete' mathematics as you seem to believe." It's a matter of presentation rather than substance, he says.

Kronecker responds with a short, polite note. He recalls earlier times—happier times. Those golden days when Cantor was a bright-eyed, stubble-chinned grad student learning at Kronecker's feet. That's the Cantor he recalls. That's the Cantor he loves.

Cantor takes that bit of nostalgic reckoning as an olive branch, Kronecker's attempt, however narcissistic, to meet him halfway. In the darkness of his mental recovery, it probably seems a beam of sunshine. A healing light. Laser love between techno-enemies. Could they, would they, now be able to put the past behind them?

Sadly, no.

As their letters cross, Cantor is once more pushing even harder to prove the continuum hypothesis. Again, his effort ends in miserable failure. He gives up for a few days, but then he has a burst of renewed energy, returns to the problem again, thinks he has the answer, and sends an excited letter to Mittag-Leffler. He is ecstatic, saying he's finally succeeded at finding a proof—and a simple one at that. *Extraordinarily simple!*

"When I've put it all in order, I will send you the details," he writes. But by the time his letter arrives in Sweden, he's already sending another message to say, *Never mind!* It's not a valid proof but yet another false rabbit hole. Still, he is undeterred. So he dives right back in. But then, a few weeks after that, he realizes he's been chasing yet another scarecrow. Nothing to share with Mittag-Leffler except the ashy taste of bitter disappointment.

Finally, in early 1885, he arrives at another false milestone and dashes off two articles for Mittag-Leffler's journal. But the reception from his friend is far from what he's expecting. Mittag-Leffler hates the papers. Too philosophical, he says—and not in a good way. Cantor needs to remember and heed the lessons of Carl Friedrich Gauss, Mittag-Leffler says, that great German mathematician from an earlier generation whom every German mathematician reveres. Gauss discovered non-Euclidean geometry in the early 1800s, but he kept it secret and locked away his work in a drawer for decades, never even attempting to publish it. It was only after Gauss died, and only because he was so famous, that his work on non-Euclidean geometry was discovered in a junk drawer of his desk. Gauss kept it close because he *knew* his work was way too far ahead of its day to publish while he was alive.

Remember Gauss, Mittag-Leffler warns Cantor. Gauss knew it was too soon for non-Euclidean geometry then. And it's too soon for Cantor's work on the continuum now. Publishing these papers now would be impetuous, Mittag-Leffler says, and "will greatly damage your reputation among mathematicians."

"I know very well that basically this is all the same to you, but it's too soon," Mittag-Leffler writes. A century too soon! Nobody will understand it now. "If your theory is once discredited in this way, it will be a long time before it will again command the attention of the mathematical world."

All that hurts Cantor. Betrayal is the bitterest form of insult. Cantor suddenly thinks Mittag-Leffler, the one mathematician whom he assumes has his back, is also abandoning him. *A century too soon*, Cantor mocks, writing those words by way of complaint to another mathematician in a letter. "I should have to wait until the year 1984."

L'INTELLECT DIVINO

True to character, Cantor is burned by the whole exchange with Mittag-Leffler and vows never to publish in his journal again. But that exchange is the least of his worries. With his mental health already shaky and his ego now deeply bruised, he turns away from math—ironically, since people are noticing his work, and set theory is finally taking off. Countless mathematicians in England, France, Germany, Italy, and elsewhere are exploring his work. But perhaps he's wise to distance himself. Others are still steadily denouncing him.

Some, like Poincaré in France, say a lot of his ideas should be thrown out. Others struggle to be enthusiastic. Or they pick and choose what they like. In a major review of Cantor's 1883 *Grundlagen*, German mathematician Max Simon dismisses Cantor's philosophical discussions. The first translation of the work into French omits the philosophical stuff altogether—another insult. "By the end of 1885," his biographer Dauben writes, "Cantor was in many respects a disillusioned man."

Sofya Kovalevskaya attends a lecture Cantor gives around that time about the philosophy of Leibniz. She writes to Mittag-Leffler that it was a disaster. "In the beginning he had 25 students, but then little by little, melted [away] first to four, then to three, then to two, finally to one single one," she says. "Cantor held out nevertheless and continued to lecture. But Alas! One fine day came the last of the Mohicans, somewhat troubled, and thanked the professor very much but explained that he had so many other things to do that he could not manage to follow the professor's lectures." Cantor promised never to lecture on philosophy again, she claims.

But if Cantor is disappointed with some of the reactions to his ideas, he is never disillusioned with the ideas themselves. Cantor never wavers in his embrace of set theory or its results, however strange they seem. If anything, after this, he turns more deeply to the philosophical and even religious implications of set theory.

Cantor begins to come off as a crackpot at times, though some experts seek to explain it away by claiming that his embrace of philosophy and religion is some sort of subterfuge—simply a way of dodging critics and silencing his inner demons. That's probably also true. But his religiosity also echoes the tragedy of his own inner struggles. (To paraphrase spy novelist John le Carré, fanatics always conceal a secret doubt.)

Cantor has always held firm to the belief that God is acting through him. He isn't driven to develop set theory because he's uncertain about it. Set theory is not a meek, defensive invention. He sees his life in heroic, even prophetic terms. He believes he's divinely inspired—not simply as the genius creator of a new field of math but as God's vessel for pouring forth the infinite secrets of the divine into the world. Cantor is certain of math's truth even when he can't believe what he sees. He has a main line to the absolute. His ideas come from *l'intellect divino* (the divine intellect)—and they are nothing short of great.

After 1885, that conviction becomes a dark equivocation into which he leans more and more while limping through failures and reeling from criticism. *Jealousy and slander of open or secret enemies!* Never is he content to sit with his head in the clouds, his nose in a book, and his elbows on the table. He *will* move forward. Alone if need be. Filled with bitter poison if he must. Paranoid, obviously. But always convinced he'll be completely exonerated eventually. He thinks mathematicians will one day embrace set theory as a "perfectly consistent, rigorous, and acceptable part of mathematics," one of his biographers writes.

"My theory stands as firm as a rock," Cantor writes a few years later. His mental health, not so much.

―――◆―――

There are three types of personal losses in life. There's the natural, trivial *force mineure* type of loss. Those are the insults we feel every day—all the minor inconveniences of life that fall beyond our control. A flat tire. A spoiled jug of milk. Stepping barefoot on a Lego block in the middle of the night.

Then there are the massive, hard, and terribly traumatic *force majeure* type of losses—and again these are things we can't control much, but they are far, far worse. The death of a parent, for instance. My own father died a few days after I started writing this chapter on Cantor, and there was too much loss-of-father sadness in Cantor's own background for me to continue. It was more than a year before I could finish the chapter. The loss of a loved one is major. Even now I find this sentence difficult to write.

All *force majeure* losses, however devastating they may be, are also authorless in a sense—in the case of natural death, delivered by the hand of God, genetics, or just plain bad luck. Coping often starts with an admission of powerlessness. Crippling tragedies. External causes. The authorless hand of the universe. Opaque divine will. Uncontrollable fortune. The mysteries of God's plan. Some sad, simple twist of fate.

This is true of Cantor. Convinced he's discovering unknown and hidden truths and that God almighty is guiding his hand, Cantor claims at one point that he's been "logically forced" to discover set theory. "Almost against my will," he says. Even though it troubles him for the rest of his life, somehow removing his own agency helps. It makes dealing with the troubles of the backlash against set theory and his inability to prove the continuum hypothesis that much easier.

But then there's the third type of loss: the unnatural loss. The despicable loss. The far more personal loss. Whether major or minor, these are the types of losses where things are taken away from us not by some twist of blind fate or act of divine judgment but through the deliberate, conniving, selfish action of another human actor: a slashed tire. A stolen wallet. A friend murdered in cold blood. Medical malpractice. These are the unnatural losses. And we move past them far less easily because we always suffer in two ways. There's the loss itself. And then there's our overwhelming sense of injustice. Of being wronged. Of knowing that a deliberate act underlies our insult—causation that adds insult to injury. It fouls our grief with anger.

Stricken with such thoughts, Cantor's mind becomes fouled in his later years. He develops what I call an Isaac Newton complex: a loathsome and pathological hatred for publishing that is informed by the paranoid certainty that your contemporaries are out to sabotage you. Either they are a bunch of backstabbing haters ignorant of your work or, far worse, they are jealous of your genius and selfishly despise you because of it. (Newton himself swore off publishing for years because of criticism of his early work on optics.)

"My own inclinations do not urge me to publish," Cantor writes in 1887, echoing Newton from two centuries before. "And I gladly leave this activity to others."

———◆———

For the next five years, Cantor is increasingly focused on new audiences. He tries to make inroads with Catholic authorities. The 1880s are a time when the Catholic church is becoming more interested in scientific discovery than ever before. The new Pope Leo XIII takes a special interest in science, especially cosmology. Science is a way forward, he claims, and he maintains an astronomical observatory at the Vatican—one whose construction he personally oversees. He fills it with the best modern equipment and keeps professional astronomers on staff.

THE GREAT MATH WAR

Cantor thinks the church has a lot to offer and that set theory has a lot to offer in return. He wants the Catholic church to become aware of his views because set theory is a way to understand the infinite nature of the divine—perhaps even the mind of God, reflected in math. *Isn't that worth considering?*

It's a hard sell.

Cantor shares his work with Cardinal Johannes Franzelin of the Vatican Council, one of the leading Jesuit theologians of his day. Franzelin writes Cantor a letter on Christmas Day 1885, saying he's gratified to receive Cantor's work. "What greatly pleases me," he says, is that it "appears to take not a hostile, but indeed a favorable position with regard to Christianity and Catholic principles." Having said that, Franzelin adds, Cantor's ideas probably could not be defended and "in a certain sense, although the author does not appear to intend it, would contain the error of pantheism."

Cantor responds in a letter just after New Year's Day 1886, assuring Franzelin that set theory is safe for Catholicism. There are only two infinities, he says—one for God and one for humans. The *Infinitum aeternum increatum sive Absolutum* (eternal, uncreated, or absolute infinity) is inaccessible to humans and reserved for the divine. Then the entirely distinct *Infinitum creatum sive Transfinitum* (the created or transfinite infinite) is open to mere mortals.

The cardinal responds politely a few days later, ignoring the bit about the two infinities, addressing one minor point at length, thanking Cantor for his thoughtful letter, and acknowledging that, as far as he can see, "no danger for religious truths lies in your concept of the Transfinite." Still, Franzelin adds, he's very busy. Please don't write to me again, he says.

INFINITUM AETERNUM INCREATUM

Cantor makes other attempts to interest members of the church. He writes to a Catholic priest named Ignatius Jeiler to convince him of the need for the church to consider his ideas. He contacts a Dominican priest in Rome who is carefully studying Cantor's book *Grundlagen* and trying to tease out its implications. "From me, Christian philosophy will be offered for the first time the true theory of the infinite," Cantor tells him.

His overtures to Catholic authorities would normally be strange behavior for a mathematician—especially one who is not Catholic—but they are the least weird part of his late-life weirdness. He also tries to prove that the English philosopher Francis Bacon is secretly the author of all of Shakespeare's plays—an elitist trope that's been around almost as long as the great bard himself. There are lots of alt-Shakespeare theories in existence—even today. The Earl of Oxford wrote his plays. Christopher Marlowe wrote them. The Earl of Derby was really the author. It's an arrogant thesis informed by the irrational conviction that no low-born artist could ever write so well.

Even as widening cracks are appearing in Cantor's psyche, he has his champions. David Hilbert in particular loves Cantor's set theory because it lends itself to something known as the pure existence proof, an invaluable tool for mathematics developed in the nineteenth century that can establish the truth of a mathematical proposition without demonstrating it. Hilbert falls in love with this approach in college. He witnesses a German mathematician named Ferdinand von Lindemann winning a prestigious chair at the University of Königsberg by using a pure existence proof to demonstrate the "transcendence" of the number pi—meaning the number is not the root of a nonzero polynomial with rational or integer coefficients. Lindemann proves that pi is a transcendental number by showing it wouldn't make sense if it weren't.

Hilbert takes a similar approach to proving something known as Gordon's theorem, which for years has been elusive and seemingly unsolvable. The mathematician Paul Gordon, who develops the eponymous theorem, could prove only a special case of it. For twenty years after that, mathematicians all over Europe are unable to extend his proof beyond that simple case. Hilbert does it in a completely novel way by using a pure existence proof in 1891, which one of his students later describes as *ex ungue leonem* (from the lion's claws). That cat has teeth!

Gordon initially responds by criticizing Hilbert's original work specifically because of his pure existence proof. "This is not mathematics, but theology," he famously says. Later, after he realizes Hilbert's proof was correct, he concedes that point with a joke by way of apology: "I have convinced myself that theology *also* has its merits."

THE GREAT MATH WAR

THE SHAKESPEARE QUESTION

In the 1890s, Cantor distracts himself by devoting countless hours to an organization called the Society of German Scientists and Physicians, which inspires him to create a similar organization completely devoted to math called the Deutsche Mathematiker Vereinigung (German Mathematicians Association). The inaugural meeting takes place in 1891.

He sets a trap at this meeting for Kronecker by preparing his brilliantly original method of diagonalization, to prove that "large infinity," the uncountable $\aleph_1$ set of real numbers, is a nondenumerable infinite set. He may have invented diagonalization solely to embarrass Kronecker. One of Cantor's biographers implies that his ulterior motive for even founding the Mathematiker Vereinigung in the first place was to seek a stage where he could force Kronecker's hand and goad him into open confrontation, like some nineteenth-century Hamlet put into a farce. Then he would proceed to embarrass him in front of everyone by countering with diagonalization and other technical arguments. Cantor's own letters bear this out somewhat. "Many who were previously blinded would have their eyes opened for the first time," he says, looking forward to the first meeting of the Mathematiker Vereinigung, where his trap is to be sprung—just as any good Hamlet would do. *Theory or not theory!*

The confrontation—if that's indeed Cantor's intention—never materializes, however. Kronecker's wife is terribly injured in a mountain-climbing accident some weeks before and dies just prior to the meeting. Kronecker can't come. He sends a short, sweet, friendly letter instead, and the assembled members of the group at the first meeting read it aloud and respond by voting him onto their board of directors. He never assumes the post, however. Kronecker himself dies a few months later.

Following Kronecker's death, Cantor's effort to build an international community for mathematics succeeds in a big way. The First International Congress of Mathematicians takes place in Zürich in 1897. By then, most people have come to fully appreciate the power of Cantor's set theory, and in the opening remarks at the conference, the plenary speaker acknowledges him. Set theory is an enormous contribution, the speaker says. This inspires Cantor to jump back into mathematics with both feet in the closing years of the nineteenth century, and he soon publishes what his biographers call his best-known and most complete publications, two papers that lay out the subject of set theory, its principles, and its implications "in an almost perfect logical form," the mathematician Philip Jourdain will write in 1912.

Those two papers will turn out to be his last great work. There are many more bumps in the road ahead. Major technical problems with set theory are emerging—logical problems known as paradoxes that seem to shake the integrity of set theory to its core. One is discovered in 1897 by the Italian mathematician Cesare Burali-Forti—the so-called Burali-Forti Paradox, which says that if you create a set of all the possible "ordinal numbers," which rank different sets in the bucket first, second, third, etc., then *that* set would contain an ordinal larger than itself, which is impossible.

Cantor himself soon finds another one, which becomes known as Cantor's Paradox (fully explored later in Chapter 5, "All Cretans Are Liars"). The disconcerting discovery of these paradoxes could add to his concern over his continued inability to prove the continuum hypothesis.

His ultimate undoing is personal, however.

In 1899, Cantor's little brother dies. Constantin Cantor is a dashing former cavalry officer with the Hessian Dragoons (they of sharp sabers and even sharper uniforms). And he's a romantic. Constantin captures the eye and wins the heart of an Italian baroness, and he retires in comfort to the wonderful island of Capri in Italy's Bay of Naples. Then in 1899, he suddenly and tragically dies, which weighs heavily on Cantor.

In November that year, a deeply troubled Georg Cantor takes a leave of absence from his university and then sends a letter to the government ministry overseeing his position. He asks if he can beg off teaching. Could he take some lesser position, such as that of a librarian, and keep his salary? He wants to get away from mathematics—quickly, permanently, and "by all means," he says in the letter. He is eminently qualified to be a librarian, he assures the ministers. And to sweeten the deal, he claims to also be in possession of certain facts and truths otherwise unknown and hidden—information that will shake the very foundations of German-English relations to their core. He claims to know *hidden* things. Information "which will not fail to terrify the English government as soon as the matter is published," he writes.

Could they send him a response within two days? he asks. He needs to know soon because if not, he plans to go to work for the Russians instead. He was born in St. Petersburg, after all. If he doesn't hear from them in two days, he will surely go to work for Czar Nicholas II. But a month later, he gives up and goes back to teaching. That's when the hardest blow falls.

On December 16, 1899, while he's giving a lecture on the Shakespeare/Francis Bacon *show-me-the-real-bard* question, his son Rudolph suddenly dies four days before his thirteenth birthday. Rudolph is Cantor's pride and joy. The boy shows tremendous promise—an extremely talented preteen violinist. Cantor studied violin himself as a child. Would he have been happier as a musician? Why did he give it up? Watching Rudolph reawakens his love for the instrument, and in his son, Georg sees what he might have been. His son plays, and Cantor is the better for it.

No father should ever have to bear such a loss. With Rudolph gone, Cantor is shattered. He spins out of control, speculating that his son has been killed by English agents. They're out to get him and instead killed his son. He sees conspiracies everywhere. He begins to decline, checking in and out of nerve clinics again and again for the last eighteen years of his life.

———◆———

When people crash and burn, they always fall for the same fatal reasons. Some climb too fast and stall after takeoff. Some never achieve enough speed to get off the ground. They crash into the trees at the end of the runway. Others get lift and soar free but then run out of gas. Others fly too high and tragically fall—like Icarus, that wax-winged Cretan of classic Greek myth. Fly too close to the sun and your wings melt.

Cantor is Icarus. He aims too high, and he flies too high as well. He revolutionizes mathematics with set theory, aspires to the divine, and he becomes convinced that his discoveries are going to move not just the minds of mathematicians but also the hearts of philosophers and theologians. And then his waxy wings melt. Everything he thought was gold has gone gray.

CHAPTER FOUR

GUNS, GOLD, AND THE GRUESOME MATH OF WAR

1899–1902

"
Ex Africa semper aliquid novi.
(Out of Africa, always something new.)
"

—Pliny the Elder

A chaplain tends to a dying soldier on a battlefield during the Boer War. Cropped from a halftone of a 1900 illustration by Frederick Judd Waugh. Public domain image via Wellcome Collection.

Bertrand Russell makes a pure-gold discovery of his own during the dog days of Cantor's troubles: love at first sight. He's seventeen when he meets Alyssa "Alys" Whitall Pearsall Smith, a twenty-two-year-old American student from Philadelphia staying with her family that summer in England before her senior year at Bryn Mawr in Pennsylvania. The Pearsall Smiths are an unorthodox family living nearby, and Russell's grandmother invites them over for tea.

Bertie is taken instantly. His grandmother, not so much. He finds Alys remarkable—smart, interesting, well educated, and beautiful. She's an outspoken feminist who advocates for some of the very things that make his grandmother cringe. Women's suffrage. Sex education. Free love. Lady Russell sees her as a loud, crude, common, and uncouth American girl. What was she thinking, inviting her family over for tea?!

When her grandson and Alys later become a couple, Lady Russell objects with all her might. When they make plans to marry, she says coldly that she cannot approve. She tries to scuttle the marriage. She skips the wedding, predictably. And when she dies soon after that, Bertie claims to be completely unmoved. "I did not mind at all," he will write in his autobiography many years after the fact.

Before getting married, Russell works as a British diplomat stationed in Paris—a plum post befitting his lofty station. But Bertie is not yet interested in politics or diplomacy, so he soon quits. After he and Alys are married, they honeymoon in Germany in 1895, and there he reaches a career decision. "I had determined to devote myself to mathematical philosophy," he says. The next year he and Alys take a trip to America. Ironically, it's in the United States that Russell first becomes acquainted with the work of contemporary German mathematicians.

Discovering the Germans is a real breakthrough for him because Russell realizes that they have done so much, shattering any illusions he has of British mathematical superiority. He is part of the last generation of "isolationist" British mathematicians who, because of the nationalist overtones in the fight over calculus almost two centuries earlier, disconnected Britain from the rest of the European math scene. At the end of the nineteenth century, the empire is still flying solo—and begging to be a backwater because of it. A "splendid but unhealthy isolation," American mathematician Einar Hille will write in 1953.

The German mathematician who will come to influence him most profoundly is University of Jena professor Friedrich Ludwig Gottlob Frege, who is famously remembered as the foremost logician of the late nineteenth century (and infamously recalled for his awful anti-Semitism in the early twentieth century).

Frege is born at a time when people are constantly searching for ever more rigorous proofs in mathematics. It's also a time of renewed interest in logic as a mathematical endeavor. A few years after Frege is born, British mathematician George Boole publishes a book that lays out strategies for deduction using math and equations and representing logic using signs and symbols in algebraic form.

That's good, Frege later shows, but not good enough.

What Frege dreams about is not simply discovering individual mathematical truths—to push disparate logical propositions to and fro from some dark corner of unsolved problems into the well-lit realm of *beyond-a-doubt* proof. What he's really interested in is strengthening the truth of the system of mathematics as a whole—all the way down to its foundations. He wants to show just how universal, connected, and interdependent mathematical truths really are. A solid foundation for mathematics is needed, he thinks. Those foundations are everything. And into them Frege pours the entire concrete of his being.

In 1879, as Cantor is reeling from his first struggles with Kronecker and Bertrand Russell is still a shell-shocked, school-aged, stay-at-home grandson sequestered in the cold halls of family expectations, Frege writes an obscure book called *Begriffsschrift* (Concept notation). Russell discovers this book twenty years later and is amazed by it. It outlines a sort of visual nomenclature for logic, a new language Frege develops by using symbols "for pure thought," as he puts it. It's a formula language. And he hopes it will replace the inadequacy, lack of precision, and ambiguity of ordinary language.

THE GREAT MATH WAR

You could say Frege is antilanguage in a way—even though he invents a new one. But his new language is meant to be different. Common vernaculars are obstacles to pure logic, he's convinced. All the nuance of ordinary language is a problem. Ambiguity. Inexactness. Subtext. Subtlety. Double entendre. Hyperbole. Pregnant meaning. Cynicism. Humor. Irony. Untrustworthy narrators. Misinformation and outright lies. All the things that make for great writing also make for terrible logic.

Even the carefully crafted writing by mathematicians doesn't suffice—whether in English, German, Latin, or other tongues. Everything that anyone has ever written about math relies too heavily on language—equations notwithstanding. And human language constrains if not distorts how we see the world. To describe something mathematically using ordinary language means accepting those constraints and allowing imprecise meaning, which Frege hopes to avoid.

Ordinary language will never allow him to express mathematical truths freely. No common tongue will suffice, he thinks. So Frege invents a new one from scratch. A language specifically for math. Precise. Free of ambiguity. Bereft of unintended meaning. A language unfettered by the baggage of language.

Why is unambiguity important? Think of it this way: When you build a house, you don't use ambiguous materials. You don't mix concrete with random amounts of cement and sand. You measure those things precisely. Exactly. You don't use structural beams cut to inexact lengths. The adage "Measure twice, cut once" reflects the great need for accuracy at the framing stage. No blueprints are ambiguous, interpretable, or ironic. No! You build a house with specifically, precisely, and accurately cut planks and unambiguous materials.

GRUNDLAGEN DER ARITHMETIK

Frege thought we needed a blueprint in math just as much as we do in architecture, and his 1879 book, *Begriffsschrift*, is that blueprint. It describes a precise, consistent, symbolic, and Lego-like logical language that allows you to mechanically piece together statements into mathematical proofs. Frege was convinced his new concept notation would "provide us with the most reliable test of the validity of a chain of inferences," he said, and "point out every presupposition that tries to sneak in unnoticed, so that its origin can be investigated."

His work is spectacularly successful, basically establishing modern mathematical logic as a subject, and many people today, according to modern German mathematician Volker Peckhaus, still consider it a revolution. Some call Frege the "founder of modern logic," though others tip their hat to the earlier British mathematician Boole, whose famous 1847 book, *The Mathematical Analysis of Logic*, gave birth to Boolean logic. But where Boole applied math to logic, Frege applies logic to math. For that reason, some consider him the greatest logician since Aristotle. In the 1960s, math historian Jean van Heijenoort, reflecting on the *Begriffsschrift*, will say simply that it is "the most important single work ever written in logic."

That being said, Frege's book is hard to approach. Its symbolism is weird. He employs a two-dimensional visualization of logic with lines and letters spilling out all over the page. It's uneconomical, complicated, and cumbersome. The pages of his book look more like tattoo designs gone bad than mathematical proofs done right. His logic may be elegant, but his shorthand is not. His visual system is also the bane of his publishers—not that Frege cares much. "The convenience of the typesetter is certainly not the *summum bonum* [ultimate good]," he says.

Frege becomes convinced that by using his system, all of mathematics could be reduced to a firm foundation of logic. That leads to another book, published in 1884, called *Grundlagen der Arithmetik* (*Foundations of Arithmetic*), which applies his system to basic concepts, like the definition of a number. Frege is the first person to show how some of the simplest objects in math, like numbers, can be derived purely from logic using just a few simple functions like "and" and "or."

"In a sense," American mathematician Frank DeSua will claim in 1954, "it can be said that Frege was the first person in history to know what a number really is."

Now, that sounds completely strange. And it is. Frege's work is weird for the same reason the cannonball or orange-stacking problem is strange—that centuries-old problem on Hilbert's list also known as Kepler's conjecture. It hypothesized that the most efficient way to stack spherical objects was to stack them on top of one another. It's befuddling because it's an intuitively easy problem to answer (any grocery clerk in a produce department could tell you the best way to stack fruit). But it's a hard problem to solve—so much so that this was one of the last of Hilbert's twenty-three challenge problems in his 1900 Paris lecture to be solved. (And when University of Michigan mathematician Thomas Callister Hales tried to publish his 250-page proof in the late 1990s, the math community was at a loss to even begin to peer-review it.)

THE ORANGE-STACKING PROBLEM

I think of Frege's *What is a number?* question and Kepler's orange-stacking conjecture as examples of a logical misunderstanding, which I call the fallacy of unimportant results: The more obvious the answer to a problem seems, the less important its solution appears to be. Why do we need a bunch of geniuses, supercomputers, and lengthy, impenetrable proofs to tell us something we already know—or that any old grocery clerk could tell us? Nobody needs proof of how to stack oranges because it's already obvious. Fewer still would ask a question like *What is a number?* In our *why-should-I-care* and *what's-in-it-for-me* modern world, few would ever care to ask the question, let alone seek rigorous proof of the right answer. *What's a number?* How about *What's a middle finger?* Thanks for wasting my time, genius!

But the fallacy of unimportant results is not Frege's biggest problem. His work, like Cantor's, is among the most innovative of the entire century. Cantor's set theory introduces a conceptual system that allows innovative new approaches to math, and Frege's formal symbolic language gives people the ability to apply the purest form of logic to its very foundations—what Bertrand Russell a few years later will call "the technical adaptability and the logical comprehensiveness that are essential to a mathematical instrument for dealing with what have hitherto been the beginnings of mathematics."

Yet, like Cantor's, Frege's work carries hints of underlying tragedy to come. In just a few years, Frege will become a brilliant failure—on the eve of his greatest success. There's a flaw in his carefully crafted system. A fundamental flaw, which Russell will soon discover. And Frege will soon see his efforts crumble to dust.

GUNS, GOLD, AND THE GRUESOME MATH OF WAR

For that reason, I like to think of Frege as an inverted unicorn, a creature known more for his hole than his horn.

JUST NORTH OF STARVATION

When Bertie and Alys return from their travels to the United States in 1899, a political and military crisis is unfolding on the world stage: the Boer War—the same one that inspires Queen Victoria's black-clad trip to Ireland in 1900. On one side of the war is the British Empire, which controls vast parts of South Africa and owns two large-footprint colonies in the region: the Cape Colony on the southern tip of the continent and the coastal Colony of Natal northeast of Cape Town. Falling outside Britain's massive colonial control in the region is a pair of landlocked countries ruled by people of mostly Dutch descent—the two independent nations of Transvaal and the Orange Free State.

The roots of the British conflict with these two countries go back almost a century to the Napoleonic wars, when the Prince of Orange fled the Netherlands for England. That allowed Great Britain, just before the Battle of Waterloo, to claim the Dutch colonies in South Africa as its own, bringing 150 years of colonial Dutch rule in South Africa to an end. The independent-minded descendants of the original European settlers there were known as the Boers (farmers) in the local Dutch-derived language Afrikaans. They bristled under the British yoke from the beginning.

That bristling boiled over after December 1, 1830, when Great Britain abolished slavery in all its colonies, including South Africa. Thousands of Boers responded by pulling up stakes, loading wagons, and migrating north to eventually establish the two nations of the Orange Free State and the Transvaal—which by the end of the nineteenth century were two of only four colonially independent countries on the entire African continent that fell outside direct European control.

Embarking on this "great trek" of the 1830s became a touchstone of Boer culture, signifying both a *get-up-and-go* streak of wanderlust and a *gone-out-to-gun* commitment to independence. It redefined the Boer life in remote and rugged terms: They were one part settler, nine parts survivalist. Children learned to survive, shoot, run, ride, herd, hunt, and hide from an early age. The great trek also ushered in fifty years of slow-moving, on-again-off-again conflicts between the British and the Boers—a series of dramas defined and sometimes ignited by the presence of rich mineral wealth, including the accidental discovery in 1867 of the fattest diamond deposits anywhere in the world on the British borders of the Boer lands.

History forgets the name of the farmer who walked the banks of the river near his ranch in the 1860s, picking up cool rocks and keeping some of the best ones in a bowl on his kitchen table. But it remembers the name of the man who recognized one of the rocks for what it really was—a massive uncut diamond. John O'Reilly was a traveling merchant who stopped at the farm to feed his animals, saw the bowl of stones, and spotted among them the largest diamond he had ever seen. He talked the farmer right out of that fortune.

News of that discovery led to more finds, and soon an even more massive eighty-four-carat uncut diamond was unearthed—later dubbed the "Star of South Africa." The lucky fool who found it traded the rock for five hundred sheep, ten oxen, and a horse. (It was later cut and sold to a British earl for a far greater fortune.)

Following that, a crush of treasure seekers stampeded to the area, and a boomtown was born. At first little more than a camp that didn't even have a name, it soon became Rush City. Then Diamond City—stupid names like that. But it eventually became formally incorporated as the imperial British town of Kimberly, just inside the northern borders of Great Britain's Cape Colony, adjacent to the Boer Transvaal Republic, and soon connected by railways to the coast.

But if diamonds are impressive to the eye, there is another mineral that's even more precious to the ear. Few words are as tantalizing to hear, as pure in the locket, or as heavy in the pocket as "gold"—that shiny yellow element of interstellar origins. No substance can be more chemically pure or psychologically impactful. It motivates people as nothing else does, *Scientific American* declares in the late nineteenth century amid the South African gold rush. "Men bow down before it the world over," the magazine says. "With it and for it we wage our wars."

GUNS, GOLD, AND THE GRUESOME MATH OF WAR

The nineteenth century is the boomiest time in history for bonanza gold strikes. Gold mints fortunes in California, Colorado, Mexico, Canada, Australia, Alaska, and many other parts of the world. Cities explode where gold is found. Merchants grow fat. Miners grow thin. And get-rich-quickers come with shovels in hand and fantasies in mind. They rake the clays of Western Australia. They climb the ropes of Chilkoot Pass. They land in old plank-town San Francisco. They wend their way to Sutter's Mill. Coyotes. Claim jumpers. Prospectors. Pretenders. Some strike pay dirt. Most claims peter out. Some prospectors settle down. Some are lost, dying grim, cold deaths in lonely, unforgiving places. Others hobble home—somewhere just north of starvation and due south of happy to still be alive.

South Africa in the 1880s sees the biggest bonanza strike ever. Gold is found in an area called the Witwatersrand (White water ridge) in the Afrikaans language of the Boers. This area is marked by a unique geological feature: a long scalp of hard metamorphic rock stretching for hundreds of miles across the continent. Over the millennia, rivers running over this scalp have formed waterfalls—thus the name. And gold, a naturally heavy element, collects in the alluvial deposits of riverbeds, making the Witwatersrand a prime place to find it.

The gold rush almost never took off, however, because the earliest gold strikes there seemed unimpressive. Unworthy of extraction. "Poor diggings" in the prospector's parlance. Famed American mining engineer Gardner Frederick Williams toured the Transvaal for two weeks in 1886, visiting and closely inspecting several mining operations up and down the Witwatersrand. His impression was *Meh!*

"If I rode over these reefs in America, I would not get off my horse to look at them," he said. There was gold, to be sure. But it was fouled with dirt and pyrite—iron disulfide, or "fool's gold." It's hard to separate gold from pyrite, and that lowers your yield. The real gold winds up sticking to the fool's gold and gets tossed away in the so-called tailings, or waste piles. A perfect gold strike has rich, glittering deposits of "lode" close to the surface. The best mines are mother lodes, where thick seams of gold are embedded in quartz crystals. That kind of gold is chunky. Heavy. Even pretty. Its nuggets look like museum pieces—and in fact, some are.

What Williams saw instead on his ride was ugly gold. Dirty gold. Fouled gold. Stupid, dirty fool's gold. Hardly worth the effort, he thought.

He wouldn't know it for months, but he'd just made the single greatest blunder in the history of mineralogy. What the Witwatersrand mines lacked in individual perfection, they more than made up for in collective volume. In fact, as it turned out, all those ugly little mines were not separate poor diggings but scratched at one and the same deposit. The Witwatersrand literally sat on a mountain of gold. Over millions of years of seasonal rains, its hard metamorphic scalp forced crashing floodwaters down and created a single long, thick, massive, and continuous reef of gold. As a strike, it hardly mattered that the gold was fouled. There was a shelf of dirty gold that was two miles wide, a hundred miles long, and a mile deep in some places—the greatest deposit the world had ever seen. What Williams thought were poor diggings was actually wealth on an unimaginable scale.

Recovering it wouldn't be easy. Williams was right on that score. Until then, mining companies had typically used an amalgamation method to extract gold from pay dirt. They spread a layer of mercury on silver-plated copper surfaces connected to batteries, mechanically crushed the ore, mixed the powdered ore with water to form slurries, and then passed that mud across the charged plates. Tiny gold particles amalgamated on the surface and could be scraped, burned, and weighed. *Et voilà—le couleur!*

But right around the time of the South African gold rush, a new Scottish innovation improved yields by adding a step where the leftover tailings were remixed with potassium cyanide into muddy slurries. Cyanide dissolves gold, chemically separating it in the liquid supernatant, and it can then be precipitated out. That tiny, poisonous process transformed mining operations in the Witwatersrand and enabled its mines to become the most profitable in the world, lifting the Transvaal out of miserable poverty and into vast wealth overnight and ultimately pushing it into war with Great Britain.

THE CITY OF THE DEVIL

Ostensibly it was a governance war. Half the gold mines in South Africa were operated by just one company, and two other companies controlled

GUNS, GOLD, AND THE GRUESOME MATH OF WAR

the bulk of the rest. These companies were backed by massive foreign investment, and the British interests were enormous—and not just the British. Many of the mine workers were American. The chief engineers of all the major companies were Americans. The Transvaal had a minuscule manufacturing base, and it was not just people the country imported from the United States but also goods. Some 90 percent of the machinery at the gold mines was imported from US manufacturers. People in the Transvaal drank American whiskey, they guzzled American beer, they drove American wagons, they worked with American tools, and they wore American clothes.

The city of Johannesburg, like its sister city, the diamond town Kimberly, was another strange beast. Jo-burg also exploded in size in the late 1880s, literally growing from nothing into a major urban center overnight. When the gold was first discovered, it was a dirty little backwater tent camp. Nobody could even remember who Johannes was. But as news of the bonanza broke, Jo-burg became a boomtown. Within six months, it was a full-fledged, wood-plank town with bona fide buildings and storefronts. Within a year, it had grown so large that you couldn't walk from one end to the other in a single day. Within a decade, it was one of the largest cities on the entire African continent.

But it also suffered from the same bad reputation from which all frontier towns inevitably suffer. Fast. Mean. Hard. Lawless. Lots of boozing. A rough place. Tons of fighting. Murder. One observer said the city could corrupt an archangel. Another called Johannesburg "the city of the devil."

Transvaal president Stephanus Johannes Paulus "Paul" Krüger was determined to keep that devil in chains. His government passed punishing laws denying voting rights to foreigners, women, and people of color. His country also maintained a strict monopoly on dynamite, an essential tool for gold mining. That added extra costs to mining operations. Shipping was also expensive, and Transvaal laws granted the government a monopoly on the railroads. And there were also huge tax burdens on individual foreigners—called *Uitlanders* in Afrikaans.

Protecting the British diaspora in South Africa became Great Britain's justification for sending troops there in 1899. The drumbeats began in the summer. War was coming, everyone knew. A refugee crisis began that fall. Johannesburg emptied. Streets were deserted. Major businesses were boarded up. A flood of forty-five thousand people fled the country before the first shots were fired that South African spring.

One person paying particularly close attention to the start of the war is Bertrand Russell. It commands his full attention, in fact. He's transfixed. A bona fide news junkie. Walking several miles a day to the train station to read the latest news from the front. "I could think of nothing else," Russell will later recall. "Philosophies seem more like children's games in comparison."

His obsession stems from the same source as Queen Victoria's: disbelief over how foully the war plays out for the British in its first weeks. At the Cape Colony town of Ladysmith, 150 British soldiers are killed or wounded and another 1,000 are taken prisoner. At the Battle of the Modder River on November 28, 1899, the British ford a muddy river filled with snakes. Some soldiers who aren't shot are snakebit. "Everything was against us," one British surgeon writes.

Three more battles sow further disaster. First comes Stormberg. After a British commander calls retreat, a third of his troops fail to hear the order. In the ensuing chaos, some six hundred soldiers are surrounded and captured. The next day, at Magersfontein, an ill-fated British advance ends in a disastrous hardscrabble retreat. Some four thousand British soldiers fall into fumbling flight. "A flock of sheep running for dear life," one witness says. The British lose a thousand men.

Finally comes Colenso, which is by far the worst. The British attempt to cross the Tugela River at its most fordable spot becomes a major blunder. The crossing is at one specific bend in the river. That funnels men, horses, and cannons to a choke point across from which the Boers have spent a week digging and occupying trenches that are now virtually invisible. When the shooting starts, people watch in horror as the British troops are suddenly exposed to a cone of sniper fire. One report claims the Boer rifles firing together sound like bacon frying in the pan. But that doesn't stop the British from charging again and again—first to cross the river and next to recover the cannons that lie abandoned on its banks. (British military doctrine of the day insists that guns should never be surrendered intact.)

"I was sick with horror," Boer general Louis Botha says after the battle, "that such bravery should've been so useless."

In England, people mass around newsstands, reading, horrified. Those three battles, Stormberg, Magersfontein, and Colenso, glue the eyes of the British public to the papers. People read shouting headlines in disbelief. How could a rough-hewn, homespun army defeat a far superior British force? Writer Arthur Conan Doyle captures the British mood in stark terms. He calls the span of five days encompassing the three battles "Black Week": "the blackest one known during our generation."

The next month sees another battle, Spion Kop, and another disaster. A young Winston Churchill is there. He describes the battle as "a bloody, reeking shambles." Hundreds of British soldiers are killed, more than a thousand are wounded, and hundreds go missing. Again, people flock to the newsstands—not just in London but all over the world.

LOGIC, INK, AND WARFARE

Never have so many journalists covered a war. As a news story, the Boer War has it all. Colorful leaders. Exotic landscapes. Heartwarming heroics. Gut-wrenching battles. Hard, cold loss of human life. Readers gobble it up. London's *Daily Mail* becomes the first newspaper in history to sell a million copies in a single day.

It soon becomes a propaganda war as well. Some European papers, siding against the British empire, present a romanticized view of Boer men as dashing commandos. Standing against oppression. Tall in the saddle. Two parts Errol Flynn and three parts Obi-Wan Kenobi.

British tabloids do the same in reverse. London papers cast British soldiers as noble, lovable Tommy Atkins heroes. Their editorial cartoons trot out one unflattering likeness of Transvaal president Krüger after another. He is an ape in cruel mockery. The big-bear Boer. A bearded beast. The "Boer Machiavelli." Hideous Krüger. Ugly Krüger. Fat Krüger. Krüger the git. *He can't even sign his own name!*

All ink stains, but patriotic ink stains deeply—and indelibly.

THE GREAT MATH WAR

──◆──

It works. In England, the mood on the street is ridiculously, riotously, and righteously in favor of Britain's war. Antiwar meetings are routinely taken over by prowar factions and broken up with chanting, singing, counterprotesting, rock throwing, and sometimes face punching. One gathering in Glasgow is mobbed by thirty thousand people. Businesses of well-known antiwar leaders are vandalized. Some antiwar activists are stabbed, others beaten. Six months into the war, more than two dozen antiwar meetings are canceled because of violent threats.

But then the war makes a turn. In 1900, just after Queen Victoria's trip to Ireland, the British bask in a stunning series of victories that change the tide of battle and promise to end the war. Ladysmith. Johannesburg. Bloemfontein. All great victories for the British. The imperial army occupies Pretoria without a shot fired. At a place called Paardeberg Drift, top Boer commander Pieter Arnoldus Cronjé is defeated and captured along with four thousand troops—10 percent of the entire Boer army. London is elated. Queen Victoria is thrilled. And Russell slowly calms down.

The relief of Mafeking is even more of an event on the home front. Mafeking is a remote, tiny, insignificant town of mud huts and tin roofs. A lonely place with nothing to see—really just a railroad stop in the sandy wasteland edging onto the Kalahari Desert. On the borders of nowhere, as British historian Thomas Pakenham puts it. But when news reaches London of its liberation and relief by the British army, it's a victory like no other. Thousands of people jam the streets of London, jubilantly shaking in a new dance move they soon call "mafficking." The party goes on all night. In some places for days. It turns into riots. Laughing. Shouting. Cheering. Singing. A newspaper headline shouts support for the khaki-clad Tommy Atkins troops the next day: *WHEN SHALL THEIR GLORY FADE?*

From there, victory seems assured. By the time Hilbert issues his twenty-three math challenge problems a few months later in Paris, Russell has fully calmed down. He no longer walks miles to fetch the latest dispatches from the train station. But something else is happening as well. A different sort of movement is taking place—a revolution in journalism and antiwar activism that will deeply influence Russell in the end and topple him off his pro–Boer War pedestal.

———◆———

In Manchester, England, Charles Prestwich "C. P." Scott, the editor of the liberal *Guardian* newspaper, is rapidly changing the press with a new approach. For him, newspaper reporting is a form of public good. Papers should not merely inform their readers, he believes, but try to guide them. For him, it's not about what readers should *know* but about what they should *think*. For him, a newspaper has a "moral as well as material existence" to influence as well as to inform, Scott says. And early on, he embraces the much-maligned antiwar activist movement and legitimizes it anew by celebrating antiwar activism as an alternative form of patriotism. It's patriotic to oppose unjust military actions, he basically argues.

And it works—at least on Bertrand Russell. In the coming months, Russell undergoes a political awakening. "At the beginning of the war, I was an imperialist more or less," he will say years later. By war's end, he adds, "the Empire has come to seem to me not worth preserving."

Russell makes peculiar claims about his peace "conversion," saying his transition from stern imperialist to staunch peacenik comes suddenly and without warning—almost like an ischemic stroke: "a change of heart, which brought with it a love of humanity & a horror of force." Precipitating this conversion, he says, is not an awakening to the horrors of war detailed in the liberal press but something completely different and unrelated: concern over his good friend Evelyn Whitehead, who suffers a sudden and unexplained illness in early 1901.

"She seemed cut off from everyone and everything by walls of agony, and the sense of the solitude of each human soul suddenly overwhelmed me," Russell will later write. He describes the feeling as akin to the ground beneath him giving way. Shaky foundations, like standing on loose sand in an earthquake. Standing by Evelyn's bedside, he claims to have had a rapid series of revelations: The loneliness of the human soul is unendurable. Nothing can penetrate the soul but love. War is wrong. Public school education is abominable. He wishes for peace. He rejects imperialism. And he has suddenly fallen out of love with Alys.

Personally, I don't buy it. This story feels like a lie Russell tells other people—perhaps even himself. There's no question he undergoes a peace conversion. But his instant-karma story is dubious. It may be an equivocation of sorts to excuse his later behavior. Russell commits barbaric acts of spousal neglect and outright marital betrayal during the years after the Boer War. He treats Alys cruelly in the coming years, ruthlessly neglecting her even as she suffers from dark depression and then finally dumping her a decade later for another woman. It's not a good look, and it's the greater part of valor for Bertie to gloss over that truth decades later by claiming to have been thunderstruck. An "Amazing Grace" moment of salvation. For an atheist, no less!

More likely, his feelings about the war gradually change, as is the case with many in England once they witness the horrors of the Boer War spilled out in the press. And most likely Bertie's feelings about Alys change gradually as well—not in some lightning-fast instant conversion. Claims of sudden salvation are always suspect—especially when the claimant is an atheist. Popular hymns and religious texts portray those *how-sweet-that-sound* moments in the storm that save you from your wretchedness. But who's to say it always happens in an instant? The truth is probably more ordinary, at least in Russell's case. His opinion of the war slowly changes. His views adapt in a gradual, not thunderstruck, way. His embrace of humanity is nothing new. It just gets stronger. He is definitely concerned for his friend Evelyn, deeply worried in that moment when she has her sudden illness—but I seriously doubt he has revelations about the inadequacy of the British school system at that exact instant. And if he falls out of love with Alys, that goes slowly as well. He has other things on his mind.

SOYLENT GREEN IS PEANO

Before Bertie's conversion, falling-out-of-love, and concern-for-humankind moment, he goes to Paris in the summer of 1900 with Alys, Evelyn, and Evelyn's husband, the British mathematician and philosopher Alfred

North Whitehead, who is Russell's close friend and collaborator. In Paris, they attend the meeting where Hilbert gives his famous future-problems-in-mathematics talk, and there Russell meets the Italian mathematician Giuseppe Peano, who has developed a logical language the same way Frege had but with better and more accessible symbols.

Peano introduced for the superset/subset relational operators a "⊂" and an inverted ⊃, that later become the common modern signs "⊂" and "⊃". In 1889, he introduced five axioms, called the "Peano postulates," that governed basic arithmetic and natural numbers. And he published a book called *Formulario Mathematico* that expressed all the fundamental theorems of mathematics using a new symbolic language of his own creation.

Russell quickly asks Peano to share with him everything he has ever written. Peano's ideas are revolutionary, not just because of those symbols but because of how he has strung them together. Peano has developed a framework he calls "ideograms," which the English mathematician Edmond T. Whittaker would describe in 1948 as formulas that break down the logical components of an argument the same way a reaction in chemistry can be represented on paper as a series of chemical changes to the compounds in question. Peano's postulates are the chemical formulas, his logical operations the reactions, and his symbolic ideograms the written statements describing the reactions as a whole—like cookbook chemistry for small molecule synthesis.

Russell loves Peano. Meeting him in Paris forever changes his life. He sees immediately that Peano has made real breakthroughs in extending logic. "I was impressed by the first that, in every discussion, he showed more precision and more logical rigor than was shown by anybody else," Russell says. He will later say Peano has a "rare immunity from error."

And there's something else: Peano is a genius at communicating his ideas. He buys his own printing press so he can personally oversee the publication of his work. He insists on making sure it's exactly to his liking. He starts a new journal called *Rivista di Matematica and Formulaire* (*Journal of Mathematics and Form*) as a venue for discussion. Peano's energy will eventually inspire Russell's own concept of "logical atomism," which is also based on the idea of building logical statements from more primitive elements. Meeting Peano throws Russell into high gear and thrusts him into math as never before. He will soon embrace a seemingly impossible goal, but one he thinks he can achieve: establishing logic as the basis of mathematics.

But first, returning with Alys to England, he sets a narrower goal for himself: finishing a book he's started called *Principles of Mathematics*. That fall, he writes ten pages a day, compiling a 200,000-word manuscript in just three months. He becomes so obsessed with finishing this book that he begins to worry he'll be run over by a streetcar, ending his life before the work is done. That dark bit of obsessed-with-my-own-death ideation aside, Russell will later describe this time as his intellectual high point. "Every day I found myself understanding something that I had not understood on the previous day," he recalls. "I thought all difficulties were solved and all problems were at an end."

He finishes the book on December 31, 1900, celebrating the achievement in a letter to a friend the same night: "I revealed a new subject, which turned out to be all mathematics, for the first time treated in its essence. I have no good resolution to make."

A few weeks later, he has his peace conversion.

UNTO THE BITTER END

Russell's conversion occurs just as London thinks it's reaching the end of the Boer War. Even though the traditional siege and set-piece battles of the early days of the war are over, hostilities have not ceased. The conflict has morphed from a traditional war into a hit-and-run guerrilla war of resistance waged by the remaining "bitter-ender" Boer commandos in the field. Nevertheless, more commandos are captured each day, and the end seems to be coming soon. *It must be!* The British army has swelled larger and larger. It's now the largest expeditionary force anyone has ever seen, with hundreds of thousands of active troops deployed in the field. In early 1901, many in Britain still think it will be over soon.

The man entrusted by the British command with wiping out the final pockets of resistance is army chief of staff and general Horatio Herbert "H. H." Kitchener, a hero of several of Britain's earlier "little" wars. Kitchener wants to end the Boer War immediately at any cost. That's not just his mandate but his goal. He wants to end the war quickly because his real goal is to take that one final step in his ever-climbing career of winning the crown jewel of plum military commands—chief of British forces in India. He thinks only a quick and decisive victory in South Africa will deliver that for him, and his urgency to meet this goal has dire human consequences, the reverberations of which will be felt for decades.

Kitchener didn't get where he is by accident. He is a brilliant military strategist, and like all smart planners, he worships at the temple of big data. While the public's mind may be put at ease by the daily captures of Boer guerrillas, the numbers paint a panicky picture for Kitchener. They suggest that the Boer guerrillas could hold out for years, if not forever. Those numbers—crunched in a gruesome math-of-war exercise—tell him he needs to try something else. So he decides on a brutal scorched-earth policy called "land clearance," which is an eye-for-an-eye, tooth-for-a-tooth approach to counterinsurgency.

Land clearance is a good example of the dangers of following a strategy based solely and foolishly on data that tracks pure process measures, with no thought given to data reflecting outcomes or cost.

I call this the fallacy of pure process data. It's similar to Goodhart's Law in economics, which says that when measures become targets, they cease to be good measures (often because they are then open to manipulation). But the fallacy of pure process data reveals a much deeper truth about human psychology. It demonstrates our inexplicable desire to manipulate ourselves into stupidly looking for our lost keys in a well-lit parking lot when we know we left them in the bar. The fallacy of pure process data leads to pathological distortions of reality where one focuses on one set of data while wearing blinders to others.

Oh, sure! Anytime we use data to drive decision-making and inform a strategic campaign, we have specific outcomes in mind. And measuring process data is almost always easier than measuring outcomes. In the Boer War, this means tracking things like the number of farms burned, the number of people displaced, the number of commandos and civilians interred, and the square miles of land cleared.

But the fallacy of pure process data says that even if our data isn't wrong, our interpretation of it may be flawed. The outcomes you achieve by following pure process data may not be the outcomes you desire. You can always follow data. But unless it leads you down a well-blazed path, you don't know where the data leads. And you won't necessarily wind up in the right place.

There are always *unintended* outcomes in life—death being the most obvious of these.

THE COST IN HUMAN LIVES

A simple glance at the data Kitchener contemplates in late 1900 suggests that the war has no end in sight. It is quickly becoming Britain's most expensive ever, and that endless economic disaster will only get worse the longer the war goes on. There are far too many commandos still out there. So Kitchener needs to try something else. He decides that removing farms sympathetic to the commandos will peel away any means of resupplying them in their ongoing guerrilla war. So he orders that any farm located within ten miles of a Boer attack be put to the torch, setting tens of thousands of British troops to the task.

Some thirty thousand houses and farms will eventually be burned, tens of thousands of civilians will be displaced, and more than 3.5 million farm animals will be killed. "Nothing but burnt farms and desolation," one soldier says, describing the wastelands of South Africa at the end of 1900. "We usually burn from 6–12 farms a day."

If you measure just that one specific variable, the number of farms burned by the occupying army during a war, then it's easy to argue for the effectiveness of land clearance. By extension, burned farms provide no food, and starving soldiers make no war. But process measures alone may fail to consider outcomes like the cost in human lives, the propaganda price to be paid for civilian death, or that damned-fool human resolve and obstinacy that cause one to fight on in the face of crushing opposition.

Land clearance never breaks the back of the Boer insurgency, as Kitchener hopes. If anything, it increases sympathy for the Boer cause in Europe, especially outside England. It steels the resolve of the commandos and their supporters. And it erodes the once solid support of people like Russell back home. Land clearance doesn't end the war quickly and more

GUNS, GOLD, AND THE GRUESOME MATH OF WAR

cheaply. It extends the war and makes it worse. The policy costs the British a whole hell of a lot, puts the government on trial morally, and helps to make the Boer War the most expensive one in three lifetimes.

Why? The main problem with land clearance is that it causes a refugee crisis. As hundreds of farms go up in flames, thousands of people become homeless. What to do with the women and children whose farms you just burned? The solution the British come up with is equally cruel. Kitchener decides to move the land-cleared refugees into pop-up camps near his military bases. As 1900 ends, he orders his soldiers to sweep all displaced civilians into these tent cities after each farm burning—whether they are willing to go or not. Kitchener soon fills his camps with some 100,000 refugees and captives, and South Africa becomes a land of barbed wire, tents, tin huts, and thin justice.

The British officially call these places "camps of refuge." But the world will soon know them by another name—concentration camps—after similar encampments dubbed *reconcentrados* by the Spanish, who established them in Cuba in the late 1890s. This name will haunt the future of humanity, and in a twentieth century chock-full of humanitarian disasters, the British concentration camps are the first.

To be sure, the British concentration camps are not on par with the apex of human horror reached by the more infamous concentration camps of the Holocaust forty years later. The Nazi concentration camps stand alone in history for their despicable brutality and crimes against humanity. No—the British concentration camps are not like that. But they are dehumanizing and savage nonetheless. They expose people to disease, starvation, and violence. And life in them is demoralizing, purposeless, toxic, and cruel.

There's nothing to do in the British concentration camps but sit around all day. People crowd eight to twelve per tent, sometimes sleeping on the bare ground. The camps are brutally hot during the day and freezing cold at night. There are not enough blankets to go around. There is no soap. Little fuel. Few beds. No clean water. No fuel to boil water. Starvation rations. Maggoty meat. Copper-flavored coffee. No fresh vegetables. Never any milk. Scarce comfort. Nonexistent privacy. The weather is hot, the flies thick, the stomachs enteric, and the water amoebic. Diseases are everywhere—typhoid, measles, influenza. Everyone talks of dying—who is and who has—and everywhere, always, the nose meets the rotten stink of death.

Once the story spills, people around the world condemn these camps as cold-blooded cruelty. Mournful letters appear in the press, some even written by British soldiers claiming to hate what they're doing. Many in England are horrified. People like Russell, who once stood staunchly in favor of the war, are face-punched by the reports of the unfolding humanitarian crisis. For Russell himself, the concentration camps are the start of a lifetime of antiwar activism.

UTILITARIAN TO A FAULT

Kitchener is immune to criticism about his concentration camps because for him, they are simply a means to an end. As awful as they are to behold, he thinks they're nevertheless a solid foundation upon which to build a decisive victory. He sees the continued existence of an armed Boer insurrection as an existential threat for both the ultimate war effort and his own personal future. And in his single-minded desire to end the war quickly and decisively, he embraces the horror of the camps. "I wish I could get rid of these camps," he confesses to a colleague in a letter, "but it is the only way to settle the country."

He's wrong, of course. His introduction of the concentration camps during the Boer War is one of the worst examples in human history of corrupt utilitarianism—the philosophy sometimes used to justify unethical behavior by insisting that the ends justify the means, one of the worst lies ever told. Good intentions are never as good as we tell ourselves. Lying to ourselves and to others about our intentions is one of the classic lies humans tell. For most people, lying about true intentions is like breathing.

But the real problem with utilitarianism is that it gives up good behavior so cheaply. Outcomes are idiopathic. End states are a common destination of a thousand different paths. Achieving a goal should never involve ignoring criminal behavior. You should never excuse human horror by claiming that the ends justify the means. Cruelty in the name of the so-called greater good is still just cruelty. Negligent homicide is still a form of murder. And an approach to ethics that ignores human horror is flawed, as Kitchener's conduct of the Boer War shows.

Six times as many people will die in the concentration camps as on the fields of battle. And in the end, the British abandon the concentration camps. They aren't effective at ending the war anyway. It's a different innovation that does that—one invented for animals but applied to people.

What turns the tide for the British in the end is barbed wire—wrapping Boer country in thousands of miles of it. They use it to surround rail lines and connect strings of homemade tin-and-dirt forts called "blockhouses." Each blockhouse is basically an earthen mound topped by a circular metal fort made by sandwiching dirt between pieces of corrugated tin nailed to wooden posts. Kitchener starts to build them all over the place, the idea being to fence in, cut off, and slow down fleeing commandos. And it works.

By the end of the war, the British will have built eight thousand of these block houses, staffed by sixty-six thousand troops and strung together by some thirty-seven hundred miles of barbed wire. They enclose fifteen thousand square miles of the Transvaal and seventeen thousand square miles of the adjacent Orange Free State in a mess of tin, dirt, and galvanized steel. Doing so allows the British to employ a shock-and-awe cavalry tactic called the steamroller formation, with troops spread out along a long line charging across the empty veldt to sweep commandos from their hiding places and leave them nowhere to run.

That is what cuts off commandos and ultimately breaks the back of the insurgency. By the end of 1901, Kitchener will more or less abandon his concentration camps and reverse his land clearance policy. A few months later, in May 1902, a peace treaty will finally be signed at Vereeniging.

———◆———

We are so often wrong in hindsight when it comes to the lessons of war, even when looking at war data. Cost. Casualties. Civilian lives lost. The amount of land lost or gained. Those things make winners and losers seem obvious, but that data may lack ground truth. There is a lesson in warfare and its conduct for the foundations of mathematics: The irony that our

THE GREAT MATH WAR

path to the goal can be completely confused, however lucid our accounting and however honest our intentions.

Data without an appreciation of the underlying mechanisms that drive the results makes it easy for people to draw the wrong conclusions. And no war exemplifies this more than the Boer War—the Anglo-Boer War, the Great Boer War—the so-called forgotten war, or whatever you want to call it. It could have been a warning of the horrors of war to come, but its lessons are largely ignored, including the greatest one: that sometimes the best outcome is not victory but avoiding war in the first place.

The same weak analysis is endemic in futurism, both then and now. Faulty ex post facto armchair analysis of any war is always flawed for the same reason flimsy futurism is flawed. We assume all the wrong things in thinking about the future. We gobble down the promise of technology and ignore the reality of implementation barriers. We assume that just because something seems possible today, it will become reality tomorrow, without really thinking about whether it makes sense. That's as true today as it was in South Africa 125 years ago.

Today, as in 1900, people are fed a steady diet of rich, tantalizing discovery, all with the promise of personal gain with none of the pain. Lose weight on drugs. Get rich on crypto. Become an influencer by saying nothing but stupid things. Beat Mike Tyson in a boxing match. Plant a flag on Mars. Cure all disease in a decade.

But the future is full of possibilities, some far darker than we can imagine. The future we are always sold is glitzy. Clean. Like an upscale steak-and-seafood house where the waiters wear those spotless long white aprons. But the future may equally well be a roadside diner. A greasy spoon. Cheap. A place you go for convenience and decor, not for the unhealthy meal itself—the kind of place that when asked why you go there, the only possible response is "It's my local."

Oh, sure! Even that future has its charms. The hipster waiter. Those fun throwback stools. The shiny chrome counters. Crackers in the sugar jar. Rice grains in the salt. Those amusing *flip-flip* jukeboxes that never work—and nobody cares. Futurism is always the same song anyway. And we listen uncritically. Johnny Cash. Juice Newton. We love it! The future is belly-filling. The future is promise. The future sizzles. It crackles. *Like bacon!*

CHAPTER FIVE
ALL CRETANS ARE LIARS

1903–1908

> "
> Those who are called mathematicians deal in matters of incomprehensible complexity and subtlety.
> "
>
> —Marcus Tullius Cicero

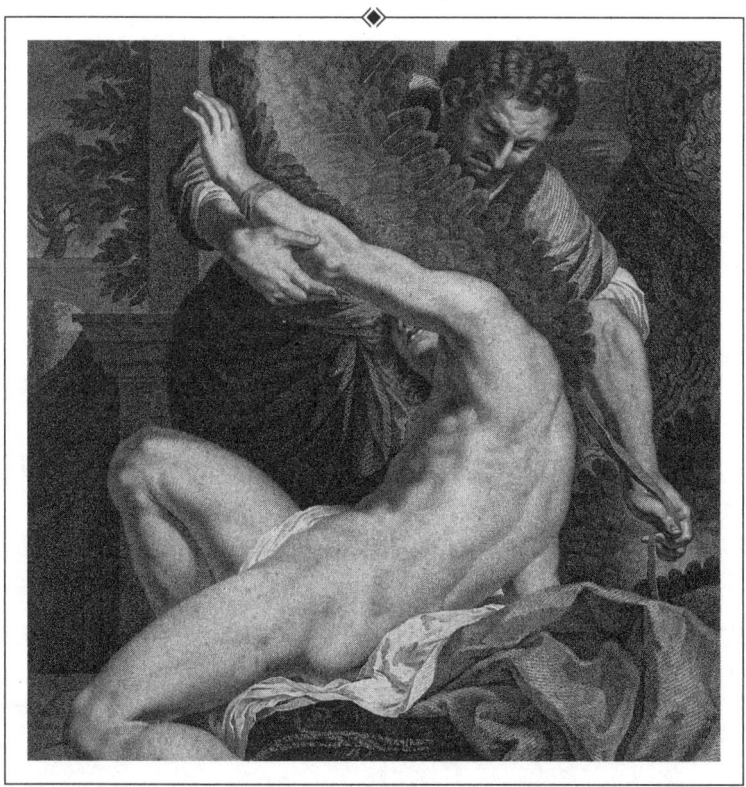

Daedalus attaching wings to the shoulders of his son Icarus. A 1779 stipple engraving by G. S. & E. G. Facius after Charles Le Brun. Cropped from a public domain image via Wellcome Collection.

The exalted vainglory of finishing a book is a dull, dry, and dispiriting affair. There are no trumpets or drums. No adoring crowds. No pomp. No parade. No hussars or high-step marches. No guayaberas or fezzes. No dragoons or dirks. Nothing waits to greet you upon final draft. Nobody is even reading your work at that point. There's just you. Alone. In the dark. Rewriting the same paragraph you've reworked a hundred times over a thousand days. When you first started your book, you couldn't wait to dive in. Now that you're almost finished, you can't wait to be done.

Russell feels this acutely in late 1900 as he puts his book *Principles of Mathematics* to bed, writing ten pages a day for months. He's burned through all his seed corn. Now he's exhausted. Nothing left to sow except a few burnt kernels of misery and despair. He's done. Finishing gives him no ecstatic joy, he says to Alys. "Merely a kind of tired relief as at the end of a very long dusty railroad journey." He's even almost bitter. "It seems to me a foolish book," he writes to a friend after its publication in 1903.

There's something else in the air. In 1902, just before his book appears, Russell makes one of the most important discoveries of his life. He finds and reads an obscure third book by Gottlob Frege called *Grundgesetze der Arithmetik* (*Basic Laws of Arithmetic*), which appeared in 1893. Russell discovers a problem in the book—a flaw in Frege's thinking. So he decides to write Frege a letter about it.

Fate plays us all for fickle pawns. Even as Russell is drafting his letter, Frege is busy reviewing the final proofs of the second volume of the same 1893 book. He's literally preparing to send his last edits to the printers as Russell's letter arrives. It's very polite. And interesting. Russell prides himself on his letter writing, which he considers a lost art—even then. (And let's just say, God save the epistolary form today.) Russell tells Frege he has spotted an error. "I am on the point of finishing a book on the principles of mathematics, and in it I should like to discuss your work very thoroughly," Russell says. "There is just one point where I have encountered a difficulty."

And then he tells him.

The error is shocking. It's not minor. This isn't some typesetting mistake or reproduction problem. It's no trivial misconception linked to some

ALL CRETANS ARE LIARS

peripheral aspect of the work. There's nothing peripheral about it. Russell has discovered a defacing fundamental flaw at the very heart of Frege's logic—a massively scarring problem at the center of masterpiece, almost like discovering *Mona Lisa* has a hidden face tattoo. We call it Russell's Paradox.

THE CLOWN CAR OF PARADOXES

Math has become infested with paradoxes in the early 1900s. In addition to Russell's Paradox, there are the Burali-Forti, Weyl, Richard, König, Greiling, Skolem, Kleene, and Rosser paradoxes. "Radical contradictions" some call them. When the mathematician Julius Wilhelm Richard Dedekind learns of some of these paradoxes, he begins to doubt whether human thought itself is even rational.

Cantor discovered one of these paradoxes himself, based on the so-called cardinal number of a set, which is a measure of the number of objects in a set. Think of it this way: A set of starting players on an NBA basketball squad would have a cardinal number of 5, a starter-plus-bench baseball roster would have a cardinal number of 26, and a set containing all the active players on a soccer pitch during a World Cup match would be 22.

One cool thing Cantor did in developing set theory was to contemplate cases where there are an infinite number of objects in your set. What happens then? Cardinal numbers designate the number of elements in a set, but if your set contains an infinite number of elements, what happens to your cardinal number then? Cantor introduced the concept of "transfinite" cardinal numbers to account for infinite sets. The transfinite number aleph-naught, $\aleph_0$, for instance, is the cardinal number he gives for sets the size of the denumerable little infinity of whole numbers.

One of the stunning assertions of Cantor's set theory is that infinite sets can vary in terms of their cardinality. Not all infinities are made the same way. There is $\aleph_0$ for the small infinity set of whole numbers versus aleph-one, $\aleph_1$, for the large infinity set of real numbers. As the mathematician Allan Calder will write in *Scientific American* fifty years later, some sets are more infinite than others.

THE GREAT MATH WAR

Cantor's Paradox starts with the concept of a "superset," which is a combinatorial set based on all the possible combinations of elements within a set. Here's how it works: Imagine you have a set of cardinality S, let's say $S = 3$ for a simple three-element set of the first three whole numbers. Cantor defined its superset as 2^S. So in this trivial three-number case, $2^S = 2^3 = 8$. More visually, your set would look like

$$\{1, 2, 3\}$$

And its superset would be

$$\{\{\emptyset\}, \{1\}, \{2\}, \{3\}, \{1, 2\}, \{2, 3\}, \{1, 3\}, \{1, 2, 3\}\}$$

Cantor was far from satisfied with trivial cases. He was Icarus, after all, flying up to the heavens. What he contemplated were infinite sets. And he didn't stop there. Straight for the sun! He thought about the set of all sets—call it $\mathcal{U}$—whose superset would be $2^{\mathcal{U}}$.

That's where Cantor's Paradox came from. Since the set of all sets includes all sets, then by definition, it must also include its own superset. But a superset is larger than its root set—Cantor proved this. That means one subset of the set of all sets is larger than the set of all sets itself. Since the set of all sets contains all sets, it cannot be smaller than one of its own subsets. But if the set of all sets contains all sets, its own superset must be one of its subsets. How can that be? That's Cantor's Paradox!

Cantor never discussed his own paradox anywhere in his writings, allegedly thinking it was not a cause for alarm. He described it to a friend, showing "no apparent concern over the paradox and its implications whatever," American philosopher Christopher Menzel will write eighty years later. Perhaps, again, Cantor's lack of strong concern stemmed from his religious convictions—his fundamental belief that set theory was God's gift to math in that it centers mathematics in the divine order of things.

You cannot shake what can't be moved, and the foundations of set theory were, in his mind, unmovable.

ALL CRETANS ARE LIARS

IT IS IF IT AIN'T, AND IT AIN'T IF IT IS

Belief and conviction are like a crutch and a salve for a limp or a wound. Oh, sure—both have their uses. They buttress the broken bone and allow it to knit. They stave off secondary infections in our cold, cruel, bacteria-filled world. But relying too much on a crutch can make your injuries worse. Limp too long and you risk further damage—a knee, a hip, your back. And if you drag a cross too far and lean too much on the solid crutch of divine knowledge while suffering the slings and arrows of outrageous insults, something's gotta give.

Cantor is not the only one who suffers. The paradox Russell finds concerning Frege's work—that *Mona-Lisa-face-tattoo* flaw—is *really* bad. It threatens to make everyone suffer. Why? Russell's Paradox emerges when you consider a very special type of set. Cantor's basic formulation of set theory, which Frege uses, holds that sets themselves can also be objects of another set. One set can be a part of a larger collection. The NBA, for instance, is a collection of basketball *teams*—each one being a set of individual players and coaches. One implication of this is that some sets actually include themselves as a member—sort of reminiscent of the Hair Club for Men commercials in the 1980s ("I'm not only the Hair Club president, but I'm also a client").

Let's take a step back for a second. If some sets are members of themselves, many sets clearly are not. Think of the collection of all prime numbers. That collection *itself* is not a prime number. It therefore does not belong to itself. Nor is the collection of all the teacups in England actually a teacup. So it does not belong to itself. Think of all the good French restaurants in Duluth (a very small set indeed). That list of restaurants is not itself a French restaurant. No bazaar is a kiosk. No sports team is a player. No army is a soldier.

Now let's get even more abstract. Think of all the things you *can't* imagine. The very act of imagining this group of things implies it's not included. Almost any set you can think of is like this. Most sets simply don't belong to themselves. Sets that are not members of their own collection are so common, in fact, that they are sometimes referred to as "normal" or "ordinary" sets. Even if you consider collections of purely imaginary objects, like the set of all things that exist only in fantasy—dragons, unicorns, talking gerbils, employee-friendly HR policies—any conceived collection of such made-up things is a defined, real thing in your mind and is therefore not a member of itself.

But although most sets don't contain themselves, some sets clearly do. The set of all sets is a good example. It contains all sets and by that definition must also include itself. Another good example is the collection of all things you *can* imagine. Among the many things that belong to this set must be the set itself, since even now you're imagining it.

Now, here is where Russell makes his major leap—one that sends him down the rabbit hole of logic for the better part of a decade, never to return. Consider, as Russell does in the early 1900s, the set of all sets that *are not* members of themselves. And then ask yourself, *Is that set a member of itself?* The answer is Russell's Paradox: *It is if it isn't, and it isn't if it is!*

WHO SHAVES THE BARBER?

The classic example Russell gives is the busy male barber who shaves every man in town who doesn't shave himself. What happens when you ask, *Does the barber shave himself?* That's a pickle. If he doesn't shave himself, then by his very job description (shaving everyone who doesn't shave himself), he *must* shave himself. But if he shaves himself, then that very same job description pretty much excludes him from the act. So does he shave himself or not? Russell's answer is that if he does, he doesn't, and if he doesn't, he does.

Never in the history of mathematics have eleven words had such a devastating effect.

Constructing the set of all sets that don't contain themselves and showing that it leads to Russell's Paradox is the most acute criticism Frege has ever encountered. Why? Who cares about stupid barbers and whether they shave themselves? What does that have to do with math anyway? It seems like a silly little word game more than anything else. Aren't contradictions commonplace in human life?

But this is more than just a clever turn of phrase, a theoretical mind game, or a puzzling contradiction. Nor is it an example of semantic or syntactic ambiguity or even simple confusion, the sort of thing that claims something could be this or could be that. This is no mere uncertainty. The logical inconsistency of Russell's Paradox throws into question the very notion of mathematical existence itself. Russell's Paradox implies that a thing can both be and not be *at the same time*. But if a mathematical object can exist and not exist at the same time, that means set theory is inconsistent. That implies that Frege's system of mathematics is inconsistent. And that could very well mean basic logic itself is flawed.

All this has terrible consequences for the mathematical mind, not the least of which is Frege's own. Logical inconsistency is like kryptonite to mathematical truth. "The worst thing which a mathematician can encounter," the Swiss mathematician Rolin Wavre will write in 1934. When a system is consistent, truths and falsehoods stay in their own lanes. You should never be able to prove that something is true when it's actually false, or vice versa. And when your system is consistent, you simply can't.

Think of it this way: We know that $0 = 0$ and $1 = 1$. Never in a thousand years using all the mathematical tools and experts who ever lived could you prove that $0 = 1$.

WITH APOLOGIES TO ERNST ZERMELO

This idea is ancient. It's really one of the most fundamental tenets of logic, dating back to the ancient Indian philosophers as well as pre-Socratic classical thinkers like Parmenides, the teacher of Zeno of Elea of dawdling-tortoise-defeats-Achilles fame. It's known as the Law of Noncontradiction. According to Plato, who gives the classic definition, the law simply says A and ¬A (not A) cannot both be true.

By uncovering a contradiction where A and ¬A are both true, Russell's Paradox upsets the apple cart in a big way. It's a rip-up and a teardown of the foundations of Frege's logic. It exposes "a time bomb ticking away in the heart of mathematics," Bertrand Russell's biographer Ronald Clark will write after Russell dies.

It's worth noting that many insist Russell doesn't deserve full credit for his own paradox. And they're not wrong. German mathematician Ernst Friedrich Ferdinand Zermelo discovers it independently around the same time, and this can be proven, since he revealed his discovery in a letter to

David Hilbert. As an aside, this has led some to insist that Russell's Paradox should really be called the Russell-Zermelo Paradox. There's a fair case for that, and I acknowledge Zermelo's parallel discovery here. But most people seem to call it Russell's Paradox, so that's the label I choose. (Sorry, Zermelo!)

But let's not leave Zermelo quite yet. There's another apology owed that guy. Don't call a paradox a "paradox," Zermelo insists. He advocates using the word "antinomies" instead—meaning an unreasonable contradiction between two equally reasonable beliefs. The word "paradox" is insufficient, he claims, since "it doesn't contain anything of an inner contradiction."

And just as you will find sources in the mathematical literature calling it the Russell-Zermelo Paradox, you will also find sources that dutifully employ the term "antinomy." But my eyes glaze over when I see that word. It falls flat. It's a failed label. A bankrupt branding. To channel Bertrand Russell's woeful lament for a moment, I hope a train hits me in the face every time I see this term. As a lifelong advocate of plain language, I can tell you why this term has never caught on. Everyone has at least some sense of what a paradox is, even if they're wrong about it, but I doubt even one in a hundred could say a single thing about an antinomy. So while I again acknowledge Zermelo's greatness, I nevertheless ignore his greater wisdom. (Sorry again, Zermelo!)

OMNI COPIA PARADOXON

Imagine the range of emotions on Frege's face when he reads Russell's letter. Intrigue. Delight. Pause. Surprise. Concern. Anger. Sadness. Shock. "Your discovery of the contradiction caused me the greatest surprise and, I would say, consternation," Frege finally writes to Russell in response. That's putting it mildly. Some sources insist Frege is thunderstruck through and through. And why not? "[Your paradox] has shaken the basis on which I intend to build arithmetic," he says to Russell. In other words, *Thank you very much; you have just destroyed the last twenty years of my life!*

Imagine working most of your adult life in pursuit of a singular goal and then, at the age of fifty-four, having some snot-nosed twenty-something pull the rug out from under you. Frege can read the writing on the wall. He acknowledges Russell's letter by adding a postscript to his third book. But he has neither the stomach nor the will to go much further. "A scientist can hardly meet with anything more undesirable than to have the foundation give way just as the work is finished," he writes in his book. "In this position I was put by a letter from Mr. Bertrand Russell as the work was nearly through the press."

Russell is truly impressed by how well Frege takes his letter. "On the verge of publishing the second volume of his life's great work, to have it all wiped away in a second by a much younger mathematician," Russell writes to a colleague. "It was almost superhuman."

The troubles with the foundations of math are not limited to Russell's Paradox. The cumulative effect of all the paradoxes together is to cast doubt on the cogency of modern mathematics. Nevertheless, the biggest one of all, the mother of all the antinomies, as Zermelo would say, is Russell's Paradox. It cuts to the heart of mathematics and has a "downright catastrophic effect in the world of mathematics," according to the American philosopher of math Wilfried Sieg. Why? Think of the rotten-apple principle. If you can derive one example of inconsistency from the fundamental logic of a system, then other inconsistencies could obviously also be constructed.

Finding a way of dealing with Russell's Paradox and fixing that foundational flaw is therefore seen as crucial—though Frege won't attempt it himself. He feels deflated. His life's work... fallen short! He writes a few more articles after this but largely abandons his designs to show that mathematics can be reduced to logic. (Although later in life, he has a brief flash and decides that geometry, rather than logic, must lie at the root of mathematics.)

Many others, however, are not content to sit idly by and let their field be plagued by paradoxes. "It seemed that the wages of the sin of biting from the tree of knowledge of infinity was self-contradiction," American mathematician Howard DeLong will write in 1970. "The dogmas of the past were inadequate, and a new approach was needed." The new approach was to use mathematical methods and develop a new mathematical logic to fix the problem. The chief proponent of this approach is Russell himself. After his exchange with Frege in 1902, he vows to find a way around his own paradox.

Cantor's set theory can serve as a solid foundation for mathematics if and only if it can be freed of paradoxes, and Russell thinks he's just the one to do it. And why not? He's tremendously influential. He will spread his influence and infuse the field with his positive energy. He's not the only one. Others will follow, he is sure. And he's not wrong. Every major approach to set theory in our modern world finds its "roots" in the response to Russell's Paradox, writes the scholar Kevin C. Clement in a 2007 essay titled "A New Century in the Life of a Paradox."

But Russell is not right either—not exactly.

MATH, PEACE, LOVE, AND HORROR

One other thing that has its roots in that global response to Russell's Paradox is Alys and Bertie's relationship. I read a lot of articulations of Russell's Paradox while writing this book, and perhaps the best one is from the late mathematician George Boolos, who in 1998 will describe Russell's Paradox most succinctly as *it is if and only if it isn't*.

That fairly well also describes Bertie's marriage to Alys. After 1901, it's on the rocks and worse. And yet they will stay in that *Remains of the Day* misery of an arrangement for years. Russell tells Alys after his peace conversion that he flat-out doesn't love her. It becomes one of those relationships that's more about staying than loving. *You couldn't call it living, for it ain't*, as American poet Robert Frost once said. He was talking about poverty in rural New England, not marriage, but it's the same frigid, fallow landscape. The Russells' marriage is a cold, half-starved thing in winter. *Good fences make good spouses!*

Bertrand Russell gets weird after falling out of love with Alys. He has a lot of compassion for humanity but is filled with personal self-loathing. And he's brimming with logic, which makes him doubly thick-skinned. Logicians never fit in easily anywhere. His friend Beatrice Webb says his faith is always in the absolute—but as an adjective, not a noun. "An absolute logic, absolute ethic, absolute beauty, and all of the most refined and rarefied type," she says. "A proposition must be true or false, a character good or bad, a person loving or unloving, truth speaking or lying." For him, there are only two types of understanding. There are beliefs obtained by inference, derived from data, and free of prejudice. True beliefs. Then there is the second type of understanding, which he says is "not obtained either by inference or by analysis of perceived facts." Another name for this type of belief, he says, is prejudice.

Russell has somewhat rose-colored glasses in this regard. Just because inferred beliefs are free of prejudice and obtained by a weighing of the facts and data does not necessarily mean they are true. It's the fallacy of pure process data again, though a slightly different flavor of it—not so much the risk of ignoring outcomes for the sake of pure process data, as was the case when the British built concentration camps during the Boer War. With Russell's mathematical philosophy, it is more the danger of inferring outcomes based on process data alone.

For instance, if you wanted to know what happened at the beach on a hot summer day, you could weigh a word cloud of collected observations like {{teenagers}, {laughing}, {crashing}, {splashing}, {waves}}. From that data, you could infer the verb "swimming." It's perfectly logical, but the inference isn't necessarily right. The data may all be true and weighed appropriately, but what they denote could still be ambiguous. Those words could be describing a group of teens who belong to a running club. They could be jogging along the beach in the surf just up to their ankles. The correct verb should be "running," then. In the same way, beliefs obtained by prejudice are not necessarily true.

———◆———

In embracing an absolute logic, Russell longs to discover a self-contained set of propositions that are sufficient to establish the logical foundations of math—his ultimate form of true belief. And he's sped along in that goal by the success of his 1903 *Principles of Mathematics*. "A clear and well-written account," one reviewer at the time says of the book.

Not everyone loves it, though. Other reviews are more scathing. "The exposition is hurried and often degenerates into bald repetition of Mr. Russell's views," another reviewer writes. "Often, too, they are unconvincing, dogmatic, or superficial." Nobody calls Russell's next book superficial.

THE GREAT MATH WAR

Even before the publication of *Principles of Mathematics* in 1903, Russell and Alfred North Whitehead decide to collaborate on a work to advance their particular philosophical and mathematical worldview, known as logicism—the idea that math is based on logic or that mathematical truth and logical truth are one and the same.

It's strange to see Whitehead throw in on the project. When they start, he's already deep into a book of his own, a second-volume sequel to his book on universal algebra, which he published a few years earlier. Whitehead is several years into his new book by the time they start on *Principia*. He's so far along, in fact, that Russell is afraid he'll say no to collaborating. But Whitehead says yes. He wants to collaborate. They're both at Trinity College, Cambridge (Whitehead is several years older). They are friends. They traveled to Paris together in 1900 and attended the meeting where Hilbert gave his twenty-three problems talk. And they are their own little set of couples—two husbands, two wives.

Once Whitehead throws in, Russell is elated. "To my great joy he has agreed," Russell says. But neither of them has any idea what awaits. Only a small corner of the universe of logic had ever been explored from ancient times, Russell will later write. Until the nineteenth century, nobody had really done anything original in logic for two thousand years—not since the days of Aristotle. But the approach of using symbols in logic, pioneered by Boole, Frege, and Peano, changes the game. Now Russell and Whitehead want to take it further. They want to fully explore "logicism" and deduce ordinary mathematics from logic.

THE *PRINCIPIA MATHEMATICA*

Conceptually, logicism claims that logic lies at the foundations of math—that arithmetic, algebra, pure analysis, geometry, and all the other areas of math always come back to it. Practically, it means that any concept, proof, or axiom in math can be defined using combinations of the most primitive predicates of logic alone (things like "and," "or," and "not").

In their collaboration, Russell and Whitehead follow Frege's example. They define the primitive notions and propositions of math using logic and then derive higher mathematical theorems with those same propositions. They marry that approach with Peano's more amenable if still somewhat arcane symbolism (as opposed to Frege's *looks-more-like-a-bad-tattoo-design* weirdness). The approach seems straightforward, even though it's not that simple. "This is no longer a dream or aspiration," Russell writes before they begin the project. Many of the mathematical underpinnings have already been worked out. "In the few remaining cases, there is no special difficulty."

No special difficulty? These are famous last words.

They start out thinking their book will be quick and easy. They think a year will be long enough to complete the job. They'll finish it by 1904. And why not? Two smart guys. One true subject. They give the new book a lofty Latin title—confusingly, *Principia Mathematica* (Principles of mathematics), which translated to English is the name of Bertie's just-published book. That makes sense in a way because they think at first that the new book will simply be a sequel to that older book. Easy enough to finish it in a year. Maybe they'll finish even sooner. But they're off by years.

"As we advanced, it became increasingly evident that the subject is a very much larger one than we had supposed," they write in the preface of the book, the first volume of which finally appears in 1910.

"Larger" is an understatement. The book is a daunting project, and not just because it takes years to complete. The book itself is massive. It fills thousands of pages and three fat volumes when they're done. The text is so heavy in symbols that the work becomes almost impenetrable. Nor does it accomplish what they set out to do. *Principia Mathematica* is a bold attempt to establish the fundamental thing many believe in 1901: that formal logic and mathematics are one and the same. But it falls short of proving that, even as it balloons into a big, thick, intimidating, fills-a-third-of-a-bookshelf unapproachable monstrosity—the kind of book lots of people talk about but few people read.

The perceived difficulty in digesting their book comes as no shock to Russell. He revels in the book's fundamental opaqueness and the fact that nothing seems obvious. He actually finds its difficulty appealing. "The fact is symbolism is useful because it makes things difficult," he writes. "Obviousness is always the enemy of correctness."

THE GREAT MATH WAR

The only thing that's obvious to Russell at the outset of the project around 1903 is that he knows he somehow needs to deal with Russell's Paradox. And he and Whitehead aren't the only ones who think so.

HOC CONTINUUM HYPOTHESIN

When he was first manifesting mental illness in the 1880s, Georg Cantor visited Heidelberg, where he called for the creation of a new mathematical society—one with an international flavor as opposed to the traditional in-country societies, which might allow foreign members but were squarely focused on nationals. He succeeded, helping create the International Congress of Mathematicians—though others did the heavy lifting for the most part.

At the group's first meeting in Zürich, the opening keynote speakers singled Cantor out for appreciation. At the Second International Conference in Paris in 1900, Hilbert gave him an even more special nod by selecting as the very first of his twenty-three problems the one remaining pickle that had dogged Cantor for years—proving the continuum hypothesis: that there is no infinite set size between small infinity, the $2^{\aleph_0}$ set of whole numbers, and large infinity—its superset $\aleph_1$, equivalent to the cardinality of the infinite set of real numbers:

$$2^{\aleph_0} = \aleph_1$$

The continuum hypothesis claims the cardinality or size of an infinite set is a two-headed Highlander—*There can be only one...or the other!*

Cantor was never able to prove it, but by making this the first of his twenty-three great problems, Hilbert hoped to inspire somebody to do so. Four years after his famous talk, in 1904, it still hasn't been done. Even so, things couldn't have been better for Cantor—at least on paper.

Some 336 mathematicians show up for the Third International Congress of Mathematicians in August 1904 in Heidelberg—the city where Cantor first proposed holding such international meetings almost twenty years before. Same city, same goal.

ALL CRETANS ARE LIARS

By now mathematical societies are sprouting up in cities across Europe like mushrooms on a wet log, and almost all of them want Cantor as an honorary member. Thus, 1904 should be a victory lap for Cantor. His organizational genius is evident. His fame and importance are growing. His community is taking off. His mathematical insight is "almost universally acknowledged," according to his biographer Dauben. That year, he receives the Sylvester Medal, the top prize of the Royal Society in London.

So when the Third International Congress convenes in 1904, people are amazed to discover that Cantor himself is there. He's almost a mythical figure at this point. This should be his triumph. His proud victory lap. He should be wrapping himself in a mathematical flag and giving all credit to God. There should be applause. There should be adoration. Adulation. Punch and cookies. But there's nothing like that to see. Things don't go well at all. And it becomes another angry episode in the long history of Cantor's shaken mental status.

THE KÖNIG CONTROVERSY

Cantor is deeply disturbed when a Hungarian mathematician from Budapest named Gyula "Julius" König presents a paper that fundamentally challenges him by attempting to knock down the continuum hypothesis. König says the continuum hypothesis is wrong, and he claims that he can prove it. Cantor refuses to believe it, of course, steady in his mathematico-religious faith. But he can't immediately spot any error in König's proof, even though he's sure there must be one. After all, God is God, math is math, and Cantor is Cantor.

König claims that the continuum hypothesis is false on technical grounds, saying the aleph-one large infinite set of real numbers cannot be well-ordered—meaning you cannot arrange the elements of the set in a logical ascending hierarchy the way you would do a finite set, like the first ten non-zero numbers of the Fibonacci sequence: {1, 1, 2, 3, 5, 8, 13, 21, 34, 55}. Well-ordering is seen as a prerequisite for proving the continuum hypothesis, so if it can't be done, the continuum hypothesis cannot be proven. That reasoning is solid, but König is nevertheless wrong. He made a mistake in his work, and the flaw is soon discovered by the same young student of Hilbert's who codiscovered Russell's Paradox. (Hello, Zermelo!)

König promises Hilbert he will submit an article based on his lecture for publication in the journal *Mathematische Annalen*, which Hilbert edits. The article never materializes, however, because König discovers his mistake. He's wrong, wounded, and filled with regret. "After a series of unhappy days, I must finally report to you that I cannot send you the article promised," König writes, calling his own failed proof "the catastrophe of the congress."

"How I regret what happened, and how I suffer from it," he says in his letter. "I find it incomprehensible how I could not see this earlier."

Zermelo, energized by the whole drama, is inspired to dive deeper. He tackles the very issue König is dealing with. König had thought he could prove that the nondenumerable, $\aleph_1$ "large infinity" set cannot be well-ordered, but his proof was flawed. Okay, fine. But failing to prove that something *can't* be done doesn't definitively show that it can. So after the congress, Zermelo decides to sit down and prove that every set, even the large infinity real number set, can indeed be well-ordered.

By the fall of 1904, Zermelo has done it. He writes to Hilbert, telling him he has a proof that every set can be well-ordered—his so-called *Wohlordnungssatz* (well-ordering theorem). "For any family of non-empty sets there exists a correspondence that associates to each of these sets one of its elements," he writes, describing his organized way of choosing, which will ultimately lead to his so-called "Axiom of Choice."

THE AXIOM OF CHOICE

What is the Axiom of Choice? Think of a school where you have at least one but possibly more students in each classroom (your nonempty sets), and you decide to order all the classrooms based on one particular student from each room. That's a trivial analogy, but Zermelo goes further, claiming that you could employ a "choice function" to do the same thing for an infinite set. (Every classroom that ever was, is, will be, or ever could be.) He uses this notion of choice and infinite ordering as the basis for working out an axiomatic system for set theory, doing more for Cantor than Cantor could do for himself.

ALL CRETANS ARE LIARS

"In future years," modern American mathematician and set theory expert Akihiro Kanamori will write a century later, "the basic scaffolding provided by Zermelo's axioms, with their schematic simplicity and open-endedness, would win out as a working foundation for mathematics." (God is God, math is math, and Zermelo is Zermelo, as it turns out.)

Some people are impressed by Zermelo's efforts in 1904. And now, years later, after a century of appreciation, many, many more are. "It turned out to be so adequate that, up to minor improvements, it is still the main basis for modern mathematics," Dutch mathematician Dirk van Dalen will write.

But if people tip their hats to Zermelo today, at the time...not so much!

When Zermelo's work is first published, it draws pointed criticism. Fire and ire. Zermelo shows that Cantor's basic assumptions are correct, but how he does it is controversial. After Zermelo's proof appears, the next issue of the same journal runs red with scratched letters from scorched-earth doubt-filled readers.

The most stinging criticism comes from Paris, from none other than the great mathematician Poincaré. He's not the only one. No German mathematician can make a dunk in those days without an armchair Franco referee whistling foul. French mathematician Félix Édouard Justin Émile Borel also objects to Zermelo's proof. It demands an infinite iterative process, which he thinks is absurd. "Any argument where one supposes an arbitrary choice to be made a non-denumerably infinite number of times [is] outside the domain of mathematics," he writes.

The Italian mathematician Peano, whose work so inspired Russell, also objects to Zermelo's Axiom of Choice. You can't set up an infinity of arbitrary choices, he claims. Impossible! Zermelo responds that Peano is too restrictive and self-limiting. He calls Peano's own mathematics "artificially mutilated" to some extent.

But by far the most profound criticism of Zermelo in 1904 comes from Poincaré. He has already publicly criticized some aspects of set theory prior to the international congress. Once he sees Zermelo's paper, he doubles down. "Although I am rather disposed to Zermelo's axiom," Poincaré writes, "I reject his proof." Poincaré's criticism is that Zermelo uses

THE GREAT MATH WAR

definitions that create a "vicious circle" because he first specifies a collection and then picks out members of it.

ZERMELO THE GOD

After Zermelo decides to respond to the criticism and address the critics, he does so by revising his 1904 proof that every set could be well-ordered. He expands his scope from there and begins working on axiomatizing all of set theory in 1905, following what Hilbert did with geometry half a decade earlier. "I intend to show how the entire theory...can be reduced to a few definitions and seven principles, or axioms, which appear to be mutually independent," Zermelo says.

Where these definitions and principles come from, Zermelo says, he will not discuss. Whether they can be proven, he adds, he'll not attempt. "But I hope to have done at least some useful spadework here for subsequent investigations in such deeper problems," he says. More ditch digger than dodger, Zermelo does just that. He publishes his revised work in 1908, and it's a huge accomplishment, as experts accept today.

Oh, sure—some parts of it will still be seen as difficult in 1908—and some even worthy of rejection. And people will still be ruthless in their assessment of him then. Russell himself will dismiss one part of Zermelo's 1908 work by calling it "so vague as to be useless." Nevertheless, Zermelo will make history.

What he does specifically is define a set of axioms that people now call ZFC. "Z" for Zermelo as well as "C" for the Axiom of Choice. (In other words, he puts the Z and the C in ZFC, the basis of modern math.) Taken together, Zermelo's axioms allow mathematicians to generate and manipulate sets and will become indispensable—winning out, in Kanamori's words.

But that's all later. In 1904, Russell is aware of but under-appreciative of Zermelo's breakthrough. Poincaré's vicious-circle criticism of Zermelo's work hits Russell's mind hard. He is by then deep into his new book with Whitehead. The vicious-circle idea shakes him. He has great respect for Poincaré, and the heaving criticism of the elder French mathematician really makes Russell think—and ultimately leads him to a basic conceptual breakthrough in *Principia Mathematica*.

ALL CRETANS ARE LIARS

THE MANUTE BOL OF ANCIENT PHILOSOPHERS

The road Russell goes down is the one first taken thousands of years before by the most famous liar in human history (Tricky Dick Nixon notwithstanding). Epimenides the Cretan is a mysterious figure in the ancient world who famously formulates the Liar's Paradox, which is something like *This statement is false.* (If it is, it isn't, and if it isn't, it is.) But that barely hints at how cool Epimenides was. Let me explain.

When I was a teenager, my father and I were both fans of the late great Manute Bol, the seven-foot-seven-inch Sudanese starting center of the NBA basketball team the Washington Bullets (now Wizards). I loved Bol for the same reason I love Epimenides. The Bullets drafted him in 1985 and paraded him out almost like a carnival freak-show spectacle next to five-foot-three point guard Muggsy Bogues—the tallest and shortest players in the NBA, side by side. But that gimmicky display was just eye candy. What I love most about Bol is that he was a unicorn—truly great in one narrow sense: He was a shot blocker unlike anyone the world had ever seen. A legend. The best ever! He still holds several NBA records for blocked shots, including a unique achievement that will most likely never be repeated in the sport: He's the only player in basketball history to have recorded more career blocks than points.

Likewise, Epimenides is the only philosopher in history who is more famous for telling lies than speaking the truth. And that's just the beginning of what's weird about him. He was said to be able to astrally project his spirit away from his body. He was said to have lived to age three hundred, an obvious lie in itself. He once supposedly slept for fifty-seven years without a break. And when he finally stirred and awoke, the first thing he said was the classic logical one-liner for which he'll be forever famous: "All Cretans are liars!"

According to the Apostle Paul in the Bible's New Testament, Epimenides's exact statement was "Cretans are always liars, evil brutes, and lazy gluttons," words that resonated for Paul and his followers in a gospelly way as a repudiation of everything they saw as detestable in human or beast: deception, corruption, excessive consumption, and the awful brutish behavior of evil people as compared to the righteous.

But Epimenides's statement took on added significance through the centuries, becoming known as the Liar's Paradox, an illogical idiom equivalent to saying, "It's true this statement is false." That makes the Liar's Paradox very close to Russell's Paradox in the sense that it's a true statement if and only if it isn't.

There's no telling how righteous or unrighteous Epimenides truly was—how much you can really say about him, or even whether he existed at all. Maybe he was a legend. A bedtime story old mathematicians would tell their little logicians. Perhaps he was an amalgamation of different stories about different people in the ancient world. His fifty-seven-year nap suggests that he may have been multiple people, a sort of ancient-day Dread Pirate Roberts, with not just one Epimenides but several—each one playing the part for a time and then selling the title to someone else when the old Epimenides retired. (True to form for a liar.)

There's actually something fitting in considering Epimenides as a collection of liars because the big breakthrough Russell makes in solving his paradox comes from thinking along those lines—not of Epimenides the man but of Epimenides the type. Thinking about Poincaré's vicious circle leads Russell to develop his "ramified theory of types," his solution to the pickle of his own paradox.

THE THEORY OF TYPES

This theory of types is one of the greatest contributions Russell makes to mathematical philosophy, and it forms a cornerstone of *Principia Mathematica*. It basically eliminates Russell's Paradox by changing the rules and avoiding it. How? Inspired by Poincaré's criticism of using the vicious circle, Russell realizes that most of the paradoxes in math emerge because of some wayward self-reference. So he seeks to do away with all that in one fell swoop by dividing logical propositions containing objects and relationships into a hierarchy of different classes. Russell imposes this hierarchy of types on sets and states what he begins to call his Vicious Circle Principle: no sets, class, or type can have members defined in terms of itself. Any given type should be exclusively defined in terms of lower types.

That means the barber who shaves everyone who doesn't shave himself is of a higher type than his clients (not in socioeconomic status, perhaps, but as a defined category—worker versus client). The barber himself is always a barber type, never a client type. And he's a barber by virtue of the fact that he shaves *them*—not *himself*. Russell's theory of types holds that the barber is not part of the shaved group because he's not their type—at least not in his own barbershop. What the barber does at home in his own bathroom is irrelevant (except perhaps to the barber's partner). Either he shaves himself or he doesn't, but either way, he's never part of the in-shop shaved set.

This avoids the paradox because you can still say the barber shaves everyone who doesn't self-shave. The question of whether that means he shaves himself becomes meaningless because the theory of types basically reformulates the barber's role: He shaves everyone *else* who doesn't shave himself. To put it more generally, the barber is held to two criteria: First, he shaves everyone who doesn't shave himself, and second, anyone he shaves never shaves anyone else in the barbershop. More technically, any expression containing a variable is itself a higher type than that variable, Russell says, thus avoiding his own paradox.

To illustrate this with Epimenides, in Russell's formulation, the type Epimenides referred to when he uttered his famous statement "All Cretans are liars" would not, under Russell's type hierarchy, actually include Epimenides himself. It's more like he said, "All *them* Cretans are liars." Thus by putting Epimenides or the barber on an elevated pedestal, in a sense, Russell avoids the paradox.

It's a strange fix, however—a *bend-over-backward-to-avoid-a-self-referential-contradiction* hack. But it works, and it accomplishes Russell's aim. His theory of types does not admit a class of all possible classes or a set of all sets but rather institutes a hierarchy of classes that cannot be members of themselves. It's a triumph of sorts because it removes the logical contradiction that is holding them back. But because Russell mucks around with basic definitions in order to find a fiddling fix, his result is also seen as a "rather awkward reconstruction of mathematics," according to the modern American philosopher Paolo Mancosu. The mathematician Kanamori calls Russell's theory of types unwieldy and unsightly: "a fearful symmetry imposed by an artful dodger."

◆

THE GREAT MATH WAR

That "unwieldy and unsightly" criticism comes from the fact that the theory of types requires Russell to introduce three more new blank-check axioms to the foundations of math to rebuild it on solid ground. The problem is that it's not clear whether these other fixes can themselves be proven with logic. Realizing this, Russell and Whitehead declare in *Principia Mathematica* that the additional axioms are not actually logically necessary—just empirically true. That causes some critics to dismiss the theory of types as a stupid game of syntax.

Whitehead and Russell impose "a ban on the construction of certain kinds of linguistic expressions," British mathematician and physicist E. T. Whittaker will say in the 1940s, "thereby avoiding all formations which could lead to logical contradictions."

Russell is aware of these criticisms but ultimately unmoved by such concerns. "The theory of types is really a theory of symbols, not of things," he says. "In a proper logical language, it would be perfectly obvious."

Having made that breakthrough and working with Whitehead for a few more years to finish the book—eight years in all from beginning to end—Russell is fork-tender and finally done. "My intellect never quite recovered from the strain," he will later say.

Queen Math is demanding, if not fickle. Several times, Russell will later claim, he contemplates suicide in the early 1900s because of mental strain—both marital and mathematical. In his autobiography, Russell writes that he would stand on the footbridge watching steam engines at Kensington Station and wonder whether he could gather the courage to fling himself to his death on the tracks below.

The love of math alone keeps him going, he claims. *Principia Mathematica* is his "chief interest [and] chief source of happiness"—though perhaps that's not solely a reflection of how highly he regards math but a dual expression of how much he hates life. (All those troubles with Alys—*You couldn't call it living, for it ain't!*)

───◆───

On that sad, terrible night when Bertie confesses to Alys that he no longer loves her, he finishes the evening in his study, toiling away on lofty mathematical ideas while Alys is left sobbing down the hall. He can hear her, but he can't help her, he tells himself. There is nothing to do, he decides. He needs to be cold "in the deliberate hope of destroying her affection," he will later write. Then he decides that he needs to withdraw even more. He'll stop making eye contact with her, he resolves—never look in her eyes again.

Math becomes his cordial cherry after that—an intoxicating distraction topping the miserable cocktail of his life. But even that's not enough. He continues to wallow. "I am unhappy beyond what I know how to bear," he writes in one of his journals. "Happiness is gone forever, my work is second-rate, and all that I care for is going or gone."

In the years to come, as *Principia Mathematica* and his marriage both start winding down, Russell dives headfirst into political activism. It becomes a sort of refuge. A crutch. A way to separate himself from his personal life for a brief, blissful recess—as mathematics once did. He becomes an outspoken critic of the ruling conservative party. He gives public lectures, writes articles, and pens op-eds. He's good too. Funny when he chooses to be. Clear when he must be. And exciting the whole time. "Morally, England is on trial," Russell once declares in typical hyperbole. Having a major intellectual like Russell stoop to petty political stump speeches seems like overkill to some. In January 1904, his own brother-in-law attends a talk Russell gives and later says to Alys, "It seems like using a razor to chop wood."

Even so, Russell's baby step into politics will soon have major consequences. The next few years will see the biggest disruption ever in Bertie's life, both personally and professionally. He separates from Alys, ruins many of his professional relationships and personal friendships, throws himself into a tumultuous churn, and falls far, far from the heavenly loft of mathematical logic and the foundations of math to the tumbledown terra firma trench of antiwar activism.

All this because of the entrance onto his stage of a new queen—his brother-in-law's best friend's wife. She's an amazing woman who will irrevocably change him the way true lovers do. His prime disruptor. The next love of his life.

This is her story too.

CHAPTER SIX

THE DAY OF JUDGMENT

1909–1911

"
*But that I am forbid
To tell the secrets of my prison house,
I could a tale unfold whose lightest word
would harrow up thy soul.*

"
—William Shakespeare

The Last Judgment: The Graves Open and the Dead Emerge. A 1615 line engraving by P. de Jode the elder after Jehan Cousin the younger. Cropped from a public domain image via Wellcome Collection.

Lady Ottoline Violet Anne Cavendish-Bentinck, like Bertrand Russell, is raised in the selfishly narrow, silver-spoon confines of an upper-crust Victorian *Grey Gardens* world of well-heeled privilege and awkward elitism.

Like Bertie's world, hers is a silk-rag society of wealth, birthright, promise, status, honor, legacy, and charm. "We were both *ancien regime*," Ottoline will later say. Her life reads more like an aristocratic period-piece drama than his, though—one that is happy, light, mournful, and stupid all at the same time: live-in servants. Extravagant clothes. Fabulous estates. Summer houses. City flats. Great halls. Gorgeous rooms. Cutting-edge art. Fancy china. Strutting peacocks. A pearl necklace that once belonged to Marie Antoinette. Lavish stays at exclusive Swiss spas and lavish Italian villas—Venice. Padua. Bologna. Ravenna. Urbino. Gubbio. Siena. Perugia. Assisi.

It's a dusty old world where even the dust glitters.

Like Bertrand, Ottoline lost her father at a young age. He was a lieutenant general. A cavalry officer. Seventh Dragoons—they of dashing, crisp uniforms and slashing, steel sabers. A photo of him as a white-haired gentleman in the 1870s shows the general seated at a desk, pen in hand, proud, imposing, yet soft. In the photo, he seems approachable, as fathers lost at an early age often are.

After her father died, Ottoline moved as a young girl with her family to an empty, rich place called Welbeck Abbey, a cavernous, cold, and hollowed-out estate on the edge of Sherwood Forest. It was a shambles of large, lonely buildings on an estate that had been run for decades by her recently deceased uncle, an eccentric old recluse people called "the burrowing duke" in honor of his most strikingly peculiar quirk: He had been obsessed with digging tunnels around his ancient estate for decades.

These were massive excavation projects, not tiny crawl holes. He dug whole underground chambers. Cavernous paths, miles long. They were no

THE DAY OF JUDGMENT

small feats of engineering either. One tunnel passed underneath a lake. Some tunnels contained full-fledged roads. Some sections of tunnel were as tall as trucks and wide enough for several people to walk side by side—not that the duke ever did. He had parallel tunnels excavated next to the main ones so that his servants and contractors could traverse those instead. "A lonely self-isolated man," Ottoline will say of him in her autobiography.

The burrowing duke had another strange quirk: roast chicken. In one of his buildings, chickens were roasted one after another, day and night—as many and as often as needed to always have fresh roast fowl on the menu when the old man rang. The servants never knew when the call would come, so they constantly cooked chickens day in, day out. Once the bell did ring, a heated truck was loaded with the fresh-roasted pollo and lashings of side dishes and pushed to the house through yet another underground tunnel along a long rail—like El Chapo sneaking cocaine into San Ysidro in the 1980s.

As a child, Ottoline thoroughly explored the strange, broken-out buildings and all the lonely, selfish caverns. She discovered bizarre hidden treasures. One building had a room stacked with cupboards packed floor to ceiling with green boxes, each one filled with brown wigs. In another room were boxes and boxes of cream-colored socks. Here was a room filled with white linen shirts, all frilly and out of fashion. Another housed box after box of fine silk handkerchiefs. There was a chest full of old velvet robes dating back to a king's coronation centuries past. In another chest, she found a little green purse with £2,000 in cash. There were snuff tins, watches, vintage statues, and many other random treasures. Once lost. Now found. In dust.

One underground passageway led to an old building where a massive sunset was painted on the roof. It had stacks of paintings dating back to King James, all removed from their frames. "Pictures that had come down from generation to generation," she later wrote. "So elegant and lovely and sentimental, with pearls in their ears and long trailing lovelocks of golden hair hanging over their collars, and of the ladies they loved, with very low-cut dresses, showing curvy breasts and embroidered skirts and exquisite shoes."

Those creepy ancient eyes were ever present. The eyes of long-gone fancy people in expensive paintings torn from their frames. They followed her where she walked. Her aristocratic past exerted its lingering gravitational pull in this way. Old money. Ancestry—like some great, faraway, forgotten galaxy in an obscure corner of the universe. The stars had long ago collapsed and gone supernova. Atomic fusion had long since ceased. The nebula was slowly cooling. And yet the galactic remnant of her family's past still had residual mass. Empty greatness. It bent the light of all the neighboring stars toward itself.

THE BEDFORD SQUARES

Years later, when Ottoline married Philip Edward Morrell in 1902, the gravitational force of her past came back to haunt her. Not everyone was enthusiastic about the match—least of all Alys Russell's brother Logan Pearsall Smith, who had been best friends with Philip since their college days. Logan hated the prospect of the marriage—and maybe Ottoline as well. Logan was "critical and suspicious of me," Ottoline wrote after their first meetings. And she was not being paranoid. Logan told Philip their friendship was over because of the marriage, and he made plans to travel in Europe so that he wouldn't have to attend the wedding.

By strange coincidence, Ottoline and Philip married at the exact same time Bertrand Russell had his so-called awakening and supposedly fell out of love with Alys in an instant. Ottoline and Russell didn't know each other then, however, and wouldn't properly meet for a few more years.

Meanwhile, Ottoline and Philip moved to London as Bertie was toiling away on the *Principia Mathematica*. They bought a place in Bedford Square in the West London neighborhood of Bloomsbury, an up-and-coming trendy neighborhood with grand Georgian houses and gorgeous tree-lined squares on the borders of urban decay: Near to trains, soot, noise, smoke, voices, vagrants, and seedy hotels.

Philip, an Oxford-educated lawyer by training, decided to run for office, declaring as a Liberal Party candidate in his home turf of South Oxfordshire. The couple campaigned tirelessly, making their rounds to this small community or that. Friar Park. Henley. Watlington Park. Ship Lake Court. Newington House—the home of the painter Ethel Sands, whose mother was friends with the author Henry James and who was in the good

THE DAY OF JUDGMENT

graces of the greatest liberal of them all, the late William Ewart Gladstone, who served as prime minister from 1868 to 1874, from 1880 to 1885, in 1886, and from 1892 to 1894.

One of the first things I read when I started researching this book was a short Gladstone obituary from the end of the nineteenth century—mainly out of curiosity, but it did inspire me to adopt the Gladstonian trait of always reading three very different books at the same time. I was familiar with Gladstone's name because when I was a teenager, I lived, studied, and worked on an estate in southern Scotland that had once been a large mansion but had become a small school. Its dining room had a great wooden table and big glass windows looking out on a flat lawn that spilled onto border hills and heather bush. Every lunch and every dinner, people sat around this table eating fabulous meals and discussing religion, philosophy, and life. Reserved for the top of the table were special place settings: cups, plates, bowls, and saucers belonging to a china set we referred to as "the Gladstones"—dishes once owned by the great man himself.

On more than one occasion, one of these irreplaceable plates slipped and cracked or dropped and chipped and sometimes broke into pieces. An artist living nearby would occasionally come to the school to mend and glue the broken pieces back together. At the top of the house was her workshop—not much of a workshop, really, just a small desk in the attic with a few plastic bins filled with sand. I once sneaked up to look. She wasn't there, but she had left the broken plates immobilized in the sand, gluing the pieces back on one by one. I can still see them there in my mind—all those half-mended plates with jagged edges and drying glue jutting out of the sand. They stuck out like broken tombstones. Monuments to a forgotten past. Chipped glory. Long-gone greatness.

———◆———

The greatest tragedy of Ottoline's life occurred after they moved to London. In 1906, she gave birth to twins. Her daughter, Julian, survived, but Julian's twin brother died a few days later. It was a shattering psychological blow, and she suffered dark spells of depression for years. The same year, Philip was unexpectedly elected to parliament, riding an anticonservative liberal wave to crushing victory. The demands on Philip's time became profound. Glad-handing constituents. Taking meetings. Attending sessions of parliament. Up, down, in, out. Ottoline had never felt so alone. "I hope he doesn't realize how bad I feel," she wrote in her diary. "I am dead—dead."

Some months later, Ottoline decided to throw herself into what she called the "great whirlpool of London," that chaotic mess of noisy machines, angry mobs, highbrow snobs, and some very interesting people. Her neighborhood was home to the Bloomsbury Group, an odd early twentieth-century set of writers, artists, and intellectuals that included Virginia Woolf. Her sister Vanessa Bell. Vanessa's husband, Clive. Virginia's husband, Leonard. Edward Morgan "E. M." Forster. John Maynard Keynes. Lytton Strachey. Roger Fry. Duncan Grant. And soon (some say) Bertrand Russell as well.

Ottoline hosted some of the best parties this group had ever seen—her house became a place where the already famous met the soon-to-be famous and the forgotten famous met the wannabe famous. Prime ministers. Painters. Drunkards. Geniuses. Henry James. William Butler Yeats. D. H. Lawrence. The famous Russian dancer Vaslav Nijinsky became a fixture there. People came and went. They laughed and drank. They stayed and danced. They contemplated and debated. Life. Art. Music. Theater. Poetry. Mind. So successful are her Thursday evenings that some call her the "high priestess of Bloomsbury."

In September 1909, Ottoline meets Bertrand Russell after Philip's best friend, Logan, Russell's brother-in-law, introduces them. Logan is a fixture at the Morrells' country cottage near Oxford this summer, and he brings his sister, Alys, and her husband, Bertie, by one afternoon to meet his old friends.

"Bertrand Russell is most fascinating," Ottoline writes in her diary. "I don't think I ever met anyone more attractive, but very alarming, so quick and clear sighted, and supremely intellectual—cutting false and real asunder. Everything he said had an intense, piercing, convincing quality."

THE DAY OF JUDGMENT

He is pierced as well as piercing. Russell is struggling that summer to oversee the final preflight proofing of the first volume of his book *Principia Mathematica*. He and Whitehead have just finished, and he's a mess of nerves. When they first started their book, Russell and Whitehead figured it would take them a year or so to finish. But it grew and grew. And now, at the end, they've spent nearly nine years and filled three fat volumes. Volume I alone is 666 pages. It uses four hundred different symbols—a typesetter's nightmare—and lays out thousands of equation-like logical propositions as complicated strings of symbols.

Russell has had terror dreams about the book. His house is catching fire. All his precious pages are burning up in the flames. Perhaps he'll melt as well. He has that sinking feeling of postliterary regret—that the loss of his pages would be no real loss and that his work doesn't matter. He's like Hercules lopping off the last slimy head of the Hydra, only to find two more heads with his own face in its place.

"DESOLATE, COLD & UNRESPONSIVE"

The publishing house at Cambridge University regards the book with crushing pain. They bleed worry. Money. It's a massive tome. Ponderous. Opaque. Filled with strange symbols. Many of the symbols are bespoke, invented solely for this work. They have never before appeared in print, and strange symbols don't print themselves. Each one will have to be meticulously hand-tooled into letterpress blocks for offset printing. The publisher knows that will add to the costs and fears the book will be as expensive to print as it is hard to read.

Reviewing the book will be harder still—and proofreading nearly impossible. The only two people in the English-speaking world who have the requisite knowledge, patience, and perseverance to digest the work are its two authors. Few people could even attempt to read it. Fewer still, the publishers fear, will want to try.

The bottom line for the publishers is this: It's a thick, incomprehensible, and impenetrable book filled with meaningless symbols that nobody will ever buy. So they punt. They won't print it, they say, unless Russell and Whitehead agree to defray the cost of the first production run by putting money up front—about £200 (more than $25,000 in today's US dollars). The Royal Society agrees to pay half this amount, leaving Russell and Whitehead to cover the rest. "We thus earned [minus] £50 each by 10 years' work," Russell writes in his autobiography.

───◆───

Reviews of *Principia Mathematica* are mixed. Even as some people celebrate Russell and Whitehead's achievement, there's no getting around the book's thickness and impenetrability. It's a "dry and desolate undertaking," the historian Dauben will later say. Artificial and contrived. "Not polished enough to be really useful," the Austrian-born mathematical logician Georg Kreisel will write in the 1970s, "not even to trained mathematicians." A Herculean folly. *Who Hydras whom?*

Other reviews are kinder. "It is hard to overestimate the importance of *Principia* as the first worked out example of how to reconstruct in detail from a limited number of basic principles the main body of mathematics," Mancosu will write in 2010.

Russell himself, however, finds the book a letdown. He confides to friends that it's almost impossible to explain what it's about. More than that, he fears it's a failure. Not because it's cumbersome, thick, and hard to read. Nor do his doubts have anything to do with audience reactions, reviews, or sales. None of that. The book actually sells surprisingly well, as far as that goes. Its failure for Russell is more foundational. He and Whitehead show that classical mathematics can be derived from set theory. But ultimately, they are never able to achieve their main goal of marrying the concepts of Cantor with the methods of Frege—of reducing set theory to logic, taming the paradoxes, and forever fixing the foundations of mathematics.

"The grand object of *Principia*," philosopher of science Ernest Nagel and mathematician James R. Newman will write fifty years later, "was to demonstrate that mathematics is only a chapter of logic." In that, Russell and Whitehead failed.

That frustration is compounded because the book has been a big part of their lives for nine years. Russell and Whitehead have been joined at

THE DAY OF JUDGMENT

the hip. They have rarely spent more than a few weeks apart. They have often worked elbow to elbow for weeks on end. Discussing. Wordsmithing. Writing equations. Fixing errors. Philosophizing. Whitehead is methodical, exacting, and patient. And Russell is—well, Russell. He's bursting with ideas. He's fearless. Clever—an artful dodger. He dashes things off and considers them publication-worthy when they are barely half formed. Whitehead talks Russell back from the edge at times, and Russell pushes Whitehead over the edge at others. Together they work. They break through. And they bash out one dizzying page after another.

The irony is that even if *Principia Mathematica* falls short in showing that all math can be reduced to logic, the book is unexpectedly something of a commercial success. It sells better than expected. The first print run sells out almost immediately. There's quickly a second. And then a third. And another. And another. "I can't think who buys it," Russell jokes to a friend. New editions appear again and again in the years to come. It's still in print today! I bought a brand-new set from Amazon when I started writing this book, and it arrived the next day.

When volume II of *Principia Mathematica* appears a year after the first, Russell has already moved on. "Odd how much passion goes into a thing and how cold it is when it is done," he says. "A vast amount of various people's solid misery is crystallized in this book." He would have done anything to finish it while he was working on it, he says, "and now it is a mere moment's interest."

By the time the third volume appears in a few years, Russell will have almost set mathematics aside entirely. Queen Math, Russell will say, has become a "cold and unresponsive love." More and more, he's deep into a political thing. Campaigning. Stumping. Writing. "I have chucked my work till after the general election, and I'm throwing myself into politics, speaking and canvassing," Russell writes to a friend.

"NATURALLY FLATTERED AND PLEASED"

At first, Logan thinks putting his sister and her husband in touch with Ottoline and Philip is a brilliant move. They hit it off well. Maybe this new friendship will spark something? Bertie and Alys and Philip and O—they have so much in common. Bertie and Alys are interested in politics. What better way for them to serve those interests than to get to know Philip, a rising star in the ruling Liberal Party? And his wife, Ottoline, is also very political, just like Alys. It seems like a double date made in liberal heaven.

And so it goes. Bertie begins to campaign for Philip in 1910, facing sometimes hostile, even terrifying crowds. People boo. Some throw stones. For his part, Bertie stands fearlessly in front of hostile crowds and takes it, talking them down with calm reason until they settle, listen, laugh, and nod. People want to hear what he has to say, even if they don't agree—especially if they don't agree. He's smart. He's funny. That impresses Logan. It impresses Philip. But even more, it *really* impresses Ottoline. She watches once as a loud, angry crowd jeers and catcalls Russell. He stands to speak, undaunted. He calms the crowd. Then silences them. They are spellbound. "Very seldom have I seen intellectual integrity triumph over democratic disorder," Ottoline writes. Who is this brave, funny genius?

Bringing Bertie and Alys to see Philip and O is one fifth-wheel introduction Logan will come to regret. "Bertrand Russell is most fascinating," Ottoline writes in her diary. He *intrigues* her. People call Russell "the day of judgement," she notes.

Ottoline is a good judge of political character. She's no casual observer of politics as the wife of a parliamentarian. She has even more political connections than Philip does. She is longtime close friends with the prime minister.

After the death of British prime minister Henry Campbell-Bannerman in 1908, the torch was passed to Campbell-Bannerman's lieutenant Herbert Henry Asquith, considered one of the most capable politicians ever to sit on the British bench. Asquith is a dear old friend of Ottoline's. They first met in the 1890s, when she was dining with her brother Henry at a London restaurant. "He took an interest in me and in my reading," she later recalls in her memoir. "I was naturally flattered and pleased." They became fast friends. He gave her books to read on her travels. She became friends with both his daughter and his second wife. And he was a solid, if distant, part of her circle after that.

———◆———

THE DAY OF JUDGMENT

Politically, Asquith was a boxer. A bruiser. A terminator. He won all the top prizes that could be had all the way through school—the Balliol scholarship, the presidency of the Oxford Union debate club. He won the coveted Craven scholarship. Diplomas. Awards. Accolades. And he was a political prodigy besides. Brilliant. A logician-meets-magician of an orator. The sentences rolled off his tongue with "hammer-like precision," according to a 1913 editorial in the *North American Review*. He was a workhorse, and he was anointed besides, handpicked by Campbell-Bannerman and personally groomed by Gladstone. Asquith was a machine. No—the perfect machine. A logical machine. "There is no appeal to passion in what he says, no loose generalities, no attempt at rhetoric, nothing over-subtle or bewildering," the *North American Review* editor writes.

If he's brilliant, however, he's not without weakness. Asquith's is that, like all machines, he is vulnerable to sabotage. His enemies call him "Squiffy" behind his back—a slur that must make him seem all the more sympathetic to Ottoline. She has faced the backstabs of her own field of "frenemy" fake friends her whole life. And long after her demise, she will fare even worse under the treatment of mean-girl historians who will insist, even to this day, that she was never a member of the avant-garde Bloomsbury Group. Not really. Even though she's at the throbbing heart of activity in the neighborhood—friends with writers and artists, host of a fabulous salon, founder of London's Contemporary Art Society, and its chief buyer just before the Great War—they exclude her. Some historians don't seem to consider her a serious historical figure at all—more of a curiosity. A person to be ignored as much as remembered. They bristle at the thought of considering her part of their precious set, the Bloomsbury Group. She is *in* Bloomsbury, they grant, but not *of* Bloomsbury.

They hate her shortcomings. She lacks a college education. She's no artist. She's not even a writer—not really. Her published memoirs, letters, and heavily edited diary entries will later be described by Canadian philosopher Nicholas Griffin in the 1990s as vague and cliché: "The letters jump waywardly from topic to topic and give every impression of a mind that wasn't properly on its task."

But if they're right about her writing, middling as it is, they're wrong about her. To all those who exclude Ottoline, I say, *Open the heavens to its own angels!* Don't take my word for it. Virginia Woolf, her good friend, once said Ottoline "created her own world." She helped to shape the time and place and to define the era. And she was there. For that alone, she should be considered part of Bloomsbury. I think Woolf would agree with me. To shape Woolf's own words about Ottoline,

> *& there used to be a great lady in*
> *Bedford Square who managed to*
> *make life seem a little amusing*
> *& interesting*
> *& adventurous,*
> *so I used to think*
> *when I was young*
> *& wore a blue dress*
> *& Ottoline was like a*
> *Spanish galleon, hung*
> *With golden coins*
> *& lovely silken*
> *sails*

BERTIE'S BALL OF FIRE

The sexual relationship between Bertie and O begins almost as soon as he finishes his book. Nine months after their first meeting that summer outside Oxford in 1910, Ottoline makes plans to see Russell again. On March 19, 1911, she gets her chance, inviting him to dinner at her London home with a few others when Philip is out of town. It's a lovely evening. They dine. They drink. They talk. They laugh. The other dinner guests both leave around midnight, but Russell stays longer. They hang out until 4:00 a.m. "Fate sometimes throws a ball of fire into one's life, stunning and overwhelming," Ottoline says.

Years later, Russell will cast this evening in prophetic terms in his autobiography. Nothing was ever the same again, he claims. "My life before 1910 and my life after 1914 were as sharply separated as Faust's life before

THE DAY OF JUDGMENT

and after he met Mephistopheles," he writes. "I underwent a process of rejuvenation, inaugurated by Ottoline Morrell and continued by the war."

To fall in love in an instant, as he claims, once again smacks of the same sort of instant-karma gratification that he claims happened with his peace conversion in 1901. But now, a decade later, his hyperbole is actually believable. Bertie pours his heart out to Ottoline. He's unhappy with Alys. All those long, cold years. Wasted years. Bereft. Pent-up. "He was consumed with desire to possess me and to cast off the unhappiness of his own life," Ottoline writes. "It was as if he had suddenly risen from the grave and had broken the bonds that held him." Bertie describes the affair as "a rejuvenation." An awakening. Far from a mere fling—for him, it's the real thing. He's been a coiled spring for ten years, and on that last day of winter in 1911, his spring has sprung.

Russell floats on cloud nine. Elated. Enamored. Expectant. "I did not have full relations with Ottoline that evening," he recalls, "but we agreed to become lovers as soon as possible." His mood is one of true happiness. "Among the few moments when life seemed all that it might be but hardly ever is," he writes. The affair is full scorching sun after an overcast decade of bitter marital misery and the "cold and unresponsive love" of Queen Math.

Biographers and historians universally agree that the affair is fundamentally important for both of them—but especially for Russell. Not only does it tip his scales and spur him to finally toss himself out of his own miserable, loveless marriage, but it also forces him to think more about politics and less about math. He's aligned with Ottoline on the social causes she holds dear, but now he has muse motivation to embrace them even more.

Biographers differ, however, about the depths of Ottoline's feelings for him. But then, biographers are full of desire and expectations for their subjects. They often see only what they want to see.

———◆———

One of her biographers insists that Ottoline is not really attracted to Bertie—at least not sexually. What she seeks instead is some form of intense and stimulating intellectual fulfillment. Russell's biographers insist that it's a heated and passionate love affair and a warm friendship as well. And they're not wrong. For the next three decades, they stay close friends—even after their *friends-with-benefits* benefits expire. From 1911 to 1938, Russell writes some 2,500 letters to Ottoline, and she sends nearly 2,000 to him. Sometimes they exchange as many as four letters a day—half a lifetime of letters that are funny, sweet, strange, and fascinating.

But at the dawn of their relationship, Ottoline tells Bertie that she has no intention of leaving Philip. He decides he's fine with that arrangement, even with her telling Philip. In fact, he is inspired by her openness and decides he should tell Alys. It doesn't go well. Alys is angry, and the meeting ends in fiery drama—as basically all *it's-not-you-it's-me* conversations do.

What is Bertie expecting? Is he clueless? Perhaps he thinks if the new arrangement makes sense to him, it should surely make sense to Alys. Things have been so bad between them for so long. She *must* see that! The real breakup was nine years ago, when he fell out of love with her during the Boer War. The fact that they stayed together a decade longer only prolonged the inevitable. What's left for them to do now except to move on?

Fools only ever see what fools want to see, and there's no greater fool than a man in the early throes of love. "England's wisest fool," as Russell will be publicly mocked in the years to come.

———◆———

Alys is bitter. She's been waiting the better part of an awkward decade for Bertie to come around. She has expected him to eventually see his way clear to settle down, not to wander off. And to fool around with the wife of her brother's best friend? Intolerable! The conversation they have—his confession—devolves into an apocalyptic argument. Awkward. Angry. Indignant. Sad. It's a nasty exeunt omnes that brings sixteen years of marriage to an abrupt, Icarus-in-free-fall, wing-melting end.

"I then rode away on my bicycle, and with that, my first marriage came to an end," Russell will later write in his autobiography.

The next day, Alys threatens to expose the affair and create the exact sort of public scandal that would ruin Russell's reputation. Bertie threatens to kill himself if she does—something he later admits was a loathsome

THE DAY OF JUDGMENT

bit of manipulation. But it works. Alys remains quiet. Then, just a few days later, her mother dies. When Russell visits her to console her for this loss, another dreadful scene unfolds. They argue. She's angry. She says so. He says more. They fight. Alys insists she needs to tell her brother, Logan. He's Philip's best friend, after all. Perhaps he should be told, Russell says. "Perhaps he had better."

But before that can happen, the proverbial cat falls out of the bag on its own.

In May 1911, a few weeks after her affair with Bertie begins, Ottoline and Philip host a fancy dinner party with Winston Churchill as the guest of honor. Henry James is there as well. But the evening is a bust. Churchill doesn't like James, and Ottoline doesn't really like Churchill. "He is very rhetorical," she writes (to put his puffery mildly).

Also at the party is Virginia Stephen (who, when soon married, will become Virginia Woolf). Roger Fry, an artist and art critic with whom Ottoline had the briefest fling, is there as well. Ottoline helped Fry a year earlier when he launched a major exhibition of contemporary French paintings in London. The show shocked a British public weaned on little more than John Singer Sargent, whom Fry considered passé. His show had dozens of Van Goghs, fourteen Cézannes, eight Manets (including his famous *Bar at the Folies Bergère*), a couple of Picassos, and eight Gaugin nudes. The show was a revolution. It torched London society, smoked Fry in controversy, destroyed his reputation for a time, but ultimately helped enshrine him as the most influential art critic of the early twentieth century. "Nothing like this had ever been seen, or hoped for, in London," one critic wrote.

At the time of Ottoline's dinner party, Fry has just come back from Turkey with Virginia. Virginia's sister Vanessa had fallen ill there, and Fry fell in love with Vanessa while nursing her back to health. Virginia confides all this to Ottoline at the dinner party, and Fry becomes furious when he later learns what they were discussing because he thinks Ottoline was gossiping about him. He worries it will threaten his relationship with Vanessa, and so he confronts Ottoline.

Two days after the dinner party, Ottoline is out shopping with Bertie. They go to stores. They have a nice long lunch. They say their goodbyes. They part ways. But when she returns home, Fry is there. Waiting. "He suddenly turned on me with a fierce and accusing expression, commanding me to explain why I had spread abroad that he was in love with me," she writes.

Fry says he heard that she said Fry is in love with her. He won't tell her who told him that, though, and before she can really respond, he storms off. He's not listening—neither hear nor there. Angry Fry. Deaf Fry. "I was so utterly dumbfounded," Ottoline says. "Whatever the cause, one of my most intimate and delightful friendships crumbled to dust that Saturday."

Ottoline pursues Fry and begs him to keep his mouth shut about her affair with Bertie (Fry knows all about that as well). But it's too late. Angry Fry speaks. Word gets back to Bertie that he's the subject of gossip that could threaten his career. Now it's Ottoline's turn to be pissed. Fry is "an untrustworthy dog," she writes in disgust, "who softly pads up to one and licks one's hand, but who will nearly always turn and bite."

And Bertie?

He is definitely worried, but only from his cloud-nine vantage. His fear of exposure is overwhelmed by his new sense of found passion and desire. While working on the *Principia Mathematica*, he claimed to have been "hardly aware of the outer world." Now he is hardly aware of anything other than his physical feelings. He's like a starving little boy from a Charles Dickens yarn—a kid who's just been handed a big box of chocolates, Ottoline's biographer the English literary critic and novelist Miranda Seymour will write in 1992. He's awkward. Enthusiastic. With fumble fingers and gripping paws. He claws with both hands at the box and bonbons. *When will the yum-yum start?*

AN UNKNOWN GERMAN

That summer of 1911 is one of the hottest anyone can recall. Ottoline will always remember it as one of her favorites. Russell will recall it as one of the happiest times in his life. Why? Because they see a lot of each other. Rendezvous in hotels. Afternoons walking in the woods. Picnics under trees. Reading together. Politics. Shelley. Ibsen. Plato. Spinoza. They give each other gifts. In many ways, Russell acts the artist who's finally found his

THE DAY OF JUDGMENT

muse. "You have released in me imprisoned voices that sing the beauty of the world," he says to Ottoline.

Having a muse gets him thinking about new book projects. He's working on finalizing the upcoming third and final volume of the *Principia Mathematica*, and he's now convinced that he's done all he can in the philosophy of mathematics. Ottoline encourages him to turn to philosophy for the everyman. He decides he wants to remake himself, and that's exactly what he does. The year he falls in love with Ottoline is the same year he becomes the popular writer he will remain for the rest of his life, publishing books and articles that will shake, delight, and enrage people and "successfully [permeate] the consciousness of Western Man," the Canadian English professor Margaret Moran will write ninety years later.

Besides basking in the glow of the new love affair, his *new* queen, Bertie has a wonderful year. He is appointed as a full-time faculty member of Cambridge University and elected to the Royal Society. He finishes another new book. And he is swept up in romance and achievement all summer. The *Principia Mathematica* is selling well. People love it! He's putting an end to a soul-sucking decade of life. His star is bright, and he is rising.

Then, late one afternoon and quite randomly, a young student walks into his office. He's an aviation engineering grad student from the University of Manchester who says he wants to discuss logic and the foundations of math, of all things. And why not? The student has just been to Germany to see Frege, who suggested that he visit Russell in Cambridge.

Bertie is amused. "An unknown German," he says to a colleague. A bit crazy perhaps. "Obstinate & perverse," Bertie writes. The boy's an oddball at the very least. What type of aviation engineer is interested in the foundations of mathematics? "A miserable creature, full of sin," Russell writes to Ottoline, adding, "Whatever he says, he apologizes for having said."

The young student begins to follow Russell around like a strange and argumentative puppy. Russell thinks he may be bereft of calm good sense. But the German engineer attends Russell's lectures. Then he arranges to meet later to discuss more. He shows up to some of these discussions with armfuls of flowers. More often he loses his cool, makes wild gestures, throws out strange assertions, and finally storms out in frustration and anger.

How tiresome, Russell thinks. A troubled engineer! But slowly he warms to the boy. He is brilliant—that much is obvious. Within a few months, Russell is telling his friends that he has met a bona fide genius, however strange. He wants to design and fly airplanes, and he's full of passion for math and logic. Could this be Russell's intellectual heir? He wonders. Could he be the person who will inherit his work and carry it on now that he has exhausted himself on the unforgiving steppes and peaks of his own personal K2, the *Principia Mathematica*? Will his troubled German engineer have the requisite energy, skill, and interest to even attempt to climb in his footsteps? Bertie thinks he will. *Perhaps he will do great things!*

Still, the boy is a rough bit of clay. Lots of shaping needed. "His disposition is that of an artist, intuitive and moody," Bertie writes to Ottoline. "He says every morning he begins his work with hope, and every evening he ends in despair."

As strange as that sounds, Ottoline likes what she hears. That all sounds odd, but...he brought flowers to a philosophical debate?! Who brings a bouquet to a knife fight?!

I think I love him, she writes to Bertie.

CHAPTER SEVEN
AIRPLANES AND SUNSHINE

1912–1913

"
Logic is the youth of mathematics, and mathematics is the manhood of logic.
"

—Bertrand Russell

Bertrand Russell in an undated, pre–World War I photo by Underwood & Underwood. Cropped from a public domain image via Library of Congress [LC-USZ62-49535].

Bertie's so-called German engineer is Ludwig Josef Johann Wittgenstein, a twenty-something who is indeed an aspiring aviation engineer. But he's not German. He's actually Austrian, the youngest of eight children from one of the richest families in Vienna—indeed the world. He basically, according to one biographer, comes from the Austrian equivalent of the Carnegies in the United States or the Rothschilds in England.

All that wealth comes from his father, Karl Otto Clemens Wittgenstein, an industrial genius who built a business from the ground up into one of the largest and most impressive corporations in Europe. But his wild success in accumulating wealth tainted Karl with those twin toxic trappings of the self-made man: unabashed ruthlessness combined with overwhelming ego. He was a dashing gambler in the pits of trade—lean in the tux and mean in the trench. "An industrialist must take great risks," he once said about his career when Ludwig was ten. "When the moment demands it, he must be prepared to place everything on a single card even at the danger of failing to reap the fruits that he hoped for."

Like many tycoons, Karl falls prey to something I call the illusion of desperate deliberateness: that stern conviction that your actions and yours alone account for all your success. It's a fallacy that ignores any forces majeure, external factors, good fortune, twists of fate, or foul chance and places the credit for your outcomes solely under the deliberate power of your choice.

Choice is as important in business as it is in mathematics, especially when you are doing new things. But in business, things like charm, wit, and luck are also major assets. For Karl, it was the whole package. He was macho and good-looking. Confident. Funny. Cultured. Intelligent. Ruthless. And possessed of a business acumen even his industrial rivals acknowledge is great. He capped it all off with a singular, compelling charisma. "The power to charm a bird from a tree," as the British philosopher, Wittgenstein scholar, and biographer Brian McGuinness will put it in 1988.

And perhaps the business where Karl busies himself the most, the place on which he lavishes the whole of his energy later in life, is his family. "Karl Wittgenstein is for his children the dispenser of all good things, the

AIRPLANES AND SUNSHINE

creator of the world of large houses, parks, and estates, which they took as a natural environment," McGuinness will say.

Silver-spoon privilege doesn't even begin to describe the upbringing Ludwig and his siblings experienced. It was more like they were wrapped in platinum, dipped in palladium, and studded with diamonds. They lived a rarefied existence. Taught only by the best tutors. Mingled only with the finest families. Chauffeured parties. Elite schools. Basically the best of everything. Legendary composers Johannes Brahms and Gustav Mahler regularly attended parties at their house—a block-sized beast of a home that people around Vienna referred to as the Palais (palace). It had no fewer than seven grand pianos and room after room of Rodin sculptures. When Wittgenstein's sister Gretl was married in 1905, famed artist Gustav Klimt painted her wedding portrait.

The family worshipped Karl like a god. But the ancient Greeks were right in their assessment of the fickle nature of gods in the pantheon—they can be monstrously cruel as well as benevolently kind. Karl was both. He was obsessed with making the right choices for his children. He wanted to see each son achieve business success akin to his own—that unique blend of engineering ingenuity and ruthless acumen. Business and engineering, engineering and business—his queen was a double major. And that was what he sold his sons.

SUFFER THE CHILDREN

Legacies are sometimes weak and fragile, however. Humans are not like seeds. You can't simply plant a son in a profession, fertilize him a few times, and watch him grow into the tree you desire. People are subject to human psychology, which sometimes resists even the most generous care. And sometimes we become pot-bound and fall into stunted growth—psychologically constricted by domineering parents who prevent us from thriving.

Children of greatness fail for the same reason that many great basketball players fail with the game on the line. To crumble under the crushing pressure of undue expectations is very human. It's psychologically normal. It's common—it's sometimes even de rigueur. And there's a lesson for all loving parents in that: Go easy! Love your children, and if you have to push them, don't do it harshly with cruel words and humiliation. Perhaps the worst thing a domineering dad or tiger mom can do is subject their children to their own selfish, never-satisfied perspective. They risk weighing down their children's lives with the punishing anchor of their own impossible-to-achieve goals. Nobody should ever have to deal with the crushing certainty that their greatest successes will always fall just short of failure. And for children of greatness, this is especially the case. The greater the parent, the more likely that the child will suffer from self-imposed psychological handicaps.

Few families in history seem to have failed to strike that healthy balance more than the Wittgensteins. Three of Ludwig's four brothers would commit suicide as adults. His older brother Hans was a music prodigy—a genius, according to his teachers. He played solo violin concertos at elite venues before even the age of ten. But that success did little to curtail the loud drumbeat of business duty being pounded into his head by his father. From an early age, demanding Queen Business drummed away any hope of winged waltz or stringed Stradivari. It was Hans's hobgoblin, really. The first word he ever spoke as a babe was supposedly "Oedipus," that ancient Greek hero who killed his father and slept with his mother. *Paging Doctor Freud!*

Hans's life took a tragic turn after he fled Vienna for the United States. He vanished off a boat in the Chesapeake Bay, in Maryland, not far from where I sat and wrote this book. He was presumed dead by drowning—and it was presumptively deemed a suicide.

A second brother, Rudi, drank poison and killed himself around the same time. A third brother, Kurt, will commit suicide at the end of World War I after enjoying the glow of his father's attention and delight for decades. Karl makes him director of one of his companies. But Kurt hates the work. Karl dies just before the war. Then the Great War ensures that Kurt will never make it home. After four long years of war, after witnessing his army's retreat from the Austrian front, Kurt will excuse himself from camp one night, saying he wants to stretch his legs. He'll wander off, take out his gun, finger the trigger, and take his own life.

Wittgenstein's fourth and final brother, Paul, somehow escapes the family curse—no doubt because of his disposition, protected by his own steely resolve. A talented musician, he's determined to be a concert pianist, regardless of what anyone thinks. That determination will be evident even when he later loses his right hand during World War I. Paul will persist in his musical career despite the injury, learning to play with his left hand alone.

──◆──

Little Ludwig probably feels the paternal pressure at first, perhaps proven by the fact that he studies engineering. His lavish upbringing makes him somewhat a stranger in the world, however—an alien. He doesn't make friends easily. His schoolmates, far below his economic station, constantly ridicule him for his weirdness. His stubborn obstinacy. His unbending assertions. His formal airs. He feels betrayed by his friends all through school, and when he finishes college, he becomes rudderless, idle, and miserable. But he soon turns boredom into passion. In 1908, he moves to Manchester, England. There he enrolls in a graduate-level aeronautical engineering program, wanting to design, build, and fly his own airplane.

And why not?

The world's first decade of powered human flight is nearing its end. It began with the first zeppelin, which flew for eighteen minutes in the summer of 1900. It was an enormous achievement but a Pyrrhic one. The airship cost so much to build and launch that its financiers soon went bankrupt. They dismantled the ship—a cigar-shaped boat called the *Count Ferdinand*—almost as soon as it landed and sold it off in pieces. The *Count Ferdinand* still fills a phantom phallic footnote in the history of aviation as both the first powered flight and the first failed airline—all in the space of twenty minutes. If ever a mechanical beast deserved an epitaph, it's *Count Fergie*. Here's mine:

> *Heavens carry Count Ferdinand*
> *For twenty minutes across the land.*
> *1,000 ft. in the air & belly up once there.*
> *"Bust that crap—and sell it for scrap!"*
> *Destroyed at the bank's demand.*

THE GREAT MATH WAR

Dreams of human flight will not crash or fade, however. In the early 1900s, they fill the air like hangar fumes. Popular magazines publish detailed wing designs, model glider diagrams, and other technical specs. Artists are commissioned to detail more and more advanced schematics. Copies of publications carrying new propeller plans fly off newsstand shelves. Would-be inventors carefully mimic and innovate upon those airfoil shapes. Lots of people are racing to build the first powered airplane.

The Wright brothers get there first, and airplanes have officially arrived. Everyone feels it—including Ludwig Wittgenstein. He wants desperately to fly, and his desire to build an airplane has a strange, unexpected, and remarkable outcome.

Initially obsessed with engines, he soon turns his attention to the basic mathematics of wing and propeller shapes. He dives into the applied math of airfoils, and from there, he discovers the 250-year-old math-heavy field of fluid dynamics. That leads him down another rabbit hole, and he discovers pure mathematics and from there the logical foundations of math.

And ultimately it leads to Russell's office, which will change the course of his life. After meeting Bertie, he gives up engineering, embraces mathematical philosophy, and never looks back. Late in life, when asked to provide a biographical essay describing himself and his philosophy for an encyclopedia entry, Wittgenstein will respond in a single sentence: "He has concerned himself principally with questions about the foundations of mathematics."

"TOO OLD TO SOLVE"

Wittgenstein formally enrolls at Cambridge in 1912. It's an exciting time at the university. The biochemist Frederick Gowland Hopkins is there, discovering vitamins and their importance in biology. Physicist Joseph John "J. J." Thomson is already famous for having discovered the electron, and he has just discovered what he thinks is a new chemical element (though later it turns out to be a heavy isotope of hydrogen). John Maynard Keynes is a student there, on his way to becoming Britain's most noted economist.

AIRPLANES AND SUNSHINE

The famous self-taught Indian mathematical genius Srinivasa Ramanujan Aiyangar is also a student.

Russell is getting more and more excited about collaborating with Wittgenstein, his new intellectual "son." And the feeling is mutual—at first. When Russell gives Wittgenstein a newly published copy of volume II of *Principia Mathematica* in the spring of 1912, Wittgenstein tells him the book is like music. He compares Russell to Beethoven. That feedback brings them closer. "He is the ideal pupil," Russell says to Ottoline. "I love him and feel he will solve the problems that I am too old to solve."

Russell calls Wittgenstein "a treasure" and praises his ideas—though not necessarily his passion. Sometimes Wittgenstein spins out of control. "I thought he would've smashed all the furniture in my room today, he got so excited," Russell writes another time. "His avalanches make mine seem mere snowballs."

He also has grave concerns that his protégé is "armor-plated against all assaults of reasoning." Wittgenstein rejects Russell's advice to write down what he thinks and lay out his arguments logically. That would destroy its beauty, he says, like crushing a flower with muddy hands.

Russell begins to be pulled in opposite directions by the twin attractions of Wittgenstein and Ottoline. She is warm, loving, and encouraging, pushing Russell to explore a softer, more accessible side. Wittgenstein is morose, angry, and off-putting, chastising Russell for "dishonesty and sloppy thinking."

"If Wittgenstein had had his way, Russell would never have written a word about religion and morals," Ottoline's biographer Miranda Seymour will write decades later. "If Ottoline had been allowed to have hers, he would have written about nothing else."

Ottoline's influence is stronger, of course, probably because it's also sexual. In 1912, Russell finishes another book, this one on the problems of philosophy. It's aimed both at professional philosophers and a lay audience, and it sells well. Some thirteen thousand copies fly off the shelves in the first couple of months. In the summer of 1912, the Fifth International

Congress of Mathematicians is held in Cambridge, and Russell plays a part in hosting it. Alfred North Whitehead gives a talk on their towering work, *Principia Mathematica*, and Bertie presides over the philosophical section of the meeting.

"I can't help reflecting that all these mathematical philosophers have different thoughts from what they would have had I not existed," he boasts. He loves the attention—especially from the Americans. He's also named president of the Aristotelian Society that year, an important milestone in his career.

At the congress, Russell again meets the Italian mathematician Peano, whose work so inspired the *Principia Mathematica*. "He has quite extraordinary nobility from single mindedness," Russell writes to Ottoline about Peano. "I have a very great reverence for him." After the meeting ends, Russell begins to think more broadly about what he should do next. He begins to envision a new project that will attempt to do for philosophy and physics what he wanted *Principia Mathematica* to do for logic and math—marry the two. "The mere thought of it makes my blood tingle," he says. "It suddenly makes a whole lot of things interesting that wouldn't otherwise interest me."

Going into 1913, things have never seemed better.

ABOVE THE PHARMACY

Across the English Channel in 1913, beside a hut on a small plot of land surrounded by trees near the outskirts of Amsterdam, a tall mathematician sits, wearing no shirt. He squats. He studies. He lingers for hours, thinking about mathematics. He soaks up the sun. He eats nuts and leaves. And at night, as soon as it's dark, he goes to bed and dreams about math. He will figure prominently in the Great Math War and the coming debate over the foundations of math that erupts in the wake of Russell and Whitehead's work after World War II.

The early twentieth century in the Netherlands is an exciting time for science. The Dutch lowlands have minted more than a few great giants: the scientists Hugo Marie de Vries, Hendrik Antoon Lorentz, Heike Kamerlingh Onnes, Johannes Diderik van der Waals, Jacobus Henricus van 't

AIRPLANES AND SUNSHINE

Hoff, and Pieter Zeeman—legends in the fields of biology, chemistry, and physics.

Hugo de Vries is often considered the world's first geneticist. He develops early theories of gene propagation mechanisms and coins the word "mutation." Lorentz and Zeeman both won the Nobel Prize in 1902 for discovering the eponymous "Zeeman effect," explaining how electromagnetic spectral lines are split by a magnetic field. Lorentz also worked out the fundamental physics of electromagnetic Maxwell equation invariance in moving frames of reference, which helped pave the way for Einstein's special relativity theory. Kamerlingh Onnes is the first person to liquify helium, and he discovers superconductivity, which will win him the Nobel Prize in 1913. Anyone who has ever studied chemistry or structural biology knows the name van der Waals. He gave his name to the weak dispersion forces between atoms and molecules in solution, which are crucial for protein chemistry and folding in biology. And van't Hoff discovered stereochemistry, created the field of physical chemistry, developed chemical kinetics, and invented the concept of chemical affinity. He won the very first Nobel Prize in Chemistry.

The state of mathematics in the Netherlands is, compared to all that, okay but not great. Amsterdam is certainly no Paris, Berlin, or Göttingen. But in the years before World War I, it boasts a few blazing suns—including this young mathematician who burns brightest of all.

Thirty-something Luitzen Egbertus Jan "L. E. J." Brouwer is reaching the pinnacle of the mathematical world. He is about to become a full professor at the University of Amsterdam. He's about to be invited to join the editorial board of the *Mathematische Annalen*—the journal led by Hilbert and the best mathematics journal in existence. And after a string of amazing discoveries and papers over the preceding few years, Brouwer is emerging as the world's leading authority in the newly energized subject of topology. He's made amazing discoveries. New methods. Bold approaches. Fixed-point theorem. Continuous transformations. Degrees of mappings. Invariance of dimension.

"Thanks to his gigantic efforts," the German mathematician Richard Courant will write fifty years later, "topology is now as amenable to rigorous treatment as the geometry of Euclid." Some even consider Brouwer one of the leading scientists of all time, though that seems a little too outlandish a claim. When I first started this book, I asked a computer scientist from the Netherlands what he thought of Brouwer, and I discovered that he had never even heard of him. A mathematician who was a friend of my late father told me even mathematicians don't know much about Brouwer. But philosophers love him, she said.

One of the reasons he's relatively unknown today is that Brouwer is something of an oddball. He has deep extracurricular interests in mysticism, art, philosophy, and new-age health. He's also uncompromising, if not a little bit insane. He pisses many people off in his lifetime. Others fall silent around him because they don't want to cross him. And that characteristic prickliness will in many ways dictate the events of the Great Math War debate over the foundations of mathematics in the coming years.

Everything that Russell and Whitehead have done on the foundations of math is only an amuse-bouche to the coming crisis. The big battles in the Great Math War won't really take shape until a decade or more after the *Principia Mathematica* is published, and they will be fought by other people, including Brouwer. In 1913, he is a strange one-man sleeper cell. A mathematical revolutionary hiding in plain sight.

———◆———

Brouwer was born on February 27, 1881, in a muddy peat marsh of a village called Overschie, where four rivers meet, half an hour's turnip-cart ride outside Rotterdam. Today, it's been incorporated as a neighborhood within the Rotterdam city limits, but in the late nineteenth century, it was a crummy little spit of cow hills, gin mills, wet meadows, and muddy black marshes.

All that sounds almost quaint in a charmingly *Masterpiece Theatre* period-piece sort of way. But consider this: The cows were fed on fermented waste—fodder from the gin mills. The smell of their gin farts filled the air, day and night. It was not nineteenth-century quaint so much as nineteenth-century glum. It was an ugly town. A dirty town. A small town. Horrible. A few miserable houses straddling one sad little street—two rows of shacks and a whole lotta shit.

AIRPLANES AND SUNSHINE

The Brouwers didn't stay there long. The family soon climbed its way north to the town of Medemblik, where Brouwer so excelled academically that he was accepted into high school at the age of nine. This would later win him recognition in the *Guinness Book of World Records* as the youngest person in the Netherlands ever to graduate from high school.

Other than that distinction, his life was unremarkable before he entered the University of Amsterdam in the fall of 1897 at age sixteen. He shone in college. According to one of his classmates, he sometimes took the chalk from the professor's hand and continued the lectures himself. Amsterdam was something of a backwater for mathematics at the time, but there were two bright spots: The university boasted the great mathematician Diederik Johannes Korteweg and the lesser but still prominent Gerrit Mannoury.

Korteweg was prolific. He worked on far-flung problems that ranged from acoustics to astronomy to probability theory to wave dynamics to voting. He was the chief editor of the collected works of Christiaan Huygens, a famous Dutch mathematician who two centuries earlier was at the center of the Calculus Wars dispute between Newton and Leibniz. Korteweg became a sort of life coach for Brouwer in his early years.

It seemed that Brouwer needed more than just academic guidance in those days. He had health problems that he had been incubating since high school—a nervous disposition, a high-strung nature, and an overwrought appearance. Early on, he embraced the new clean-living trends of the time—strict diets, sunbathing, fasts, vigorous exercise, and spa visits. He became a health freak, according to his biographer Dirk van Dalen, "with a strict and even eccentric way of living."

One thing that motivated this lifestyle change was a bad sinus infection—nastier and harder to treat in the days before antibiotics. His doctor prescribed a bizarre food cure, telling him to eat seven goose eggs and a half pound of steak daily. Okay, steak and eggs, yum. But seven eggs a day? And goose eggs? Yuck! If you've never had one, they're huge, larger than chicken eggs, and they have a dark yolk with a strong, off-putting gamey flavor. Consuming a single goose egg is a feat, not a feast, and eating seven of them a day is a horrendous, lysozyme-choked punishment. If the prescription was intended to make him better, at least he could be assured it couldn't have gotten much worse.

When he sought a second opinion, a different doctor prescribed a simple, spartan vegetarian diet. Brouwer embraced it for life. Seasonal veggies. Lots of herbs. Foraged leaves. Stinging nettles. Dandelion greens. Fruits and nuts. Milks and grains. He combined this with an austere program of early-to-bed, early-to-rise clean living punctuated by rigorous exercise—sometimes extreme exercise. He skated one winter from Amsterdam to Rotterdam. There and back, all in a day. It was a 130-km (81-mile) round-trip ice-skating century that nearly killed him. At one point, sick in bed for days afterward, he summoned a city official to draw up his last will and testament.

He was a maniac at times!

THE POISON MIXTRESS

After Brouwer finished his PhD qualifying exams around 1904, two major things happened in his life. First, he bought a tiny piece of land in that small town outside Amsterdam called Blaricum, where he built his hut. Second, he met his wife.

Blaricum was part of an area known as *het Gooi* (the throw). It was a small, poor village—desperately scratch-poor, according to one source—where fallow farms were tended on sandy soils. Years before, the area had become overrun by artists, health nuts, and people living in communes, including a colony called Walden, founded at the end of the nineteenth century based on romantic notions of Henry David Thoreau transposed onto the Amsterdam flats. It had faltered and failed because of bad management, but some of the colonists had stayed. The woodsy, sandy, lake-dotted land had clean air and was near the city. Brouwer loved it. He called Blaricum his "faithful home," a short train ride from Amsterdam, "where the dunes are close by," he said to a friend.

On his property, Brower built his little hut as a private retreat where he could enjoy the air, sunbathe in the nude, sleep under the stars, and live according to his strict, austere tenets. His vegetarianism made him fit right in with the post-Walden colony stragglers, whom the local townsfolk called the "grass eaters."

AIRPLANES AND SUNSHINE

The same year he built his hut, he married Reinharda Bernadina Frederica Elizabeth de Holl, who went by "Lize." She was a divorcée who was eleven years his senior and the single mother of a preteen daughter. She was the survivor of a horrible marriage, even by nineteenth-century standards. She had been married to an abusive former army doctor sixteen years her senior who had insisted that they have no children but had taken no family planning precautions. When Lize became pregnant, he forced her to have an abortion. Then he promptly impregnated her again. This time she left, giving birth to a girl in 1893.

Brouwer met Lize through an old high school friend who studied in Amsterdam and became a pharmacy assistant in the city. He reconnected with this friend after moving to Amsterdam, and she told him about a woman named de Holl who ran the pharmacy where she worked. Madame de Holl had a daughter (Brouwer's eventual wife, Lize) and a granddaughter. All three lived above the pharmacy.

Intrigued, Brouwer scaled the roof of the building across from the pharmacy, eye level with the apartment above, in order to catch a glimpse of Lize. He apparently liked what he saw, and he arranged to meet her. They courted. Lize was skeptical at first, but she warmed up to him. He proposed. She accepted. And they were married.

The whole climbing-the-building episode seems a bit creepy and stalker-like today, but for Brouwer, it may have simply been a chance to climb. Once, at a Canadian conference after World War II, Brouwer at age seventy-two will disappear from a picnic and later be found high up in a tree. This will surprise everybody, according to later accounts, but it really shouldn't. He loved climbing—always had and always would.

Brouwer borrowed money from a friend and bought his mother-in-law's pharmacy. It was an odd thing for a twenty-something grad student to do—especially since the sales contract required him to pay his mother-in-law a significant annual annuity. The pharmacy became the third pillar of his life, after Blaricum and the university, and he and his wife ran it for years. Lize was also a vegetarian. She loved experimenting with homemade herbal concoctions and dispensing a range of herb-infused homeopathic treatments at the pharmacy. She became known for boiling up cauldrons full of herbs, which would later earn her the nickname *poison mixtress*.

LIFE, ART, AND MYSTICISM

While Brouwer was working on his dissertation in 1906, his PhD adviser sent him a write-up of Hilbert's twenty-three problems. After reading it, Brouwer responded that he had already solved three of them. It was a boast. Even partially solving one of them would have been enough for a thesis—and indeed, well into the twentieth century, people would still be getting doctoral degrees on the strength of having contributed even incrementally to such efforts.

But Brouwer had other fish to fry. Around the same time, he published a small book of personal philosophy titled *Life, Art, and Mysticism*. His aim was to distill his thoughts into a practical mystic's guide for a moral, if austere, life. That sounds okay, but the book was marred by hard-to-stomach, serve-your-master misogyny. He characterized women chauvinistically as illogical creatures who should figuratively bow to men. He described his ideal "real woman" as pale, supple, and weak. She should have dull, dreamy eyes, he said, but don't be fooled. In truth, she's a temptress! "Man should avoid, ignore woman," he wrote, "but the woman should live in the man." *Insignificant. Powerless. Worthless. Sacrificing everything to her man.*

Say what?

The book seemed bizarre even then. Today, it's worthy of harsh cancellation. His friends didn't like the book. Brouwer's PhD adviser was aghast. "I thumbed through it, but it is not the reading that I wish or that is good for me," Korteweg said. Other than that, the book went mostly unnoticed, receiving only a single scathing review in one local paper. That affected Brouwer deeply. He wrote a letter to a friend describing life as a magic garden full of flowers and evil gnomes. "They stand on their head, and the worst is that they call out to me that I must also stand on my head," he said. "I am afraid of them."

One thing *Life, Art, and Mysticism* did do was hint at his later work on the foundations of mathematics. While in graduate school, Brouwer had discovered Hilbert's 1899 book *Foundations of Geometry*. He had studied

Cantor's set theory. He had explored Russell and Whitehead's work on the logical foundations of mathematics. And he had become deeply influenced by Poincaré's view that logic could never be the basis of mathematics. In fact, Brouwer took it one step further and flipped the script on Bertrand Russell. Rather than admitting that math is derived from logic, as Russell advocated, Brouwer argued that the opposite was true. "Mathematics is independent of logic," he said. "And logic depends on mathematics." Both statements appeared in Brouwer's PhD thesis draft.

Logic relies on language and is therefore linguistic, Brouwer thought. Surely that must mean mathematics comes first. "Theoretical logic does not teach anything in the present world," he said in a letter to Korteweg. "And one knows this, at least sensible people do." Math by necessity should be exclusive of language, he added.

He sent chapters of his thesis to his adviser in late 1906, and Korteweg was appalled once more. It was way too much like his student's just-published *Life, Art, and Mysticism*. "Really Brouwer, this won't do," Korteweg said. "A kind of pessimistic and mystic philosophy of life has been woven into it, that is no longer mathematics, and has also nothing to do with the foundations of mathematics." There was lots to love, Korteweg wrote, but the jarring bits of bizarre philosophy disturbed him. "In my opinion your dissertation can only gain by removing it," he said.

In the end, they came to an agreement. Some parts were excluded. Some sections were reordered. Brouwer finished his thesis and graduated. Korteweg, while turned off by Brouwer's mysticism, nevertheless saw his potential as a mathematician and championed his student tirelessly. Few in Amsterdam appreciated that they had in their midst one of the great mathematical geniuses of the day, but Korteweg did. He vowed to help his student.

Obtaining a PhD in the Netherlands in the early 1900s did not make one immediately employable. There were far too few academic positions. In fact, when Brouwer graduated, there were only four math departments in the entire country, each with only two professors. And Brouwer had no experience. So after he finished in 1908, Korteweg secured him a position at the university as a privatdocent—a low-level instructor, basically. Brouwer was not excited about this entry-level position. "Of service to nobody," he said to Korteweg.

THE GREAT MATH WAR

THE TOPOLOGY TRAILBLAZER

Brouwer quickly climbs the academic ladder, however, making a name for himself in the emerging new field of topology. Topology starts with the basic geometry of lines and shapes—and more complicated objects like cycloids and conics—and looks at what happens to their properties when you stretch, scrunch, or otherwise transform them (actions technically known as homeomorphisms, or mappings). What Brouwer and the other topologists discover is that some properties of geometric shapes are invariant to transformation. Change a shape as you will, and the song (or geometric property) remains the same.

He's a trailblazer, developing many of the tools future topologists will use. His work is highly acclaimed and makes him a household name among mathematicians. One of the first papers he publishes has the curious distinction of being the first math article ever printed with color illustrations. Another paper, "On the Mapping of Manifolds," published in 1911, is groundbreaking for technical reasons. "The importance of this paper can hardly be overestimated," one of his biographers writes. "It contained virtually all the tools of the new topology."

Brouwer succeeds in his career through topology, but what he's really interested in are the foundations of mathematics. Few know this. Fewer still can guess how radical his views are, including Hilbert, who is one of Brouwer's early influences. They are friends before the war. Brouwer is influenced by Hilbert's framing of the twenty-three math problems, and Hilbert publishes some of Brouwer's influential early work.

Once, when Hilbert was vacationing along the Dutch coast, Brouwer met up with him. At the time, Hilbert was reeling from a huge personal loss. His good friend Hermann Minkowski had died just a few months before after developing severe appendicitis. That was a terrible blow for Hilbert, but it would ultimately have the unintended consequence of bringing him closer to the younger generation of mathematicians—something that would forever shape the future of his field. One of the first of this new generation to bond with Hilbert was young Brouwer.

Hilbert and Brouwer met at the resort town of Scheveningen. They climbed the dunes. Swam. Took walks. Talked math. And after listening

AIRPLANES AND SUNSHINE

to Brouwer's ideas, Hilbert revised one of the sections of his famous *Foundations of Geometry*. Brouwer was over the moon. He wrote excitedly to a close friend that Hilbert was the greatest mathematician alive—the "first mathematician of the world"—and that finally meeting him was like a fresh young apostle meeting his prophet. "It was a beautiful new ray of light through my life."

One should beware of getting too much sun, however. You risk getting burned.

———◆———

Not everything is peaches and cream in Brouwer's life. He still has financial woes. The pharmacy is a money pit, and he has to borrow money from his brother to keep it afloat. Running the business day-to-day, Lize stays in the apartment above the pharmacy most nights, while Brouwer travels to and from Blaricum as needed. Sometimes he stays in the city overnight. Sometimes he commutes home and spends the evenings alone with his stepdaughter. That's where the trouble starts.

Brouwer and his stepdaughter, Anna Louise Elisabeth, will become locked in a decade-long difficult relationship, for which he bears much of the blame. He has what I consider a heavy-handed, harsh, and brutal approach to raising children, even in those spartan, Protestant, spare-no-rod days of yore. Brouwer is only a few years older than Louise, but because of his position as head of the household (and don't forget the blatant sexism he displays in his book), Brouwer treats Louise like a child. He's demanding. Domineering. He berates her. He calls her "my silly daughter." He forces her to eat by herself when Lize is home, and when she's not home, he rules her with an iron fist. He slaps her if her attention wavers. He sees her as lazy and stupid. She deeply irritates him, and she's the one who suffers for it.

All this comes to a head in 1914 when he decides to enroll twenty-year-old Louise in a laundry education school, where she will learn how to be a laundry teacher. At the beginning of the course, Brouwer asks the school for a list of fellow students. He selects two from the list to be her study mates. One of them, he says, she should bring home to meet him. That's how Correi Jogejan, or Cor, as she's called, comes into their lives. She is most suitable, he deems.

Louise can't stand her. Cor is fun-loving. Affable. Cheeky. Bold. Flippant. Funny. Outspoken. Engaging. Good with people. Great at working a room. She is basically everything Louise is not. Cor, Cor, perfect Cor. She's also beautiful, which throws even more attention her way. It's tough to be overshadowed, especially in your own household.

Brouwer works out an arrangement with Cor's parents in 1914 that allows him to basically adopt Cor and move her into the household. Initially, Cor's parents are thrilled at the prospect of the distinguished Brouwer bringing their daughter under his wing, but they soon sour on the arrangement. As the months roll on, they eventually cut ties with the girl, who has by then become Brouwer's personal secretary. Cor will take his notes, type his letters, copy his manuscripts, and become every bit the invaluable assistant he needs. More than that—she becomes his friend, his confidant, his companion. She travels with him. They are constantly together, which is something Lize learns to accept.

Louise, however, finds her former classmate's presence much harder to bear. She grows to loathe Cor. After all, Cor has Brouwer's eye, but Louise gets only his ire. She soon moves out and into a convent, determined to become a nun.

CHAPTER EIGHT

BOSTON LOVES BERTIE

1913–1914

"
When Mr. Apollinax visited the United States
His laughter tinkled among the teacups.
...
I heard the beat of centaur's hoofs over the hard turf
As his dry and passionate talk devoured the afternoon.
"He is a charming man"—"But after all what did he mean?"—
"
—T. S. Eliot

The produce center at Boston's Quincy Market crowded with wagons and horses a few years before World War I. Cropped from a public domain image from H. C. White Co. via Library of Congress [LC-DIG-stereo-1s13133].

Back in England, things start going downhill for young Wittgenstein in 1913. He storms into Russell's office one day, agitated. He berates Bertie for how he uses the terms "finite" and "infinite." Russell is in the middle of jotting a note to Ottoline at the time and literally writes, "Here is Wittgenstein just arrived, frightfully pained by my [recent] article, which he evidently detests."

Russell tries to calm Ludwig down. It's not an isolated incident. He's worried about Wittgenstein's health in general. He advises him to go riding. Catch some sun. Eat better. See a doctor. He invites Wittgenstein to watch a regatta one afternoon at Cambridge. Alfred North Whitehead's son is rowing. It's an exciting race, but Wittgenstein doesn't see it that way.

"He suddenly stood still and explained that the way we had spent the afternoon was so vile that we ought not to live, or at least he ought not," Russell says. Wittgenstein wails. Blah-blah this. Lowbrow that. This isn't Beethoven, he laments. What's he doing here? He has accomplished nothing this afternoon. Maybe he never will. Life sucks. Everything is the devil. Berate. Bemoan. Bellyache. Beef. Boat races are evil, he says. Awful. This whole day is disgusting. "Nothing is tolerable except producing great works or enjoying those of others," Russell recalls Wittgenstein finally shouting.

Larger cracks soon appear in their relationship. When he first meets Wittgenstein, Russell finds him argumentative, tiresome, bullish, and even a tad foolish—logical to the point of ludicrousness and hard to the point of cracking. He's so obstinate that he refuses to admit any prior knowledge he cannot prove. "I asked him to admit that there was not a rhinoceros in the room, but he wouldn't," Russell writes.

———◆———

Russell gives Wittgenstein advice, telling him his domineering way of talking and questioning others contributes to his being "generally considered a bore," especially when it comes to questioning his "intellectual betters"—which, if you go back and read the preface, is the term my mother will inscribe in a book to my father more than fifty years later.

Wittgenstein is somewhat immune to that criticism. But he soon has some criticism of his own.

In May 1913, Ottoline travels to Switzerland, and Russell resolves to finish his new book in the time she's gone. He is envisioning a treatise on the nature of human knowledge. He sits down and writes at a breakneck pace. Ten pages per day for weeks. By the end of the month, he has 225 pages done. "What I am writing now is amazingly sincere," he says to Ottoline. "There is absolutely nothing 'clever' anywhere, except possibly a few words on the very first page, which I shall alter someday."

Sadly, he never does.

A few days after his letter, he writes another note detailing a toxic conversation with Wittgenstein. "We were both cross from the heat," he writes. "I showed him a crucial part of what I had been writing. He said it was all wrong, not realizing the difficulties—that he had tried my view and knew it wouldn't work. I couldn't understand his objection—in fact he was very inarticulate."

Russell's book basically becomes stillborn after that. Wittgenstein's "notoriously obscure" criticism evaporates Russell's confidence. "I feel in my bones that he must be right, and that he has seen something that I have missed," Russell says to Ottoline. It "has rather destroyed the pleasure in my writing."

He keeps going, though. More meekly. Eking out another hundred pages or so by early June. And then he up and abandons the manuscript, feeling the pangs of failure. Wittgenstein writes him a letter articulating all his issues with Russell's philosophy, and then he departs from Cambridge. He's losing interest in Russell as a collaborator, thinking there's nothing more he can learn from him. Wittgenstein writes a letter to another colleague basically saying he's over Cambridge. And why not? There's nobody there "who is not yet stale."

Later in life, Russell will claim in his brand of classic over-the-top autobiographical hyperbole that receiving Wittgenstein's letter and hearing his criticism was "an event of first-rate importance in my life," affecting everything he did afterward. "I could not hope ever again to do fundamental work in philosophy," he says. "My impulse was shattered, like a wave dashed to pieces against a breakwater. I became filled with utter despair."

THE GREAT MATH WAR

That's over the top, of course. He never abandons philosophy. He doesn't even abandon his maligned book. He merely repurposes it into a series of lectures a few months later, delivers those as a guest lecturer at Harvard University, and publishes the transcripts in a book called *Our Knowledge of the External World*, which is a brilliant if somewhat overlooked piece of work.

The real failure is Wittgenstein's. Who knows how interesting Russell's original book could have been if he had kept going at his original pace? Wittgenstein could have given him normal, ordinary constructive criticism instead of subjecting him to a selfish, self-loathing, contemptuous screed. What did Wittgenstein have to gain by being so nasty and negative? Why not be more supportive? Russell has certainly done that much and more for him. That's why this episode is emblematic of their relationship. Russell gives and Wittgenstein takes. "Russell was better off before meeting Wittgenstein," according to Kevin Klement, a modern American philosopher and board member of the Bertrand Russell Society.

Russell never fails to credit his friend and student. Wittgenstein benefits greatly from Russell's guidance. But he himself is all take, take, take. Within a few months of meeting Russell, he changes his career goals from building an airplane to rebuilding logic. He never looks back, and Russell is there in those crucial early years. He bends over backward to support Wittgenstein, going far beyond what one would expect of an official adviser, let alone an informal one. He tries to manage Wittgenstein's moods. To lift his confidence. To encourage him to put his thoughts on paper. And when all that fails, he enlists the help of others. Russell is the Batman to Wittgenstein's Robin. The Kirk to his Spock. The Bonnie to his Clyde. The Simon to his Garfunkel. And the Magic Johnson to his Kareem Abdul-Jabbar.

NOTES ON LOGIC

Wittgenstein has a strange and stingy disposition in relation to his own work, so it's not surprising that he dismisses the work of others without a second thought. He is a compositional lazybones, to begin with. With the possible exception of E. E. Cummings, never has any writer in history become more famous with fewer words. Some would apologize for this by insisting that Wittgenstein is a perfectionist. But I say, hooey. With Wittgenstein, the perfect is not just the enemy of the good. The perfect is an evil psychopath who grabs the good by the throat, throttles the good into silence, and slowly chokes the good's life away while laughing and eating good's dinner. Wittgenstein and Russell are fundamentally different in this regard. Russell is a verbal fire hose of helpful tonic. Wittgenstein is a fickle trickle of foul poison. His work is confusing, difficult, unapproachable, and reminiscent of Thomas Hobbes's view of humanity—*nasty, brutish, and short.*

One of philosophy's great ironies is that a few months after Wittgenstein essentially kills Russell's book in 1913, Russell breathes life into his protégé's work. That fall, Wittgenstein meets with Russell and attempts to explain everything he's thought, written, and done since they first met. He also tells Bertie he's reached a decision. He wants to move away to a desolate spot in Norway. To live alone—a simple, cold, spare, monastic existence in a cabin. That will allow him to work freely, far from the crowds. Russell tries to talk him out of it but is met with derision.

"I said it would be dark & he said he hated daylight. I said it would be lonely & he said he prostituted his mind talking to intelligent people. I said he was mad & he said God preserve me from sanity," Russell writes to Ottoline. Fearing his student is about to disappear from the face of the Earth, Russell tries to convince him to at least write *something* before leaving for Norway. He talks through draft ideas with Wittgenstein. That doesn't work. He hires a shorthand expert to listen to Wittgenstein and capture the discussion, but that's a disaster—the transcript is confusing to the point of incomprehensibility. He tries to cajole Wittgenstein into reworking the transcript by hand. He won't.

"After much groaning, he said he couldn't," Russell writes to Ottoline. Only the most perfect ideas should be captured, Wittgenstein says. Nothing less is acceptable.

Finally, during one of their back-and-forths, the assistant of one of his colleagues wanders into Russell's office to borrow a book. Russell leaps at the opportunity. Please come back, he begs, and take notes. She agrees. So as Russell sits, prompting and prodding, Wittgenstein speaks. The assistant takes notes. The transcript is typed up and finished into a manuscript that becomes *Notes on Logic*, Wittgenstein's earliest work and one of the few things he ever actually publishes in his life. (Despite its vaunted pedigree, the book is almost unreadable. It's more like a series of weird philosophical fortune cookie messages than anything resembling a coherent essay.)

A few days later, Wittgenstein leaves for Norway. In an exchange of letters around Christmas 1913, Wittgenstein says he needs to break ties with Russell completely. "We really don't suit one another," he writes. "I shall be grateful to you and devoted to you with all my heart for the whole of my life, but I shall not write to you again and you will not see me again either."

The silence lasts only a few weeks, but their relationship never recovers. In January 1914, Russell and Wittgenstein have a more serious falling-out, which Russell admits to Ottoline is his fault—he was too harsh. But the damage is done. Wittgenstein is now convinced that they have "enormous differences."

LOVE IS ONE HELL OF A PRISON

Problems are percolating with Bertie and O as well. He is exhausted, having burned the candle at both ends for the first half of 1913. Teaching. Dealing with the third and final volume of *Principia Mathematica*. Writing his new book at a breakneck pace—to only then be psychologically shattered by Wittgenstein's withering criticism. On top of that, he and Ottoline have a huge blow-up fight. He later writes her a letter to apologize, speaks of suicide, and ends with the pleasantry "I love you out of hell."

"Wittgenstein affects me just as I affect you," Russell despairs to Ottoline. "I get to know every turn and twist of the ways in which I irritate and depress you from watching how he irritates and depresses me."

Truth be told, things have been souring for awhile. By 1913, the leather of their romance has worn thin, and she's grown weary. At times she finds him annoying and at other times exhausting—even intolerable. "I would give my right hand to be free of Bertie," Ottoline writes in her diary. His bad breath. His tactless aloofness. His awkward movements. His narrow chin. His clumsy, forceful hands ("the paws of a bear"). His nonexistent grace. His overwhelming self-absorption. His commanding instructions. His intense need to possess her. His skewering and oppressive criticism.

"Poor Bertie," she says. "I feel sometimes when I am with him as if I were in prison." She feels smothered. "He expects me to be entirely at his disposal, morning, noon, night, and becomes very angry if I am not," she writes.

By the end of 1913, their relationship is all but on the rocks. They make plans to meet in Rome just before Christmas in 1913, but in their correspondence, Bertie drops a bombshell. He says he will meet her and then afterward see a younger German woman he's interested in. Awkward! When Bertie arrives in Rome, he makes his way to the Sistine Chapel, where they've arranged their rendezvous. There he finds Ottoline—with her husband, Philip. Double awkward! She's jealous. Bertie is furious. He tells her he feels betrayed. He rushes off. They see each other again upon returning to England in early 1914. They meet. Argue. Heap invective. Have a gloriously bitter fight. He later says that he feels nothing, as if he were an actor on a stage.

That's why I suspect they were truly in love—albeit strangely and unconventionally. Love is not always wine and roses. It's sometimes more like a survival knife. It has a smooth, curved blade that cuts clean to the heart—and an angry, jealous, jagged edge that only rips the skin. Emotional wounds come from love's jagged edge.

In March 1914, Bertie leaves for a lecture tour in the United States. Just before he goes, Ottoline writes him a letter. She could not be more intimate with anyone else on Earth, she says. She is with him—and he with her. "The tie is too strong ever to be broken," she writes. "You really have made a huge change inside me and are really important to me—I don't mean in a personal way, but in building up something fine and good and religious inside me."

He's energized by the letter, and their love is rekindled in an instant.

That's the strange thing about love—it's both resilient and defiant. It can be refreshed over and over, never once needing to justify itself. True love is offered not as protest ("*But* I love you"). Nor as aspiration ("I *want* to love you"). Nor as a solemn confirmation ("*Yes* I love you"). Nor as necessary update ("I *still* love you"). All those are wanting. Love needs no introduction. Warrants no preposition. Takes no adverbs. And precludes all grammatical modification. (Unless you marry a magazine editor.) True love stands alone—together.

◆

Bertrand Russell is at the peak of his powers in the spring of 1914, riding high on a white-capped wave of reputation and prestige, when he arrives in the United States. And that's the moment he truly arrives. Courted by Harvard, he's nominally there to deliver a series of guest lectures, but in ulterior mode, the university hopes to cajole him away from Cambridge to be its "chief professor." And Harvard pulls out all the stops. Wining. Dining. Theaters. Art shows. Garden parties. Lunches. His escorts show him ivy-covered halls, gorgeous private homes, and all the extravagant airs a fancy university can muster. The mood is electric. The crowds dazzled.

The Nation calls Russell "easily the most commanding figure" among a new generation of philosophers. Some say he's the most discussed thinker since Aristotle. Hundreds pack the lecture halls to hear him. And he doesn't disappoint. He has repurposed the book Wittgenstein trashed, and to great effect. He could not have been more prepared.

Russell wants to believe in America. A land of opportunity. A fertile ground free of pettiness, snobs, and prejudice. When he gets to Cambridge, Massachusetts, he receives a warm welcome from the faculty. Harvard is at the center of the so-called new realism movement in American philosophy. According to one expert, Russell's *Principia Mathematica* is the Holy Bible of this movement. People are falling over themselves to meet Russell. Gathering. Gushing. Lining up. Swooning.

And Russell?

He enjoys the attention, of course. But he's dismissive of the new realism movement in its entirety and unimpressed by what he sees on his visit in particular. He heaps scorn on the greater Harvard community as well. Filled with "rich, over-eating, selfish, feeble pigs," he says.

Russell calls his American colleagues barbarians. Stupid. Superficial. Dull. He dismisses Harvard president A. Lawrence Lowell as a "deadly" bore. Summing up his experience and his impressions, he says, simply, "This place is hell."

As far as Boston goes, Russell finds the city intolerable. His first impression is that it's ugly. His second impression is that it's worse. "Endless incredibly muddy streets of squalid wooden houses, almost always with the tram shrieking down them," he writes to Ottoline. He's amused by the ever-present spittoons and the "hard, efficient un-meditative men coming and going and talking in horrible American voices," he says. "It seems to me Boston is the worst place in America."

There are some bright spots. He enjoys interacting with certain students—in particular a third-year fellow named Thomas Stearns "T. S." Eliot, who is considered one of the best and brightest at the university (surprise, surprise). Bertie loves T. S. But more than that, he loves T. S.'s wife, with whom he will later have an affair.

HOME RULE REVISITED

Back in England, Ottoline receives Bertie's letters with relish. They make her laugh. Things are tense at home, not for personal reasons this time but for political ones—there is nothing short of a credible prospect of a civil war in Great Britain, which is consuming and threatening to overwhelm the British government.

A few years earlier, a deeply divided British government needed to push through a budget over a promised House of Lords veto, and so Prime Minister Asquith's government entered into a marriage of convenience with members of parliament from Ireland. Asquith readily agreed to their one demand for caucusing together: that he move forward a "home rule" bill for Irish self-governance. And why not? Irish home rule had been a Liberal Party dream in Britain for decades. It was the last great unrealized legislative agenda of Asquith's mentor William Gladstone, who tried and failed to push through Irish home rule in the 1880s. Maybe now it was time, Asquith thought. That set up the greatest internal conflict his government will ever face. In 1912–1914, there's much division in England and Ireland over home rule.

THE GREAT MATH WAR

While Bertie basks in Boston, civil war within Ireland is beginning to feel inevitable. Ireland is divided between the nationalists (or republicans) demanding home rule—who will settle for nothing less—and the unionists (or Ulsters) opposing it. Belonging to the empire is sacrosanct for the unionists. They see themselves as British. They want to be British. They are British.

Those divisions are mirrored in London, and each side has its supporters. In 1912, as the debate was heating up, a compromise proposal suggested carving out and excluding from home rule the largely pro-union Northern Ireland counties of Antrim, Armagh, Down, and Londonderry. But Asquith decided against it. He thought the threat of civil war was not real. So he rolled the dice and called what he perceived to be a bluff. People would accept home rule, he reckoned. *They must!*

———◆———

What Asquith failed to see, however, was a great big Union Jack flag unfurled at a unionist rally in Ireland in 1912 as the compromise was being debated in London. The flag was more than a symbol; it was an expression of Ulster admiration for the empire. It was, in fact, the largest British flag ever sewn. And there's a political lesson that every politician should learn: Never ignore a big-ass flag!

The threat of a British civil war feels real in 1913 and inevitable in 1914. A pro-independence group calling itself the Irish Volunteer Army (a forerunner of the Irish Republican Army) forms in November 1913 and within two weeks has recruited some four thousand volunteers. In early 1914, Ulster factions begin arming themselves. By the time Asquith finally starts to take the threat of civil war seriously, it's too late. In April 1914, twenty-five thousand rifles and three million rounds of ammunition are smuggled into Ireland through the port at Larne in County Antrim. The guns are divided up and hidden—so well, in fact, that some are rediscovered only during house-to-house searches in the 1970s.

The Ulster factions had been a toy army to that point, with antique flintlock rifles and desperate slogans. Now they're a heavily armed militia doubling down on their demands. Larne isn't just a paramilitary success—it's a political statement that nobody is in the mood for compromise.

Cue the parades—large ones. Armed paramilitaries. Marching formations. The green and the orange. Propaganda campaign blitzes. Newspaper articles. Op-eds. Handbills. Posters. Pamphlets. Photos. Films. Banners. Buttons. The catchphrase "No surrender" is first coined in this contested time. Anticipating war, newspapers send foreign correspondents to Belfast. They take up residence and wait for war to break out. Any day now, it seems that it will. Some people begin attending Red Cross classes, learning how to dress wounds and set broken legs.

PITY POOR HELEN

As the Larne gunrunning plays out, Bertie finishes his lectures at Harvard and departs for the Midwest for a few more stops. At one, he pays a house call in Chicago to the doctor E. Clark Dudley, whose daughter Helen Dudley is an aspiring poet. Russell met her a few years earlier while she was studying Greek at Oxford University. Helen knew Russell was coming to the East Coast for his guest lectures at Harvard, and she wrote to him upon his arrival in Boston, inviting him to stay at their place if he could make his way to Chicago. Will he come?

Will he ever!

With characteristic speed and intensity, Russell sweeps in and is swept off. He seems to fall for Helen completely, and he soon makes her fall for him. They initiate an intense physical relationship, dive headfirst into romance, and appear ready to commit to something much deeper. They sleep together, with Helen's sisters standing guard in the hall while she and Russell snuggle between the sheets. Russell begins to make plans with Helen. Will she come back to England? *Will she ever!* What does she think about marriage—assuming he can divorce Alys? *Boy howdy!*

When he leaves Chicago, he writes a letter to Ottoline to say he has something rather important to tell her. At the same time, she's writing him a letter to say she wants their friendship to be platonic moving forward. Their letters cross, and once she reads his words about Helen, Ottoline's reaction is severe. She wishes them well but says she wants to break her friendship with Bertie. She's cold. "If you very greatly desire to see me," she says in June 1914, "I will do so." But better for both their sakes if they don't, she adds.

That alarms Russell. In his autobiography, he will later reflect how even then, deep in Helen's thrall, he sees Ottoline as his true lover—someone so delightful that to leave her would be impossible. That sets up a very strange love triangle when he returns because Ottoline still has feelings for Bertie as well. Of course she agrees to see him when he returns, and of course they fall back into each other's arms—more physical and sexual than ever before. They enter into a new phase—one "of blazing intensity which obliterated the image of Helen Dudley," one of Ottoline's biographers writes.

———◆———

Pity poor Helen! As Alys experienced before her, Bertie's attention, once gone, is *really* gone. Helen has the misfortune of traveling separately so that Bertie arrives back in Britain weeks before she does. But by then, he is deep in a lavishly loving renewed tête-à-tête with Ottoline that summer, which has him falling into her lasciviously loving arms. That causes Russell to lose all interest in Helen. The merciful thing would have been to cable her in America to tell her to forget it. Don't come. But by then it's too late. Helen's ship has sailed—literally. She's on her way to London with her father in tow, full of expectations and joy.

As Helen sails, Ottoline swoons. She writes in her diary, "I feel tremendously alive, and very happy." Bertie echoes the feeling in a letter to her a few days later: "I cannot tell you how full of happiness I feel or what complete joy it is now when we are together. It all seems so easy and natural—there is not a trace of the constraint which had grown up."

And by then Bertie is basically like, *Helen who?*

CHAPTER NINE

1914:
THE SINS OF AUGUST

"
What fun it all was that summer. Everything seemed easy and light, as if the atmosphere had something electric and gay in it.
"

—Ottoline Morrell

Archduke Franz Ferdinand and Sophie, Duchess of Hohenberg, with one of their children in the months before the assassination. Cropped from a public domain image via Library of Congress / George Grantham Bain Collection [ggbain-17181].

The summer of 1911, when Bertie and O first hooked up, was nice. But no summer is more glorious than that of 1914. It's not just nice— it's the *perfect* summer. The finest one in living memory. Sunny. Hot. Sexy. Tinder hot, even! All over Europe, people bask in glorious sun and smiles, baskets of flowers, and hope for the future. It's a summer of love for Bertie and O. Their passion is unbridled.

These are the last sane months of peace in Europe. They come at the end of an era of political unrest—not just in Ireland. These are the golden days of political assassinations; on average, one head of state around the world has been murdered every year in the two preceding decades. Four kings, three American presidents, one empress, two heirs to a throne, and God knows how many minor politicians and courtiers have been killed— shot, stabbed, poisoned, blown up, thrown from a high window, or otherwise murdered in separate acts of terror.

Notable assassinations include the president of France (1894). The shah of Persia (1896). The president of Uruguay (1896). The prime minister of Spain (1897). The president of Guatemala (1898). The empress of Austria (1898). The president of the Dominican Republic (1899). The king of Italy (1900). US President William McKinley (1901). The king and queen of Serbia (1903). The prime minister of Greece (1905). The prime minister of Bulgaria (1907). The prime minister of Persia (1907). The king of Portugal (1908). The prime minister of Egypt (1910). The prime minister of Russia (1911). The prime minister of Spain (1912). And the president of Mexico (1913).

The summer of 1914 sees the most famous of all: Archduke Franz Ferdinand Carl Ludwig Joseph Maria, heir to the throne of the so-called dual monarchy of the Austro-Hungarian Empire—a throne Franz Ferdinand will never hold.

———◆———

Pity poor Franz for the horrible way in which he and his wife, Sophie, Duchess von Hohenberg, are treated up to that point. They love each other dearly. The royal court does not feel the same. Franz is not well liked. Sophie is despised. She has a title, but she's not royalty, and members of the court treat her with contempt. They call her names and make her endure all manner of other insults. At state dinners, she has to sit at the lower-status "kid's table" of minor courtiers and not the high table with her husband.

Getting away from court is always nice, and what better chance than when Franz Ferdinand is invited on an official state visit to Sarajevo? The trip falls on their wedding anniversary, and there could be no more wonderful way to spend it together than in that gem of a city surrounded by splendid mountains. The Paris of the Balkans! It's a chance to take in the crisp air and say "Ahhh" together.

But in the long, tragic history of unfortunate timing, theirs is among the worst. The trip falls on June 28, the anniversary of the First Battle of Kosovo in 1389, which some at the time consider the beginning of ancient Serbia's decline and its downslide into five hundred years of Turkish rule, a fate sealed with the capture of Constantinople in 1453 by Mehmet the Conqueror.

Franz Ferdinand pays no attention to this history. Why would he care? He's on an official visit, so it's all furs and finery for him and Sophie. Feather-plumed uniforms. Loud motorcades. Church visits. Orphanage tours. Military inspections. High-step marches and honor guards. The night before the assassination, Franz Ferdinand and Sophie celebrate their anniversary with a fancy feast. Hosted, toasted, and feted—sitting side by side, as they should. Cream soups. Egg soufflés. Viennese waltzes. Jellied fish. Roasted meat. Wines from Bordeaux. Sherries from Spain. Speeches. Strauss solos. Cheese courses. Cognac. Candy. Cigars. All-around fun. *A night to remember!*

The next morning is Sunday, June 28, 1914. It's a glorious day—the most beautiful day of the nicest summer in living memory. Franz Ferdinand and Sophie take a short train ride from their hotel and then ramble into town in a car with the top down. There's a photo of them that morning. The archduke sits in the back of the convertible. He shakes hands with an official. Sophie holds flowers, smiling.

Photography is a strange art. Next to murder, it's the only form of expression where even rank amateurs can achieve professional-looking results.

THE GREAT MATH WAR

A SUNNY DAY IN SARAJEVO

Gavrilo Princip, that "thin spidery youth with the piercing black eyes," as he was once described by historian Edwin Kiester, proves that good planning doesn't always matter for achieving outcomes. He's like that amateur photographer who somehow accidentally captures a moment in a single frame that encapsulates an entire era. He is a scrawny, weak, undernourished, preyed-upon little boy who grew up dirt-poor, even in the poorest part of his country. He was one of nine children, six of whom died as infants. His childhood was spent stuffed into a two-room house with no chimney. On rough land outside, the family tilled in endless toil—for which they paid exorbitant rent. Gavrilo hid inside books, dazzled himself with stories of history and songs of greatness, dreamed of being a poet, and longed to start a revolution. When he went to Sarajevo to attend college, he joined a secret society—really little more than a bunch of disaffected teenaged malcontents with a few revolvers, a handful of grenades, and a stupid plan.

The assassination of Archduke Ferdinand and Duchess Sophie is sometimes called "the perfect crime"—two shots, two kills. But it's really more like dumb beginner's luck. The assassination succeeds only because it at first fails.

Sarajevo is a water town. A river, the Miljacka, runs right through it. When the archduke and duchess and their motorcade cross the river, the bridge is narrow, the crowds are thick, and several would-be assassins wait. Before Gavrilo can even make his way through the crowd, one of his fellow conspirators finds himself close to the archduke's car. He has a shot, and he takes it. He hurls a grenade. But after the bomb is thrown, the driver sees it. He steps on the gas. Sophie ducks. Franz Ferdinand swings. He hits the grenade with his arm, knocking it away. It bounces off the back of his car and explodes beside the car behind them.

The assassin tries to run, but it's a crowded street. His way is cut off. So he jumps a barrier, hoping to swim away in the river. But it's midsummer. The water's only up to his knees. He's quickly captured. Hauled away. Beaten. Shouting, "I am a Serbian hero!"

The other would-be assassins don't even come close. One is too nervous, standing near a policeman. He's afraid to make his move, so he melts away into the crowd instead. Another assassin is ready to throw a

1914: THE SINS OF AUGUST

grenade but finds himself too far back. Not a chance of hitting anything. Another coconspirator discovers himself filled with regrets. He has a clear shot, but he's paralyzed by what he sees. The archduke's wife. Sophie, innocent Sophie. Lovely Sophie. Why murder her in the cross fire? Why make her his collateral damage? He won't—and he doesn't.

All those failed attempts! But if poorly thought-out strategy provides faulty foundations for a murder, it doesn't matter in this case. Strange fortune favors Gavrilo because after speeding off in his motorcade, the archduke soon comes back.

The bomb thrown by the first would-be assassin exploded and injured one of the passengers in the next car. Once safely away, the archduke insists on going to the hospital to visit the wounded man. When word reaches his entourage that the prospective assassin is caught and in custody, they feel it's safe enough. So they make the decision to drive to the hospital with the top of the car still down.

But the streets are unfamiliar. The entourage takes a wrong turn. They stop to figure out which way to go, and they come to a halt just five feet from where Gavrilo stands. It's too crowded to throw a bomb, so he draws his pistol. He steps onto the car's running board. And he fires two shots into the open carriage. The first hits Franz Ferdinand in the throat. He reacts violently and slumps onto Sophie's lap. "In God's name, what is happening to you?" she screams. The second hits Sophie in the gut. Franz Ferdinand sees that she's been shot. "Don't die, don't die!" he cries. *The children!*

———◆———

Nothing reveals the fundamental issues in a political era so much as an assassination. It's a tragic crime that could have been handled through simple police work and legal justice, but because of the foundational flaws of European politics, it turns into a global crisis. Dark fascination with the murders grows. People see larger plots. The striking success of the assassination attempt makes it look like the work of precision professionals—immaculately planned and well financed. The London *Times* quickly claims a larger conspiracy. Theories abound. The Germans are blamed. The Serbian government is blamed. The Austrians themselves are blamed. The novelist Thomas

THE GREAT MATH WAR

Mann says a secret international cult of illuminati Freemasons did it. The wheels turn. The machine sparks. Soon all of Europe will smell the sweet scent of saltpeter amid an unrelenting grind—the clang, clang, kerchunk, kerchunk of war. *Only monsters make sounds like that!*

An Austrian investigation in the days after the murders definitively and correctly concludes that there is no evidence of Serbian government involvement. "On the contrary," the official report reads, "there are reasons to believe that this is altogether out of the question." But this report is set aside and ignored. Instead of the murder being a tragic end, it becomes a terrible beginning. It sets a spark atop the pool of gasoline. In July 1914, France is bitter, Germany is overconfident, Austria is paranoid, Serbia is scared, Russia is humiliated, Belgium is terrified, Great Britain is indignant, and Europe is doomed.

Fifty-two months of pure horror follow. Austria bombs Belgrade, Germany backs Austria, Russia backs Serbia, France backs Russia, Germany declares war on Belgium and France, Great Britain declares war on Germany, and all roads lead to a common doom.

TO EAT HIS CAKE AND HAVE IT TOO

All the dark clouds clear from Bertie's mind even as they gather over Europe as a whole. Some biographers, apologizing for Russell, say his motivations for how poorly he handles his personal life are driven by his responsibilities as a high-profile antiwar activist. They say he wants to take a public position on the war, and they say that it's to that noble end that he sidelines Helen. He is, after all, still married and cannot risk the scandal of an extramarital affair leaking out to the public.

Ottoline's version is much the same. She claims in her memoirs that Bertie is worried about a scandal with war breaking out, convinced it will interfere with his antiwar activities. Ottoline goes so far as to say he won't even allow himself the luxury of thinking of his own personal happiness at a time like this—what with the war and all. *While the whole world was in pain and ruins!*

Helen, in this narrative, is less love-triangle victim than innocent bystander, albeit a tragic one. "She came, poor girl," Ottoline says, expecting love and "panting with high hopes." Bertie's version is simply that the unfolding events of the war are so shocking and distracting that they "kill" his passion for Helen. As a result, he writes, "I broke her heart."

But all these apology narratives fall flat. They are all spun years after the war is over, ignoring the fact that when Bertie shoves Helen aside, it's still just the dawn of the days of war. His antiwar activities don't really start in earnest until months later. The real reason for Bertie's change of heart regarding Helen is more likely his renewed romance with Ottoline. Worst of all, Russell's own narrative glosses over and offers no explanation for why he never entirely breaks ties with Helen. They don't even stop being intimate. Some sacrifice!

Bertie's failure to clue Helen Dudley in to what's happening is something of a crime of passion. She sets sail, expecting to rush into Bertie's loving arms. Her overwhelming fantasy—that they will settle down and live happily ever after—comes undone even as Europe comes unhinged and she and Bertie continue to get undressed. "She stayed in England, and I had relations with her from time to time," he later says matter-of-factly.

The simple truth of all this is more likely that Bertie wants to eat his cake and have it too. He has always been more interested in Ottoline. And she is suddenly passionate for him as well. "I felt happier with him at this time than ever before," Ottoline says.

◆

Helen arrives in London with her father—a dull, pompous American lawyer, in Ottoline's words. (He's actually a surgeon, if truth be told.) Her father soon leaves, but Helen stays. She needs somewhere to live, however, and Bertie suggests that Ottoline host Helen as her guest at Bedford Square. Is he crazy? Never in the history of dating has such a thing been done. To suggest that your new girlfriend, whom you want to dump but with whom you are still sleeping, should move in with your old girlfriend, with whom you are also still sleeping—but secretly so? That's insanity! What was he thinking? Did he expect to wrangle a threesome out of it?

Ottoline isn't thrilled at the prospect of hosting Helen at all. She finds her odd, "languid and adhesive, sympathetic but insensitive and phlegmatic, and entirely without self-assertion." Nevertheless, she agrees, saying, "What could I do but acquiesce?"

So two weeks into the war, Helen moves into the spare bedroom of Ottoline's house in Bedford Square. Bertie is already deposited in a flat nearby. And that's when things get really weird. Helen is full of expectations. She shows up with suitcases full of cheap, frilly dresses. "When I saw with what hopes and preparations for a honeymoon she had come, I could not help feeling extremely sorry for her," Ottoline says. "She was like a beaten and caged animal."

Ottoline begs Russell to help her by telling Helen he's in love with someone else—but for God's sake, please do not tell Helen it's her, she adds. Helen, meanwhile, proceeds to tell Ottoline everything, thinking she has a sympathetic ally whom she can trust (oblivious to the dotted-line hypotenuse opposite her angle). But now it's Ottoline's turn to be shocked. She doesn't know the whole truth either. She knows very little of what happened in Chicago. And she has no idea that Russell is still intimate with Helen. Now Helen tells her everything. She even lets Ottoline see a stack of Bertie's letters.

———◆———

Ottoline is shocked—not so much by what Bertie says in the letters as by how he says it. He uses the same language with Helen that he has used with Ottoline. From the dates of the letters, she can see that even as she and Bertie were basking in newfound love as never before and enjoying reinvigorated intimacy earlier that summer, he was writing passionate letters to Helen back in America as well. He was telling Ottoline that his relationship with Helen was over before it had begun, but here he was wooing Helen all the while with familiar-sounding sweet words. His duplicity continued after Helen arrived in London. Russell *is still* sleeping with her, Ottoline discovers.

Is he just lying to both of them, or—worse—hedging his bets? "I feel very keenly the disappointment in Bertie," Ottoline says. "He used all these extravagant terms of devotion to me such a short time ago, and now they are all gone, and he professed to her all the things he professed to me." All this is made more difficult by the fact that now, living with Helen, Ottoline has to play it cool with her guest. She's limited in what she can say.

Then there's the incident three weeks after the war starts when Ottoline is with Bertie at his nearby flat and Helen comes looking for him there. She stands outside, beating on his door, not knowing Ottoline is inside but suspecting Bertie is. "I imagined I heard her crying and panting outside," Ottoline says. Later Helen tells Ottoline she knew Bertie was there the whole time. She could hear him breathing, she says. Creepy!

Ottoline is forced to choose. Despite her mixed feelings after reading Bertie's letters, she chooses Bertie. She's tired of Helen anyway. A few weeks living with her is quite enough. She finds Helen too annoying. Too American. She smokes too much. She talks too much. And she constantly wants to chat about her "situation," as she calls it. B and H. H and O. O and B. Ottoline has had enough, and having reached that point, she sends Helen packing. Off to a rented room in Chelsea. Later in the war, she meets Helen again and says she still finds her "a queer creature."

THE PUNCH-DRUNK WAR

Even more queer from Ottoline's perspective is the unfolding war. Ottoline doesn't understand it. She and Philip are solidly antiwar from the start, and so is Bertie. But most people seem excited to support the war effort. They become "drunk with a mysterious primitive emotion," Russell writes.

In the hours after Britain declares war, Ottoline walks the London streets in gloom. The bars have just closed. Patriotism is on parade. Crowds everywhere. Troops. Bands. Flags. Throngs. People are marching. Children play-fight with sticks. Kill the Kaiser. The cheers go on all night. Hardly anybody knows what horror the war will bring. They rush toward it anyway. Nobody walks at the beginning of a marathon.

The paradox of World War I is that it's seen as both inevitable and unnecessary at the same time. It never had to happen, but people at the time tell themselves that it did. The foundations of Europe are threatened, people think. Surely war will settle that. A means to an end. People don't see the war for its potential horror, as they should. They focus instead on war as some sort of soft-glazed, sugary Charles Dickens Tiny Tim fantasy. It's

THE GREAT MATH WAR

A Merry Christmas to us all; God bless us, every one! type of war. A necessary and quick war. *Home before the first leaves fall!*

There is a tangle in the European mind at the time. Some social Darwinists think war is natural—even necessary. A famous German general published a book in 1910 declaring war to be a straightforward expression of "the natural law, upon which all the laws of nature rest." *The law of the struggle for existence!*

People repeat the *war-as-necessity* lie over and over until they are fully convinced. H. G. Wells coins his famous phrase "The war that will end wars," not out of some sense of tragedy or horror, as the phrase will later come to be colored, but from a place of genuine enthusiasm. He expects Britain to crush Germany once and for all—and for the Great War to make the world a better place, one safe for a British-style democracy.

◆

At the same time, it's amazing that the war ever happened at all. International trade is so important in 1914 that economic concerns might easily have overridden social or political ones. Many at the time see war as woefully economically unviable—something to be avoided at all costs because no matter who wins, globalization guarantees that everyone will lose. Why should the vast machinery of commerce give way to the destructive machinery of warfare? But these voices are few. And quiet.

Most see the war in patriotic terms: that gentle "God, king, and England" pride, like a warm blanket on a cold night. Some feel an urge toward violent attack—a *never forget, never forgive* form of patriotic pride, like a T-shirt for sale at a southern US truck stop after 9/11. An eagle crying. Its talons tearing. Afghanistan. Iraq. *Git 'er done!* For others, the war offers an intoxicating appeal to honor. Glory. Righteousness. The allure of adventure. Some in London even proclaim that the war is being fought for honor, that intoxicating fallacy. Whatever the rationale, war has incredible capacity to cover human suffering with the illusion of a just cause.

Russell doesn't buy any of the justifications for the war. He sees it as a mistake, plain and simple. "War is a mad horror," he says to a friend. He's shocked that many of his close friends disagree with him. Even some of his

1914: THE SINS OF AUGUST

pacifist friends, who initially oppose the war, soon fall under the patriotic sway. Some even volunteer for military service. The city of Cambridge fills with troops staging for deployment to Belgium and France. Almost all his students at Cambridge are swept away. His school empties of undergraduates. They rush to join the officers' corps, the cream of a generation sent to their deaths. Things get hostile for Russell at Cambridge, especially as casualties climb. "The older Dons got more and more hysterical, and I began to find myself avoided at the high table," Russell recalls. "I feel very much as if I had been dropped from another planet into an alien race," he writes to Ottoline.

RECKLESS AND BLIND AND ILL

The war evaporates the awkwardness between Bertie and Philip, giving them a common cause. And it draws Bertie and Ottoline closer as well. "But for her, I should have been at first completely solitary, but she never wavered either in her hatred of war, or in her refusal to accept the myths and falsehoods with which the world was inundated," he will later write in his autobiography. He begins attending her Bedford Square salons. Drinking and dancing. Talking and laughing. Furniture messed up. Hair behaving wildly. Gray walls and yellow taffeta curtains. A huge trunk of costumes for all to wear. Cigarettes ground into the carpet. More dancing. Drinks spilled. Pots of potpourri filling the air with their thick scent. Even Bertie's getting into the act. The philosopher is dancing!

Other relationships suffer and end. Russell's dear old friends the Whiteheads are prowar. They think it's the right thing to do. Evelyn Whitehead writes Bertie, decrying the menace of Germany, heralding Britain's mobilization, and bursting with pride over the fact that her son is about to enlist. "I feel as if my relations with the whole family could never quite be the same," Russell writes to Ottoline after reading this.

Even more shocking, his protégé enlists in the Austrian army. Wittgenstein was living in Norway in 1914, and he left that summer intending to spend the season in Vienna. He tries to depart from Austria as the first mobilizations are starting, but he is unable to leave. So he decides to volunteer instead. Even though he's exempted from the draft for health reasons, he is able to enlist—a decision his good friend David Pinsent calls "extremely sad and tragic."

Wittgenstein doesn't see it that way. "Now I have a chance to be a decent human being, for I'm standing eye to eye with death," he records in his diary. "Perhaps the nearness of death will bring light into my life." One of his sisters speculates that he joins out of "an intense desire to take something difficult upon himself and to do something other than purely intellectual work."

———◆———

Wittgenstein is posted to the Austrian First Army, which is sent to the Eastern Front, crossing into Russia in the early weeks of the war in what his biographer Ray Monk will later call "one of the most absurdly incompetent campaigns." Apparently the Russian and Austrian armies both act on bad intelligence, marching willy-nilly across each other's border to meet the other in different spots, and they miss each other completely. It's a disastrous beginning of the war for Austria's First Army as its headlong charge into Russian territory stretches thin its supply lines and forces the troops to limp one hundred miles back in retreat. They lose more than one-third of their 900,000 person force in the campaign.

Over the next eighteen months, Wittgenstein will agitate for an assignment with the infantry on the front lines but will be held back. Finally, in March 1916, he will be sent with Austria's Seventh Army to the Eastern Front just outside Romania, facing Russian forces. There he will finally see action, face down his fears, and open himself to the possibility of death and the promise of enlightenment, according to Monk. "Only death gives life its meaning," Wittgenstein will write.

He is given night guard duty amid heavy shelling, and he relishes the duty—even before he takes it on. "Only then will the war really begin for me," Wittgenstein records in his diary. And that isn't all. During the worst of the fighting in 1916, he actively thinks and writes about logic and records what will become his life's greatest work—the book *Tractatus Logico-Philosophicus*, which he will publish shortly after the war.

1914:THE SINS OF AUGUST

In the war's early months, Russell has no idea what has become of his former student, and he is despondent. He is convinced that Wittgenstein will die during the war. "He is reckless & blind & ill," Russell says to Ottoline.

Ottoline also faces the loss of her own friends. She recoils both from friends and family who support the war and those who fail to share her overwhelming sense of horror over it. "I find it very hard to see [them]," she wrote. "It is almost impossible to talk to them without quarrelling, and I feel a disgust for them." Case in point: Prime Minister H. H. Asquith. He is one of her oldest friends, since long before the war. Yet Ottoline writes in the fall of 1914 that he's basically a murderer who "doesn't really feel the war or anything else very deeply." She coldly tells Asquith's daughter to start looking for new friends.

CHAPTER TEN

RELATIVITY + TURNIPS

1915–1916

> "
> My friend, you would not tell with such high zest
> To children ardent for some desperate glory,
> The old lie: *Dolce et decorum est
> Pro patria mori*
> [Sweet and fitting it is to die for one's country].
> "
>
> —Wilfred Owen

Portrait of David Hilbert. Courtesy of the artist Anna Gorban. CC BY 4.0.

At the start of the war, David Hilbert is consumed with physics problems, as he has been for years. He teaches statistical mechanics. He is working hard on trying to derive the mathematics that correspond to theories on the molecular and electronic structure of matter. And he teaches just about every other topic in applied math that is trendy at the time.

For two years before the war, his focus has been on the kinetic theory of gases, a mathematical framework that envisions a gas (quite accurately) as a cloud of fast-moving atoms and molecules bouncing around a vessel and colliding with each other. Hilbert organized a symposium at his university on kinetic theory, and it was a who's who. Max Planck was there, talking about quantum theory. Peter Debye was there as well. He had just been appointed professor of physics at Göttingen, and he was talking about equations of state and heat conduction. Also there was Walther Nernst, famous for his Nernst equation for calculating reduction potentials in chemical redox reactions. And Hendrik Antoon Lorentz, the codiscoverer of the Zeeman effect, who spoke about kinetic theory and electron motion. And those were just the Nobel laureates in the room!

There was an air of desperation in the room as well, emanating from Hilbert most of all.

He had watched in horror over the previous few years as theories about matter had become more and more complicated. They increasingly relied on complex expressions of statistical mechanics, the mathematical framework that allows a cloud of gas to be treated not as a monolithic airy entity but as a huge ensemble of independently moving molecules. This idea had become very advanced in the prior few decades thanks to the work of Austrian physicist Ludwig Eduard Boltzmann and the rigorous treatment of the subject by the American mathematician Josiah Willard Gibbs.

Hilbert was stressed over the overcomplication of science, however. A quantitative treatment of gas clouds that relies on probability distributions of massive ensembles of particles may be exhaustive, but it's also exhausting. It may be accurate and descriptive, but it's also excruciatingly complex. Statistical mechanics is the opposite of easy. It's hard to compute, harder still to comprehend, and almost impossible to teach. A simple splash of puke would give more conceptual clarity to the minds of most people than a rigorously treated statistical ensemble. And what flies in the face of simplicity defied everything Hilbert stood for. Simplicity is like a religion for him, and in the years prior to the Great War, he becomes convinced of the need to rethink everything.

Make no mistake—Hilbert's interest in applied mathematics doesn't stem from an inherent preference for applied analysis over pure and abstract ideas. It comes from a place of duty and conviction. He serves Queen Math. Mathematics has a lot to offer other fields like physics, he thinks, and what will finally crack the molecular secrets of matter is mathematics. And not just any mathematics—his mathematics. His methods. His approaches. He is convinced that with hard work, and given sufficient time, he'll be able to reveal how matter is structured, form the basis of a unified theory for all physical forces, and do for physics what his formal axiomatic treatment does for geometry: nothing short of knitting up and completing the subject.

Only a mathematician can do this, he thinks. "Physics is much too hard for physicists," he once famously says.

RELATIVITY LOVES COMPANY

The Greeks employed logical proofs, and they were the first to develop the axiomatic method, brought to full flower by Euclid in his famous book *The Elements*. Hilbert was hardly the first person to embrace axiomatics when he wrote *Foundations of Geometry* in 1899. Nor was he daringly original a dozen years later, when he first started thinking of what it would look like to axiomatize an applied math topic like the kinetic theory of gases. He was following a tradition that was literally thousands of years old. Archimedes himself had tried to do something similar with gases.

But Hilbert was unique at the beginning of World War I for having successfully, single-handedly axiomatized an entire subject—geometry. He didn't intend to stop there. As he thought about applied physics, his mind drifted to the pure: the possibility of axiomatizing anything—from basic arithmetic to the most arcane areas of applied math. All of mathematics could be axiomatized and systematically developed, he will soon believe, and that could be the solution to the flaws in the foundations of math.

THE GREAT MATH WAR

But with his intense focus on physics and applied math in the years since his 1900 lecture in Paris, Hilbert was blissfully unaware of the fact that others were thinking about the foundations of mathematics as well. For years before the war, Brouwer was thinking about it too. Hilbert had no idea.

Hilbert was also at first unappreciative of what Russell and Alfred North Whitehead tried to do in *Principia Mathematica* and how they were seeking to tackle the foundations of math. His first interest in their book doesn't pop up until it's too late. He dives into the work in December 1914 while he's supervising a PhD student writing a dissertation devoted to mathematical logic. Hilbert is impressed with the *Principia Mathematica*, and he thinks of how much he'd love to invite Russell, the twenty-something kid he met in Paris all those years before, to Göttingen. But because of the war, the possibility of even exchanging letters let alone traveling across enemy borders to meet in person will be impossible for years.

Perhaps it's just as well. He soon becomes distracted by another subject—also applied but far more baroque. In the summer of 1915, physicist Albert Einstein visits Göttingen for a week. It's his first trip to the city. He's not yet famous—not in the way he will be in just a few years. But within math and science circles, he's already a rock star. He gives six guest lectures on general relativity, the physical framework for thinking about mass, space, and gravity in 3D space. It describes things like gravitational waves and accounts for how gravity bends light, warps space, and alters the experience of time in different moving frames of reference.

It's a mind-blowing subject, and Einstein's lectures are a huge success—an "event," as Hilbert describes them. No one in the crowd is more excited than he is. Hilbert latches on to the subject and forms a deeply intense and immediate keen interest in it. He can't believe what he's hearing. Identifying a challenging problem in search of a mathematical solution is like winning the lottery for him. And this is an unexpected megajackpot: a physical theory that's incomplete, ripe for analysis, and hard in search of its underlying equations. Relativity is just begging for his attention.

Hilbert enthusiastically dives into finding the equations of general relativity after listening to Einstein—and Einstein himself is thrilled to know it. That's great. Nothing like a little friendly competition in the race to discovery. "To my great joy, I succeeded in convincing Hilbert and [others] completely," Einstein recalls. Relativity loves company!

Interesting others in relativity is why Einstein came to Göttingen in 1915. For most of the previous two years, he has been struggling with how to represent gravity as the geometry of space-time. Mathematically, this demands a set of generally "covariant" field equations with which to describe gravity in relativistic terms. But as of the summer of 1915, those equations still need to be derived.

THE "UNOFFICIAL" GENIUS

Hilbert is boosted by a junior recruit he makes that year—hiring into a low-level, no-pay postdoctoral position perhaps the most overqualified unpaid intern in all human history: a researcher named Amalie "Emmy" Noether, one of the most brilliant mathematicians of her generation.

Noether is of an age to have benefited from Sofya Kovalevskaya's efforts to crack open the doors of German universities to women, even though those openings remained diminishingly narrow and treacherous at first—and often still depended on special dispensations, approvals of committees of men, and pernicious case-by-case favor granting. "A favor which can be denied at any moment," Kovalevskaya complained in her day.

Long after Kovalevskaya's trailblazing occurred, those doors remained locked for most women. And those who did follow in her wake still had to fight every inch of the way toward a degree, facing door slams at every turn. Germany was a particularly slow adopter of women in higher education. Women were allowed into French universities in 1861, English institutions in 1878, and Italian schools in 1885. But in much of Germany, they were excluded for years longer.

And women in Germany were not just passively discouraged from pursuing higher education degrees—they were actively prevented from doing so. The eminent (and eminently sexist) German historian Heinrich von Treitschke once warned that "surrendering" universities to the invasion of women would be "a shameful display of moral weakness." He thought little of men and even less of women. Sexist trolls like von Treitschke succeeded in keeping women out of German universities only until the end of the nineteenth century, but succeed they did. As late as 1898, the University of Erlangen's academic senate declared that admitting women would "overthrow all academic order."

Things improved somewhat when German laws changed—first prior to World War I, when new laws at the turn of the century allowed women to enroll, and then after the war, when women were granted the legal right to hold faculty positions. Even then, women were often forced to suffer male reluctance to colleague up.

Perhaps nobody's career illustrates this roller coaster more vividly than Noether's. She was a brilliant mathematician, one of the best in her generation, but she faced awful obstacles at each successive step of her excruciating climb. Even as she rose in the coming decades, many men who were far less capable passed her quickly, rainbow-trailing their way to the backslapping top while she was stuck on the lower rungs. Unpaid and underpraised.

In 1900 changes to German law allowed Noether to *attend* the University of Erlangen—becoming one of two women among 984 male students—but still not actually *enroll*. She did so well on her finals that the university granted her a degree, though still without allowing her to matriculate (basically affording her the same accommodation Kovalevskaya had received three decades before). Then, in 1904, the laws changed again. Women were now, finally, officially allowed to enroll for classes and degree programs. Noether was at the exact stage in her career to take advantage of this, ready to enroll in graduate school, so that was exactly what she did, finishing her dissertation in three years.

And then... nothing.

———◆———

For nearly eight years after graduation, until Hilbert hired her in 1915, she continued to do research and give lectures, but only in an "unofficial" capacity with no actual appointment and with no salary. Even that ungenerous situation was tenuous. Had she been anybody else, she probably would not have been able to stay at the University of Erlangen at all. But she was able to lean heavily on her father, who had been a prominent mathematician and professor at the university for decades.

RELATIVITY + TURNIPS

One interesting aside here is that Emmy Noether's father was Max Noether, a prominent math professor in Germany in the late 1800s. Because of that pedigree, Emmy is often held up as evidence of the simple-minded notion that mathematical ability is genetic. There is really no scientific basis for believing this—no math gene or genes that we know of—just a lot of anecdotal cases of mathematicians who happened to be related, like the Bernoullis or the Noethers. But there are many more counterexamples. George Boole was the son of a shoemaker. Carl Friedrich Gauss was the son of a bricklayer. Isaac Newton was a farmer's son. Nor was Max Noether himself from a long line of mathematicians. He was a unicorn as far as that goes.

There could be a much simpler explanation than genetics for why people like Emmy excel in math. Her case may have been nothing more than a strange twist of fate combined with plain old hard work. Her father, Max, had polio as a teenager. Flaccid paralysis affected him for the rest of his life, and he was driven to mathematics in part because his physical condition steered him toward a desk job. Max's father ran an iron business in Mannheim, and Max would probably have inherited that business had he never contracted polio. Instead, he became the first person in his family to earn a college degree—as well as a doctorate in mathematics.

Would Emmy have become a towering figure in math if she had been heir to an iron forge instead? It's impossible to say, but what is clear is that she was hugely helped in her career by her father's prominence—not genetically helped but environmentally. He nurtured her to become like him, and the rest was history.

Max's colleague at Erlanger was Paul Gordon, a cantankerous, impulsive, brilliant, and odd individual. He had the rare ability to carry out long, complex mathematical calculations in his head but almost no patience for casual two-way conversation. He was a great speaker but a poor listener—said to seem half asleep when someone else was talking and half crazy when he himself had the floor. He also talked to himself when nobody was around—and sometimes when they were. And when he talked, he did so more with his hands than his mouth. Abrupt gestures. Gesticulations. Unmatched intensity. Gordon was a man of few words, many violent gestures, and even more violent equations.

Noether studied under Gordon in grad school and finished her PhD in 1907, becoming the first and only student to be graduated by him in his entire career. Her thesis was on the theory of invariants, which was Gordon's main area of interest and expertise—they called him the "king of invariants," in fact. Being the king's protégé landed her a job at Göttingen eight years later. Invariants were as key to Einstein's theory of general relativity as salt and butter are for mashed potatoes. Invariant theory allows you to model how geometric properties will remain unchanged under transformations. General relativity, as Einstein saw it, posited general covariance, which held that the laws of nature were invariant with respect to transformations.

What Einstein lacked at the beginning of 1915 was a set of equations to capture this—something both he and Hilbert begin hotly pursuing after his visit in the summer of 1915. It is Hilbert's good fortune, then, to have recruited Noether. As he starts looking for the invariant mathematical formulations of Einstein's general relativity, she will be invaluable.

"WE ARE NOT A BATHHOUSE"

And yet, once again, oppressive academic sexism rears its ugly head in Germany. When she first arrives, Hilbert is forced to argue in front of a committee in favor of Noether's appointment. They want to disqualify her on the basis of gender—as idiotic as it sounds to "disqualify" someone who is perhaps the most qualified person in the world in her area at the time—even more so than Einstein himself on the topic of invariants. But Hilbert's colleagues object to her sex, not her work. It's unseemly to allow women to mix with men, they say. To this, Hilbert famously responds, "We are a university, not a bathhouse."

The committee is unmoved. "What will our soldiers think," one member of the faculty says, "when they return to the university [from the war] and find that they are expected to learn at the feet of a woman?"

Because of such selfish, sexist dogs, Hilbert is forced to hire Noether as an unofficial guest lecturer without pay. This is extraordinary, as she's reached a point in her career and a level of success in her work where many men with the same qualifications would have been full professors by now—and indeed, many were.

RELATIVITY + TURNIPS

Hilbert isn't satisfied. Later that year, as the race for relativity is proceeding and he sees how invaluable and brilliant Noether is, Hilbert again tries to secure a salaried position for her, changing his strategy and appealing directly to the government ministry overseeing the university. He shows letters of support from top faculty. The answer is still no. He asks if they could maybe grant her a paid position without an official title. Again, they say no. Hilbert says he's worried that she'll be stolen away by another university. To this, the ministry responds with sexist relish. That's fine, they say. The other university can keep her. Neither Noether nor any other woman will be allowed to officially teach at a German university, they say. Ever.

The race for relativity, meanwhile, reaches its climax in the fall of 1915. Einstein and Hilbert, separately trying to work out the mathematical underpinnings of general relativity, take it down to the wire. It's actually insane how close the race is. Einstein wins by mere days, presenting two papers on general covariance in relativity to the Berlin Academy on November 11 and 25, 1915. Hilbert delivers his own paper describing basically the same results to the Royal Society of Science in Göttingen on November 20, 1915, smack in the middle of Einstein's two papers.

Some minor squabbles over priority follow. Some at the time say Hilbert stole Einstein's work. Others claim it was Einstein who borrowed Hilbert's. For his part, Hilbert is gracious and acknowledges Einstein's priority. He calls general relativity "one of the greatest achievements of the human spirit ever" and places Einstein in the scientific equivalent of the leading role in a centuries-spanning three-act play that began with Pythagoras, continued with Isaac Newton, and culminated with Einstein himself. *Veni, vidi, vici!*

But this is all small comfort to anyone. As relativity goes high, the war to end all wars goes low.

YOUR COUNTRY NEEDS YOU!

From the start, Britain faces an enormous people problem in World War I. Unlike France, Russia, and Germany, which all have some form of mandatory military draft at the outset, Great Britain's army has long held to a tradition of purely voluntary service—and it is piddly in size compared to its neighbors. Britain's army in 1918 is only one-fifth the size of Germany's, making recruiting more soldiers a priority from the start.

The face of this recruiting effort is Field Marshall H. H. Kitchener, the same Boer War general of land-clearance and concentration-camp infamy. Early in World War I, the military rolls out an iconic advertising campaign highlighted by his mustachioed mug, a white hat, and a stern look. He faces the camera. Points right at the viewer. Big finger, bigger mustache. *Britons*, the poster shouts, *your country needs you!*

"He is not a great man. He is a great poster," Asquith quips upon seeing the ad.

Bertrand Russell is recruiting as well. His strong and resounding peace activism emerges during the war. He gives lecture after lecture to growing crowds. He's a great speaker too—though some say people come to see him just for entertainment value. He draws upon his mathematical and philosophical roots in his writings—especially in the framing of his arguments. "Of all the evils of war," he writes in a magazine during the war, "the greatest is the purely spiritual evil: the hatred, the injustice, the repudiation of truth, the artificial conflict."

It is an "unwise and wicked war," Russell writes. "Young men embarking in troop trains to be slaughtered on the Somme because generals were stupid." Strong statements like these open up new audiences (including, decades later, my own father), but they also earn him enemies among the majority of Britons who support the war, including many of his close friends and colleagues.

If it were a little war, his words would matter little. But this is the Great War, and people take great umbrage when they object to his statements. The scale of destruction in World War I is unlike anything anyone's ever seen. The Battle of Verdun, from February to December 1916, sees more than 700,000 casualties on both sides, including 300,000 killed in battle. The Battle of the Somme in 1916 sees 57,470 casualties and almost 20,000 deaths on just its first day. No-man's-land becomes an open grave as far as the eye can see, and the battle eventually accounts for some 420,000 British, 200,000 French, and 450,000 German casualties.

RELATIVITY + TURNIPS

All that death causes untold suffering on the home front. People suffer in countless little ways and one inconsolably major one: the grievous loss of more close loved ones, friends, and neighbors than they can count. Fathers. Sons. Brothers. Cousins. Coworkers. Besties. Men die by the tens of thousands—and many, many women serving in Europe and Asia are killed as well. Everybody in Britain knows someone who has died or is dying in France and Belgium. And all those losses make for no small amount of confusion and consternation among Russell's friends and colleagues when they read or hear what he's saying about the war. His sudden shift from writing about mathematics to questioning the war and Britain's need to continue it strikes some as bizarre, even inappropriate—and is made all the more strange by the fact that he continues to write and discuss the foundations of logic and math from time to time.

But Russell isn't just transforming. He's transcending. He becomes different things to different people—to some an activist. To others just a logician. To still others something of both—a mathematical philosopher interested in activism or a peace activist interested in mathematics. To many of his activist readers, his mathematical writings are confusing, if not incomprehensible. Definitions. Methodologies. Principles of knowledge. Philosophies. Set theory. Alien ideas from an alien visitor from another planet. For his mathematical readers, his ethical writings are often easy to understand but hard to stomach.

◆

A war of words begins to be waged on the pages of the publications he writes for—a fierce battle of outrage and ire, some opinions falling softly in favor of Russell and others ripping harshly at his work and words. They object to his invocation of British evil. How can you call your own side evil when you are yourself fighting the devil in the field? "Either Mr. Russell is 'astonishingly naive and uninformed,' or he is calmly insolent, or he is merely enjoying some good-natured amusement," one writer says in 1915.

Detractors aside, few protesters in human history have been as active or effective as Russell. Even the normally cynical and blisteringly acerbic writer and historian Lytton Strachey is impressed by Russell's war speeches and writings. "I don't believe there's anyone quite so formidable to be found just now upon this earth," he says.

But Strachey is not the only one watching. And others have more hatred than admiration. British authorities take note and draw up a map for Russell, "marking out the territorial limits within which he might speak and beyond which he must not travel," one contemporary writer says. Russell is unmoved by the threat. He writes to Ottoline in 1916 before setting out on another lecture tour, "I don't care what the authorities do to me, they can't stop me long."

One of the main things he protests is the British draft. Great Britain still doesn't have enough soldiers by the end of 1915. The recruiting campaigns succeed somewhat in filling Britain's ranks, but they continue to fall far short. So efforts begin to institute a draft. Support for a draft is strong within Asquith's administration, though he himself only reluctantly goes along with it. One historian says Asquith feels like he has a gun to his head, fearing his government will crumble if he resists. Once he's on board, however, the well-oiled gears of the British wartime propaganda machine turn and grind to make it happen. In the second half of 1915, some fifty-four million posters promoting the draft are printed. Eight million letters are sent directly to consumers, and twelve thousand community meetings are held.

Finally, on January 23, 1916, Britain's so-called Military Service Act passes. Philip Morrell and a few others in parliament try to stop it, but to no avail. The draft becomes mandatory for all single men under age forty-one. A year later, it's expanded to cover all married men in that age range as well. Philip and his allies do record one major win: They get language into the law allowing people to claim conscientious objector status, a hugely unpopular idea in Britain generally. Some high officials in the British army say all conscientious objectors should be taken out and shot, even though, relatively speaking, few try to claim such status. During the years of the Great War, only sixteen thousand British men register as conscientious objectors—a drop in the bucket compared to the nearly five million who serve in active duty.

RELATIVITY + TURNIPS

Still, conscription plays a major role in the Great Math War to come because it takes Bertrand Russell out of the picture. Conscription gives the efforts of the peace activist movement more structure and meaning. Protesters quickly pivot from antiwar to antidraft activism. For Russell in particular, conscription makes permanent the change he's been cultivating for years by becoming a social commentator. He is swept up and swept away from the foundations of math, never really to return.

But first, he goes to jail.

CHAPTER ELEVEN
COLLECTIVE INFLUENZA

1917–1918

"
If mathematical thinking is defective, where are
we to find truth and certitude?
"

—David Hilbert

Detail of a 1914 recruiting poster showing Lord Kitchener with his finger pointing at the viewer. Alfred Leete/Victoria House Printing Co. Cropped from a public domain image via Library of Congress / George Grantham Bain Collection [LC-DIG-ppmsca-37468].

When Russell finds himself in antiwar meetings early in the war, he tends to think of the ones he goes to as either dull, futile, or worse. "It seems like eight fleas talking of building a pyramid," he writes to Ottoline after one such meeting in late 1914.

You can't really blame him for such cynicism. He's a firsthand witness to a pacifist farce soon after Britain declares war when he attends a "neutrality committee" meeting that ends suddenly when a thunderclap outside shakes the building. What was that sound? People wonder. A boom. An echo. A gasp. Then a hush. Quiet murmurs. What was that? Anger. Shouting. A bomb! It's a German bomb! "This dissipated their last lingering feeling in favor of neutrality," Russell writes. Thus ended the meeting, and thus ended the committee.

Spy fever takes hold in England soon after that. Some fifty thousand German citizens are living in Great Britain when war starts, and right away, police arrest twenty-one people alleged to be spies. Just days after war is declared, Britain passes the Aliens Restriction Act, which mandates that Germans and other foreign nationals register with local police. They face detention and deportation, are barred from owning guns or radio transmitters, and are forbidden to live near coasts or military bases. People begin to imagine spies all around them: store clerks, waiters in restaurants, even local barbers and hairdressers. (Who spies on the barber who doesn't spy on himself?)

One newspaper resurfaces a report from 1907 claiming that ninety thousand German spies are living in Britain and have stockpiled weapons in every major city. Within weeks, the London police have received nine thousand (largely false) reports of suspicious Germans. Germanophobia is ever present after that. Newspapers print panicky letters from readers. People demand action. And in the first two months of the war, more than ten thousand people are arrested.

When the steamship *Lusitania* is sunk on May 7, 1915, Germanophobia boils over. The timing of the sinking coincides with the publication of a report documenting German war crimes in Belgium. That spurs letters to the editor. "The British people were now persuaded that they were engaged in a crusade against evil," British historian David French will write in 1978.

———◆———

Every white-knuckle, hand-wringing worry imaginable then plays out. Speeches fall in parliament. Petitions circulate among the masses. There are riots. Ransackings of German shops. Protests. In one march, a large group of businessmen crowds downtown London and demands the immediate arrest of the entire German population in England. A few days later, the Asquith administration agrees to arrest all remaining Germans.

Anything even remotely German becomes suspect. Schools stop using the term "kindergarten." The Asquiths have a German governess who is quickly sent away. King George V issues a decree changing the royal surname Saxe-Coburg-Gotha to the more Anglo-friendly Windsor, distancing them all from their thick German roots. The king himself is a descendant of George Louis of Hanover, who was elevated to king in Isaac Newton's day. Queen Victoria's mother is a German princess, and Victoria herself marries a cousin, Prince Albert, who's also German. History records that she's raised by a German nanny, hearing and speaking only German until she's three years old.

Some resist this anti-German backlash, believing that not every German could possibly be a secret agent. Ottoline begins working with an organization named Friends of Foreigners, which focuses mainly on German men who live in Britain (many of them for years) who are married to English women. But her focus soon shifts after she and Philip move to a farm near Oxford. Like many with antiwar sympathies, they become enmeshed in fighting against the draft. And when mandatory conscription becomes law, they work to support their friends intent on dodging the draft by claiming conscientious objector status.

Philip defends conscientious objectors who are called to appear in front of military tribunals. And he and Ottoline turn their farm into a "Mecca for pacifists," as some call it, where people claiming to be conscientious objectors can work in lieu of being drafted. In the new law, people who claim this status have to find employment in approved jobs that help the war effort. Farming is considered essential work, so the Morrells are able to keep a number of conscientious objectors employed and out of the war.

In her memoirs, Ottoline complains that it's sometimes hard to find jobs for these "essential" workers. Many have no practical experience, hail from socioeconomic stations far removed from manual labor, and are more used to intellectual or artistic endeavors. "Farm workers have seldom sounded so lazy and incompetent as the conscientious objectors of Garsington proved to be," one of Ottoline's biographers writes, referring to the name of the farm.

PUGS, PEACOCKS, AND POTPOURRI

Garsington is a place of art and theater as well. Of prance and pomp. Somewhere to see and be seen. Ottoline paints one room bright red and another dark green. In the garden, there are peacocks. In the house is a pack of pugs, including one named Socrates. Lots of drapes. Curtains. Cushions. Carpets. Unexpected rooms with potpourri scents that, much like her, are exquisite, overwhelming, and strange.

She's known for an almost theatrical extravagance in her garb. It's one of the things I love about Lady O. Once she dresses like a boy ripped from a Cossack folktale, according to a biographer. Tight silk trousers. A mannish medieval tunic, down to her knees. Embroidered patches. Bright red boots. Another time she parades around London wearing a huge hat with a large feather and her once-worn-by-Marie-Antoinette string of pearls. Garsington doesn't just reflect her. It *is* her.

Wartime life on their farm is rich. Here is Aldous Huxley sleeping on the roof between the gables and the chimney stacks. Here is Belgian war refugee Maria Nys meeting Huxley, her future husband. Here is Lytton Strachey polishing passages of *Eminent Victorians* in the garden. Here is Russell in the nude, skinny-dipping in the pond and running into Asquith. "The quality of dignity which should have characterized a meeting between the Prime Minister and a pacifist was somewhat lacking," he writes. Here are the poets William Butler Yeats and Robert Graves hanging out

COLLECTIVE INFLUENZA

by the pond. Here is Virginia Woolf, writing of scented cabbages and pale yellow silk. Here is an Easter party where preproof copies of T. S. Eliot's "Love Song of J. Alfred Prufrock" are handed out. "Hot from the press like so many Good Friday buns," one person says.

———◆———

On this scene lands David Herbert "D. H." Lawrence, a young writer Ottoline discovers at the beginning of the war through his book *The Prussian Officer and Other Stories*. She throws a dinner party for him and E. M. Forster, and she and Lawrence hit it off right away. D. H.'s wife, Frieda Lawrence, however, is suspicious of Ottoline from the beginning and gossips about her behind her back—even with Ottoline's close friends. She storms out of Ottoline's forty-second birthday party. She rips up letters D. H. is writing to Ottoline.

Ottoline is no less uncivil in her assessment of Frieda. Irritating. Domineering. Lazy. She's convinced that Frieda is a rusty anchor made flesh, weighing D. H. down just as he's ready to spread his wings and soar. He is "crushed," Ottoline writes. "Unhappy." *Following her like a whipped dog!* "Poor Lawrence," Ottoline says. "Frieda will wear his nerves out in time." Ottoline says Frieda is D. H. Lawrence's "dark abode." She is dark energy. Like some distant, nonluminous galactic remnant full of dead stars and dark matter. Nothing to see, and we know it's there only because of how it bends the light. She encourages D. H. to leave his wife.

The relationship between Lawrence and Russell is even more volatile. Ottoline introduces the two legends to each other in 1915, and at first things go well. They are full of ideas—collaborating on philosophy, strategizing how to protest the war—but it doesn't last. First they're pink with joy. Then yellow with boredom. Then red with pride. Then purple with envy. Then black with anger. Then slimy with mistrust. It doesn't take long for them to turn bro hug into backstab.

"You are too full of devilish repressions to be anything but lustful and cruel," D. H. says to Bertie in a scathing letter. "You simply are not sincere." No, he says, it's more than that. *What you say is a lie.* No—it's worse than even that. *You are the lie!* D. H. says he'd prefer to talk peace with a warmonger than Russell, a thousand times over. Better to be a murderer and admit it than to be a liar. He calls Russell "the enemy of all mankind."

"It is not the hatred of falsehood which inspires you," Lawrence says. "It is the hatred of people, of flesh and blood." *Let us become strangers again!* Frieda is to blame for the whole falling-out, Ottoline thinks.

CHROME YELLOW ANEMIA

Ottoline is someone who understands backstabbing, especially when it comes in gossip form. An unkind word. A sharp cut. A jagged edge. That awful serrated rip. It's the world in which she lives.

During the war years, gossip about her abounds behind her back. One friend calls her "venomous, dangerous." Another says she's cruel. One compares her to a hungry tiger, waiting to claw some unfortunate lover. Another dubs her "Omega Muddle," a reference to Roger Fry's Bloomsbury neighborhood art studio, showroom, and storefront called the Omega Workshops. Some say she's superficial. People mock her looks incessantly. They savage her intelligence. They highlight her awkwardness. They even make fun of her chronic health problems. And her writing—good God, her writing! They hate her writing. They destroy her for her writing. Her experience illustrates painfully how sometimes your friends are not your friends.

In some ways, these are the spoils of an active social life. She surrounds herself with writers and artists, so what does she expect, really? They wield verbal knives that rip and rend. Many of the writers who know Ottoline go on to depict her, directly or indirectly, in their works. Her personality is captured in J. A. Revermort's *Cuthbert Learmont*. She inspires the character of Evelyn Murgatroyd in Virginia Woolf's *The Voyage Out*. She is skewered as smug Lady Caraway in Walter Turner's *The Aesthetes*, smothered in bandages. A pack of fifteen pugs in tow. "Crooked as the Tower of Pisa," he writes.

Juliette Huxley, the former tutor of Ottoline's daughter, describes the Garsington house in *Leaves of a Tulip Tree*. H. G. Wells parodies her farm in his 1915 novel *Boon*. Ottoline herself is parodied as Lady Rusholme in

Gilbert Cannan's *Pugs and Peacocks*. She is sheepishly mocked as Priscilla Wimbush in Aldous Huxley's *Crome Yellow*. Years later, Huxley publishes "Those Barren Leaves," which has a much more direct and skewering portrayal of her as someone with a prominent chin, saggy cheeks, sloppy lipstick, and overarching hypochondria. She gets an obscure nod as the creature Lady Septuagesima Goodley in Osbert Sitwell's *Triple Fugue*.

——◆——

All those depictions are for the most part benign, even if occasionally mocking or cruel. A much nastier and by far the most stinging literary depiction of her comes from D. H. Lawrence, who portrays Ottoline as the hideous and garishly grotesque snake goddess Hermione Roddice in his novel *Women in Love*. Lawrence gives intimate and familiar details of her guests and depicts her as an old hag in a dirty dress. Obsessed with sex. Deficient. Broken. Immoral. Possessive. Fundamentally flawed. "Gnawed as by a neuralgia," he writes. He depicts a scene where her maid brings out a chest of costumes for the guests to wear while her husband plays strange songs on a pianola.

Ottoline considers the book to be genuinely cruel and a total betrayal. D. H. was someone she had championed, nurtured, and fought for. She had stood up for him during a court case after his previous book *The Rainbow* was published. The book was panned in the press and pilloried in public. One reviewer compared it to an epidemic disease, only worse. There was a huge outcry over the novel because it depicted lesbian sex and had a scene where one character mocks a soldier in uniform. (Hard to say which was more outrageous in 1915.)

A high British official successfully sued to have every copy of the book collected and destroyed. The books were banned in some places and burned in others—put to the torch by London's official public hangman, no less. In November 1915, when Hilbert and Einstein were finishing their race for relativity, Lawrence gave Ottoline his last copy of *The Rainbow*, telling her he never wanted to see it again.

That was not the end of it. When police seized copies of his novel, they found some of Lawrence's artwork, also deemed to be lewd. A hearing was held to decide what to do with the paintings. Should they be set ablaze too? In court, when the hearing took place, Ottoline stood up, impressively tall and wearing a big hat. She pointed a withering finger at the magistrate. *YOU are the one who should be burned*, she said, *not the paintings.*

In the days leading up to that trial, D. H. and Frieda stayed at Garsington. This arrangement drove a further wedge between Ottoline and Frieda. Ottoline lamented of Frieda, "If only we could put her in a sack and drown her."

Nevertheless, the fact that she hosted them illustrates how much Ottoline did or tried to do for the Lawrences. So, months later, when Ottoline discovers the awful, cruel, skin-ripping skewering of her persona in *Women in Love*, she is shocked. D. H. refers to her as "that old carrion" and ruthlessly steals scenes from her life, weaving in bits from her actual childhood. Unhappy episodes. Formative moments. Personal anecdotes. Things she trusted him with. She's furious. Once again, she blames Frieda. "All the worst parts were in Frieda's handwriting," Ottoline says.

She shows the book and the letter to her friend Aldous Huxley, and he's horrified too, Ottoline records. Philip is pissed as well. He seeks out Lawrence's agent and threatens to sue the publisher for libel.

Ottoline has a falling-out with Bertie as well. They had begun to drift apart at the beginning of the war after she read all his letters to Helen Dudley. Later, when she moves to Garsington, he sees less and less of her. Just after Christmas in 1916, Bertie writes her that he wants to end their relationship so he can be free to pursue other romances. She doesn't know it then, but he's already deep into a thing with an actress and writer named Constance Malleson, who goes by the stage name Colette O'Niel. Bertie finally tells Ottoline about Colette in 1917, after he's been seeing her for a year, and Ottoline is stunned. "It only showed me how far apart we had drifted," she says.

But her shock is due to get much, much worse.

THE GHOSTS OF THEIR CHILDREN

Philip in 1916 is distant and superbusy. He's renting a bachelor's pad in London, spending nights there during the week and sometimes weekends

COLLECTIVE INFLUENZA

as well. A member of parliament and a leader of the peace faction during wartime, he deals with crisis after crisis. By the end of 1916, he's exhausted. It's been a terrible year.

First, the draft went into effect. Then there was the armed Easter uprising of Irish republican nationalists in Dublin, thrusting the unsettled and postponed question of Irish home rule past the point of no explosion. Then in June, the *We-Want-You* poster boy of the recruiting effort as well as number-one war secretary, H. H. Kitchener, dies when his boat is sunk by a German mine off the coast of Scotland. Everybody is distraught because of that. Kitchener has been the main architect of Britain's war effort. Then the Battle of the Somme begins in July. For the rest of the year, hundreds of thousands of people die in this battle—sometimes more than ten thousand British soldiers in a single day—all to nudge the front forward a few meager miles.

Public opinion has begun to sour on the Liberal Party, and the popular press eviscerates its leaders over the cost of war. "Daily vindictive, merciless attacks in the columns of the newspapers," in Prime Minister Asquith's words. There is a growing chorus of calls for new leadership on Downing Street. Adding to his woes, Asquith is ill prepared to lead the war effort, having previously left all the major wartime decisions to Kitchener and Winston Churchill. In the summer of 1916, Kitchener is dead and Churchill is gone, having resigned after the disastrous Gallipoli campaign. There is nobody left to guide Britain through the war except Asquith's colleague David Lloyd George.

Asquith relinquishes the war reins to Lloyd George, which is a bit like Archduke Franz Ferdinand handing a loaded gun to Gavrilo. Lloyd George solidifies his own power in that position, and he orchestrates Asquith's downfall in December 1916 when he and several members of the cabinet threaten to resign in a clever coup that forces Asquith's own resignation a few days later. "Who Killed Cock Robin?" one British newspaper reports, referencing the old nursery rhyme and the head-spinning way Asquith's ouster went down.

All that death and drama weighs heavily on Philip at the end of 1916. Ottoline assumes it must explain his distance that fall. He's busy. London is a mess. The war effort is desperate. The country is in chaos. She sees him less and less, and even when he's home, he buries himself in the odious duties of the farm. But she's wrong,

Ottoline is crushed to discover in 1917 that Philip has been having affairs—not just one but two—and both with women she knows. One is her former personal maid, whom she prophetically wrote in her diary in 1916 she didn't trust. "I do all I can to make her happy, but she really is very unkind to me," she said. "I don't know why." Now, in 1917, she does.

At the end of 1916, the maid had quit and moved to London. She was already pregnant, apparently, and in London, she continues to see Philip. Also in London is another love interest of his—a woman who works as a secretary at *The Nation* who begins picking up a few hours on the side doing secretarial work for Philip. She also occasionally works as a secretary for Ottoline, and she even visits Garsington in the fall of 1916 when Philip brings her to stay there for a weekend. They have their own thing on the side, and in early 1917, she is pregnant as well.

The truth comes out soon thereafter when Philip is spotted at the Palladium in London with one of his pregnant mistresses. Having been spotted and knowing his goose is cooked, Philip decides to confess everything to Ottoline. She is thunderstruck. Two mistresses. Both with child. And, worse, her pregnant former maid is threatening to go public. The ensuing scandal will ruin them. End his political career. Philip is "terrified," one biographer says. He has to be admitted into a nursing home for a time and for the next two years suffers intermittent nervous attacks.

Philip and Ottoline decide to buy the maid's silence, making regular payments to her for the next five years until she gets married in 1922. Ottoline is forced to sell her heirloom family jewels to stay solvent and pay off the maid. Worse, the affair dredges up her own infant son's unfortunate death a decade before. Now she must contend with the fact these two other women have both given birth to healthy boys. "It is an awful obsession," Ottoline says. "The ghosts of their children."

"I feel as if some black evil cloud had descended onto this place," Ottoline writes in her memoirs. It "has blackened all the happiness and joy, eclipsing color and sunshine."

---◆---

COLLECTIVE INFLUENZA

Meanwhile Russell, forced to resign from Cambridge, spends almost all his time on peace activism. Writing. Speaking. Glad-handing. Things are rough for him, though. Several people are arrested and imprisoned for distributing an antiwar pamphlet Russell has written, and he declares himself the author in order to draw the prosecutors away from them. London's chief magistrate summons him before the court and orders him to pay a fine. Russell refuses. The magistrate has his books seized and vows to auction them unless Russell pays his fine. Again, he refuses. So the city sells the books—but several of Russell's friends buy them back for him.

A far worse punishment awaits him after Christmas 1917, which he spends at Garsington with Ottoline while she is at her darkest. There he hatches an idea that soon lands him in deep trouble. With the United States just having declared war and now beginning to mobilize its troops to staging areas in England and France, he wonders: Since American soldiers are routinely used by the US government to break strikes in the United States, won't they also do that in Britain? Will the UK government use American troops as strikebreakers to shut down peaceful labor protests? Will American troops bully English citizens?

After Russell publishes this idea, it lands him on trial for sedition in early 1918, accused of making "statements likely to prejudice His Majesty's relations with the United States."

His statements were flippant, but the British government perceives them as dangerous. The war's end feels tantalizingly close at the beginning of 1918, when Russell's trial takes place. America is shipping millions of soldiers to Europe. The US army balloons from 190,000 active-duty soldiers at the beginning of 1917 to 3,665,000 by the end of 1918—most sent to the Western Front. Russell's silly little conspiratorial screed about Britain potentially using US troops as strikebreakers appears as hundreds of thousands of Americans are being drafted. Trained. Trucked to the Atlantic. Sent up the gangplanks. Steamed across the drink. And vomited out onto the front lines. American soldiers are disembarked to the tune of ten thousand men a day in cities like London. Liverpool. Le Havre. Glasgow. Bordeaux. Bristol. Brest.

That river of men and machines will end the war, British authorities are convinced. They'll take no chances rocking the boat and risking negative public attention now that America's tap of troops is turned fully on. No way, nohow! They'll do anything to clamp down on dissent. And that's exactly what happens. Russell is accused of sedition for "disaffecting the troops."

Russell seems not to care. Even as his trial is going on, he gives a series of lectures in London on his philosophy of "logical atomism," describing the basic units of logic from which higher knowledge can be built as akin to atoms used to synthesize molecules. Those lectures are brilliant, if bizarre, given everything that's happening. Today, the transcripts show little sense of Russell's impending doom.

"EXTRAORDINARY DIFFICULTIES"

Unbeknownst to Russell, David Hilbert is becoming more and more enchanted with Russell's ideas. After his grueling race with Einstein for general relativity, he's ready for something new. Mathematical logic is that thing. The same month as Russell's trial in 1918, Hilbert's student Heinrich Behmann is finishing his PhD thesis, which focuses on Russell and Whitehead's books. Hilbert likes the thesis, but he's starting to sense cracks in Russell's ideas. He's struggling. "At the very top of Russell's theory—as highest axiom of thought—stands the so-called axiom of reducibility," Hilbert notes. "This axiom, including the related theory of types of Russell, poses extraordinary difficulties for its comprehension."

This is a change for Hilbert because for months, he's been lecturing on logicism, even endorsing it at one point. Russell knows nothing about this, of course, because of the war. But now, as the war is winding down, Hilbert's brief flirtation with logicism is starting to fall apart. He is quick to acknowledge the value of what Russell and Whitehead have done. The *Principia Mathematica* remains a remarkable achievement in Hilbert's eyes. But he thinks it may be barking up the wrong tree. Hilbert begins to doubt whether logic can actually provide a solid foundation for mathematics.

After all, if Russell and Whitehead didn't succeed in asserting solid foundations of math based on logicism, probably nobody would. They worked on their book for years. Two of the most brilliant minds of their generation. And still it falls short. Who, then, will succeed with logic alone? Nobody, Hilbert thinks. "The aim of reducing set theory, and with it the usual methods of analysis, to logic, has not been achieved today and maybe cannot be achieved at all," he will say later. Hilbert soon begins to map out another approach, a more rigorous program based on axiomatics—like what he did for geometry.

"To conquer this field we must, I am persuaded, make the concept of specifically mathematical proof itself into an object of investigation," he writes. "Just as the astronomer considers the movement of his position, the physicist studies the theory of his apparatus, and the philosopher criticizes reason itself."

———◆———

It's just as well that Hilbert cannot contact him. Russell has changed. The scholar Hilbert thought about inviting to his university at the beginning of the war is not the same man who emerges at its tail end. Russell has lost his academic career and has walked away from many of his friendships. He is about to be thrown in jail, tried, and convicted. He will lose an appeal and be sentenced to six months in prison.

After former prime minister Arthur James Balfour intervenes on Bertie's behalf, Russell is allowed to serve his time in the more comfortable confines of the penitentiary in Brixton, where he spends the final months of World War I. Remarkably, this gives him time to write yet another book, *Introduction to Mathematical Philosophy*.

Ottoline visits Bertie in jail often. She brings him armfuls of flowers. Bags of books. Perfumes. Scented soaps. Little bunches of herbs. Within these care packages she hides her own sweet form of Easter eggs. Smuggled notes. Concealed letters. Tiny pieces of paper with intellectual thoughts. Philosophy and thyme. Gossip and sage. Anything to afford him relief from his boredom at being locked away. He in turn often "recommends" a book to lend her on each visit and tucks secret notes for her inside the pages. This marks the start of a new phase in their relationship. They will forevermore be friends with no illusions of anything more. Good friends. Great friends. True friends. The best of friends. Ottoline raises money for Bertie—broke once he is released from prison—and moves him into one of the cottages at Garsington after he's free.

THE GREAT MATH WAR

The war is grinding to a halt when Russell emerges. Four years and four days after it began, the French, British, and now American troops finally succeed in pushing back the German lines, and efforts toward peace begin. The Austro-Hungarian army surrenders on September 29, 1918, the Ottoman army surrenders on October 30, and the German armistice follows two weeks after that. With feelings of isolation and dread, Russell wanders around London. Jubilant crowds. Joyous celebrations. Cheers everywhere. He sees a man and a woman—complete strangers—embrace and kiss. "I felt strangely solitary," he reflects in his autobiography, "like a ghost dropped by accident from another planet."

In the United States, the war ends even more weirdly—four days too soon. On November 7, 1918, American newspapers report prematurely that the armistice has been signed. People dash out into the streets. They sing, dance, and celebrate. Waving hats. Flying flags. Parading. Laughing. Hugging. Crying. Wrapping themselves in the thick wool of *thank-God-it's-over*. In New York City alone, some 155 tons of ticker tape is chucked out the window of all the buildings lining Fifth Avenue to drift and spin onto the cheering crowds below. The celebration is premature, and a few days later, they do it all again.

Europe is less jubilant. The Great War has turned boys into men, the living into the dead, trenches into graves, and tri-tips into turnips. Nothing like it has ever been seen in terms of scale, scope, or severity. Europe is left saddled with impossible debt. Globalization is dead. And everyone is tired.

British military historian Gary Sheffield will sum it up in 2007 as the perfect storm: "Industrialized societies fueled by nationalism, organized for total war, and producing armies of unprecedented size with war-making material to support them."

In 1918, The US army alone buys 96 million pairs of socks, 84 million pairs of underwear, 85 million T-shirts, 30 million pairs of shoes, 21.7 million blankets, and 22 million overcoats—along with 5.4 million gas masks, 2.7 million steel helmets, and 227,000 machine guns. The official statistician of the US army estimates that the country spends $1 million every hour of every day it's in the war, some $22 billion total, a national debt twenty times larger than its prewar debt. And that's only the United States, a relative late starter in the war. Great Britain blows through $38 billion. France $26 billion. Russia $18 billion. Austria-Hungary $21 billion. Germany a whopping $39 billion.

In human terms, the costs are even more staggering. Some armies have seen casualty rates of 20–25 percent. Modern estimates put the body count at 10 million, with another 10 million civilian deaths—more than 12,000 people killed per day in total. Another 21 million are left wounded, crippled, maimed, disfigured, blinded, or scarred by PTSD.

Now that it's over, the hard slog of waging peace in Europe begins. Peace should have been built on all those lessons of incalculable, incomprehensible, and unacceptable cost. Instead, a selfish peace follows a selfish war.

CHAPTER TWELVE
ALL EYES ON ALBERT EINSTEIN

1919–1920

> "
> *Wovon man nicht sprechen kann, darüber muss man schweigen.*
> (Whereof one cannot speak, thereof one must be silent.)
> "
>
> —Ludwig Wittgenstein

Hercules fighting the nine-headed Hydra. Cropped from a public domain image of an etching via Wellcome Collection.

Absolute human greatness is not easily quantified, even if you have good, objective data with which to do it. Who is history's greatest sports star? Pele or Gretzky? Michael Phelps or Sonja Henie? Usain Bolt or Babe Ruth? Simone Biles or Serena Williams? Aleksandr "the Russian Bear" Karelin or Björn Borg? Muhammad Ali or Nadia Comăneci? Those are impossible questions to answer.

Even if you simplify the question by looking at a single sport, it's still tough. Who is the all-time best professional basketball player? Russell or Bird? Chamberlain or Jordan? Kareem Abdul-Jabbar or Cynthia Cooper? LeBron James or Lynette Woodard? Caitlin Clark or Dirk Nowitzki? Stats will get you only halfway to an answer because you are talking about different people who played different positions with different styles at very different times.

What if you compare GOATs across completely different fields and periods of time? Who had more impact on humanity: Marie Curie or Pablo Picasso? Queen Victoria or Julius Caesar? Malala, Madonna, Socrates, or Shakespeare? It's ridiculous to even consider such questions. And at the end of the day, even our most powerful comparisons fall short. What I call the fallacy of rich data says that the more data we use in an analysis, the more valid our results appear to be. But data can be (and often is) wrong. And even if it's right, our interpretation of it can be wrong. No comparison is ever truly objective.

---◆---

So how do we start to assess greatness? Perhaps our linear way of thinking about it is all wrong. Perhaps greatness is not a ladder, each rung higher than the next, but more like a cloud kingdom—an infinite, opaque expanse where giants roam, thrust by accident or having arrived by some deliberate climb up one of many Jack's beanstalks of fame. Superstar status, according to the sociologist Max Weber, sets individuals apart from others as if they are "endowed with supernatural, superhuman, or at least specifically exceptional powers or qualities."

ALL EYES ON ALBERT EINSTEIN

The example Weber used at the time he wrote those words was famed German chancellor Otto von Bismarck, but he may as well have been talking about Albert Einstein, who enters the 1920s with an almost divine superhuman aspect.

It's hard to say exactly when many famous people achieve superstar status, as it almost never happens overnight. But some claim it's easy to say when Einstein achieves his: on or after May 29, 1919. That's when English astronomer Arthur Stanley Eddington, observing a solar eclipse from a small island off the west coast of Africa, gathers the first observational proof of Einstein's theory of general relativity. The eclipse reveals the displacement of a star behind the sun, its faint light traveling millions of miles through space and then getting bent by the gravity of the sun before falling on Earth.

One of Einstein's colleagues telegraphs him right away. Einstein is so excited, he sends a postcard to his mother. "Good news today," he writes. After the observations are published a few months later, newspapers around the world pick up the story. "Einstein awoke in Berlin on the morning of November 7, 1919, to find himself famous," one of his biographers will write—though it's worth adding that some scholars find this notion of overnight fame oversimplified.

Fair enough, but few would argue that Einstein's fame becomes an unstoppable force after 1919. By January 1920, general audiences are paying good money to hear public lectures on his work—precious marks for impenetrable remarks. Einstein himself complains he is becoming so hounded that he can hardly get any work done, telling a friend in the summer of 1920 that he's "pestered by excessive adulation." Some later historians will agree, finding that all the lionization of his legend clouds the analysis of his work.

ALWAYS CHASING RAINBOWS

Lionization is perhaps an understatement. We still marvel at Einstein's mind today. His name is synonymous with overwhelming genius. And genius, it should be said, is always seen as essential, never emerging. Thus, after he died seventy-five years ago, his brain was poked, prodded, and preserved in the hopes that it would one day yield its secrets. Every few years, another of these speculative n-of-1 studies tries to account for his genius on the basis of some formalin-fixed paraffin brain section observation. Perhaps it was his one-hell-of-a-huge corpus callosum?

"If society should happen to become infatuated with a man, believing it has found in him its deepest aspirations as well as the means of fulfilling them, then that man will be put in a class by himself and virtually deified," French sociologist Émile Durkheim once said. "Opinion will confer on him a grandeur that is similar in every way to the grandeur that protects the gods."

That's definitely the case with Einstein. The public in 1920 is fascinated by relativity, even if they don't understand it. But its very opaqueness doesn't matter. The impenetrability of relativity only makes Einstein and his subject seem even more significant. Greatness is great because it's out of reach and unattainable, not in spite of that fact. Nobody does flips on a balance beam like Simone Biles, passes the ball like Magic Johnson, shoots clutch shots like Michael Jordan, or dances in the boxing ring like Muhammad Ali. We admire their greatness because of its almost alien quality. We love them for their *out-of-reach-ness*. Thus, where 1920s audiences fail to comprehend relativity the theory, they focus on relativity the man. Article after article. Book after book. Photo after photo. Talk after talk. The man, the man, the man: Einstein.

A WORLD OF JUSTICE

One major outcome of Albert Einstein's skyrocketing reputation is that it has a huge impact on the Great Math War. When Einstein is appointed an editor of *Mathematische Annalen*, the journal where David Hilbert and L. E. J. Brouwer both work, it's good for Einstein and good for the journal. But it also puts him squarely in the middle of the coming fight between Hilbert and Brouwer, whose own reputations are steadily growing.

By the end of the war, Brouwer has lost touch with almost all his international colleagues. The war forced him into relative isolation. Publications dried up across Europe. Meetings were largely canceled. And Brouwer more or less stopped doing research in topology. Instead, he became increasingly focused on his other love: the foundations of mathematics. Once the war is over, he feels that he can finally get back to the all-important intellectual pursuit of serious mathematical concerns over the next decade.

In the United States, we glorify the 1920s as the romanticized past—a roaring time. Back from the war. All the boozy excess of prohibition, art, fashion, and economics. Women's bathing suits shed their miserable tunics and stockings. Skirts get shorter. On the streets of New York and London, short-haired women and long-haired men begin to appear. Makeup loses its offensive, even immoral stigma in high society. Cars crowd the streets. It's a time of sparkling silver-screen stars. Charlie Chaplin. Douglas Fairbanks. Mary Pickford. It's an age of jazz, exemplified by the hit song "I'm Always Chasing Rainbows." Things are looking up.

Brouwer is especially looking forward to this back-to-normalcy fantasy. He longs to get back to the sort of community he misses. Letters. Conferences. Meetings. Feedback. He was denied all those things while living as the lonely genius shut up in the Dutch oven of neutrality. Now that the war has ended, he's anxious to renew his professional ties. One of the first things he does after the Armistice is to send Hilbert a letter wishing him well. "May the healthy heart of your fatherland overcome the present crisis," he writes. "And may the German lands soon prosper to unknown bloom in a world of justice."

Several months later, Brouwer writes Hilbert another letter recalling their walks together a decade before. Those sand dunes. That Dutch beach resort. Oh, when they first met! It seems a lifetime ago. The war changed everything. Nothing will ever be the same—except maybe their friendship. Brouwer tells Hilbert he cherishes their closeness and prizes all the knowledge the older man has imparted to him.

In October 1919, a faculty position at Göttingen opens up, and Hilbert offers it to Brouwer. The University of Berlin makes Brouwer a separate offer. Suddenly he's the new "it boy" of mathematics. He shrewdly leverages both offers to improve his position in Amsterdam, increasing his salary to the highest pay allowed under the law and winning a sizable stipend to buy books and journals. He turns both German offers down, thinking the postwar economic malaise in Germany will increase the cost of living and devalue university salaries. He's not wrong. Germany's economy is wrecked and worse. Its empire has dissolved. Its army is defeated. Its navy is sunk. Its colonies are gone. And its people are exhausted.

THE JOYLESS REPUBLIC

The Weimar Republic rises from the ashes of the war, a shadow of Germany's former self. It's a place of breadlines and chaos. "A joyless Republic," the Canadian historian Lewis Hertzman will say in 1960. A "Republic without Republicans." A kingdom deprived of emblems. A ruined monarchy. A humiliated land of poverty that's been promised peace, handed social upheaval, weaned on war, and guaranteed political violence. A prostrate kingdom. "Monarchy in its deepest humiliation." That's the Weimar Republic. "Born in defeat, lived in turmoil, and died in disaster," the 1920s-era Austrian journalist and novelist Joseph Roth writes. It has the "whiff of fragility, of scandal, of doom about it."

But even in those dark postwar days, Göttingen is a bright spot. There are more students coming every day, and many of those who do arrive feel like they have arrived. For the rest of their lives, they'll brag about having been there. The university is becoming a hotbed of modern physics, home to the emerging field of quantum mechanics. New students know this is the place to be.

The university's fame is thanks in no small part to Hilbert. He lures smart grad students like "the sweet flute of the Pied Piper," one of his former students says, "seducing so many rats to follow him into the deep river of mathematics." People from all over the world come to Göttingen. Hundreds jam the halls to hear his lectures. They fill the rooms to capacity and feel honored to be there. Sometimes they stand in the aisles or sit on the windowsills just to get in the room.

Success breeds success. Göttingen rises high in the early twentieth century. This is extraordinary when you consider its competition. The other places comparable in terms of prestige are all located in major cities—Paris, Berlin, London, Zürich, Moscow. The general rule is the more towering a city in terms of culture, the taller its intellectual peaks. Göttingen breaks that rule. It's situated in a sleepy backwater. According

ALL EYES ON ALBERT EINSTEIN

to someone who studied there in the 1930s, the best restaurant in Göttingen was in the train station. A minuscule city in size, it's a Himalayan peak in stature.

Once the war is over, Hilbert takes a deep dive into the foundations of math. Neither he nor Brouwer has any idea the other is working on the same thing—albeit taking radically different approaches. Hilbert, for his part, is soon focused on little else. He wants to spend the last decade of his career on the subject, returning to the approach he took in his 1899 *Foundations of Geometry* and reinventing math from first principles. He thinks math can be constructed by simply defining and adhering to a consistent set of axioms.

He wants to establish the ability to develop "absolute" proofs that don't rely on human intuition or some concept or understanding of 3D space or even external reality. Geometry, for instance, has a connection to real-world shapes and forms and 3D space itself baked in at every level. But Hilbert thinks the fundamental foundations of math need not be so reality based.

He thinks we should be able to construct proofs purely symbolically and completely bereft of meaning—except that which we specifically add. Mathematics can still be an applied subject, he thinks. After all, Hilbert has spent almost his entire career as an applied mathematician. He sees the need for this more than anything else. But in that case, mathematics would have meaning *when* it's applied. For the fundamental, formal treatment of math, reality is not needed at all. It's simply a matter of consistency. Where there's consistency, there's completeness. Where there's completeness, there's proof. Where there's proof, there should be simplicity. And where there is simplicity, there will be clarity. That's the future, he thinks.

THE MAGIC WAND

Hilbert wants math to be both complete and simple, which in his eyes means any axioms he derives should be few in number, independent of each other, and complete all together. And his "formal axiomatization" of math will do what Russell and logicism could not: solve the foundations of math once and for all and show math to be "an indivisible whole," Hilbert says, "an organism whose ability to survive rests on the connection between its parts."

"Axiomatics is the rhythm that makes music of the method, the magic wand that directs all individual efforts to a common goal," Hilbert writes in one of his notebooks—a metaphor true to form for a man people call the Pied Piper. It's a bit of a mixed metaphor, but that's okay. He's getting it, grooving it. Feeling the beat!

Brouwer's approach to math is almost the opposite of Hilbert's, and the basis of the foundational crisis of the Great Math War, as I call it, is largely that they both believe their competing worldviews to be mutually exclusive. I'll talk about Brouwer's far-more-outside-the-box approach in a bit, but first it's worth pointing out there are a few things on which he and Hilbert strongly agree.

One is Russell's failure—though for different reasons. Hilbert sees Russell's work as immensely heroic but thinks a new approach is needed nonetheless. An axiomatically defined fundamental approach to math should be its foundations, he thinks, not logic. Brouwer agrees, but much more aggressively, that logic is not the answer. He sees Russell's whole attempt to establish the foundations of math based on logic as pointless. For him, logic is built on mathematics, not the other way around. Trying to establish the foundations of math based on logic, Brouwer says, is "an unproductive, sterile exercise."

Second, Brouwer and Hilbert both agree on the inadequacy of human language. Actually, Russell, Whitehead, and Frege would all agree with this as well. Ordinary language fails to unambiguously capture mathematical rigor. The whole reason Frege developed his *Begriffsschrift* (concept notation) was that he didn't trust ordinary language.

Hilbert and Brouwer both see the wisdom of looking askance at ordinary language, but they both go even further. For Brouwer, language isn't merely something to distrust and avoid. He sees the question in human evolutionary and anthropological terms, believing mathematics came before language. We wouldn't have language if we didn't have math, in his view. The very acts of speaking, gesturing, and listening rely on our intuitive ability to discern the abstract meaning attached to a snort, sign, or pantomime.

◆

For Hilbert, it's less of a chicken-or-egg question than a laser-focused practical one: What should we do about ordinary language in math? That's another thing he and Brouwer agree on. They both want to chuck language from the equation—though they'll go about that very differently.

Hilbert's idea, as I said, is to create a perfect *artificial* language for deduction. He wants to strip ordinary language away entirely, cut mathematics to the bone, and leave it as nothing more than what he calls "meaningless marks on the page." To put language aside, in other words. Separate the symbols of math from their meaning. That way you can establish and apply rules unclouded by meaning. And then let the axiomatics and the rules be your copilot.

While Brouwer agrees with some of that, he ultimately thinks Hilbert's approach is folly for the very thing Hilbert considers its strength: language. Hilbert thinks he's divorcing math from language with his meaningless-marks-on-the-page approach, but to Brouwer, symbolic language is still just language. He thinks it won't be possible to construct a mathematical language free from contradiction because at its core, a symbolic language will always be a human language. "Still as far away as ever," he will write later. Brouwer coins and popularizes the term "formalism" as a sort of poisoned-tongue petty slander to dismiss Hilbert's approach.

Some names we take for ourselves. Others are given out of love or respect, like a title retained postretirement. Coach. Colonel. Ambassador. Mr. President. Other names are skewering—either offered in jest or given to us out of anger, hatred, contempt, or condemnation. "Formalism" leans toward the latter. Brouwer thinks it's a stuffy label for a stiff approach and an even stiffer neck.

Brouwer's own approach of "intuitionism," on the other hand, he loves. That's the name he claims for himself, borrowing a term from an earlier philosophy. It "highlights the existence of pure mathematics independent of language," he writes.

This gets back to his almost anthropological perspective. Brouwer sees mathematics as a primitive human ability—something ingrained in our individual consciousness and imprinted on the human species by the long hand of evolution. The purest expression of this is his attempt to strip math to its most primitive essentials. Math isn't some formal thing we construct to describe the world but part of our essence, our existence—one of the very things that makes us human.

COGITO, ERGO SUM

This idea of constructive math as an innate human ability is not entirely original with Brouwer. Philosophers trace its origins to seventeenth-century French mathematician René Descartes, who rejected long-held cultural and religious appeals to authority and insisted that the rational human mind grasps the immediacy of its surroundings and from that intuits knowledge. That's the basis of his famous phrase *Cogito, ergo sum* (I think, therefore I am), which, amazingly, four hundred years later is not inconsistent with modern neuroscience.

Brouwer agrees with Descartes to the extent that he sees mathematics as essentially a way of viewing the world—but in a human primitive sense as opposed to more of a formal human learned way, as Hilbert sees it. But Brouwer takes it further than Descartes or anyone else ever has.

Brouwer also moves beyond a very famous older form of intuitionism developed by the philosopher Immanuel Kant a century before. Kant was a genius except in one simple-headed way: He saw time and space as absolute. His idea that we intuit space and time through the "immediate awareness" of our senses is spot-on, according to modern views of cognition. But he was done in by his conception of absolute space and time. By the 1920s, people were actively dismissing the absoluteness of time and space, however true it seemed to the senses. Einstein's rise to world fame was based entirely on the light-bending proof that time and space are *not* absolute but rather relative. Time dilates. Lengths contract. Space stretches. Gravity gives rise to spooky weirdness. And Einstein's theory of relativity rigorously violates space-time absoluteness nine ways from Sunday. By the 1920s, science was further from Kant than anyone could have imagined.

Another, even more fundamental change to humanity's worldview is coming. Astronomers will soon gather evidence, based on viewing galaxies farther out in space than ever before with larger and larger Earth-based telescopes, that the universe is not what we have thought it was since the dawn of time. It's not fixed and endless but finite and expanding. That leads immediately to the leap that the universe has no steady-state existence without beginning or end but instead has a narrative arc. An origin. A beginning. A big-bang birth. Our present middle. Someday to end. *Kant who?*

GIMME THAT OLD TIME 'TUITION

Kant's quaint views on intuitionism become collateral damage in light of Einstein's relativistic revolution and the emerging concept of the big bang. "An old form," Brouwer says of Kant's intuitionism in the 1920s, "now almost completely abandoned." He argues that what came before was not really intuitionism at all but something else. Preintuitionism. Neointuitionism. Semi-intuitionism. (Quasi-intuitionism?)

Call it what you will, but Kant shines through Brouwer. After the end of the war, Brouwer returns to the basic Kant-inspired approach he pioneered in his strange book *Life, Art, and Mysticism* twelve years before. Like Kant, he takes the view that individual consciousness and the basic human interaction with the external world through our senses and reason are the only ways to obtain true knowledge. One "intuits" meaning about the world, he believes, which is where the term "intuitionism" comes from.

This, by the way, is not just the single, most basic defining aspect of Brouwer's belief—it's the major point of departure where his worldview differs profoundly and departs sharply from Hilbert's view and almost everybody else's. Why? Because it leads Brouwer to the conclusion that in mathematics, existence can never be taken for granted. It must be *established* "as a natural function of his intellect, as a free, vital activity of thought," Brouwer's foremost student, the Dutch mathematician Arend Heyting, will write a few years later. "For him, mathematics is the production of the human mind," Heyting will say of Brouwer.

In Brouwer's view, foundational mathematics should be purely *constructive*—conducted in a process where you build mathematical objects and proofs as a cognitive activity. For him, math is less concerned with logical consistency and formal proof than with building such mathematical objects through reason. He holds that humans have a natural, primordial, innate intuition for counting and for natural numbers. *That* should be the basis for mathematics, he thinks—not anything based on corrupt language or pure logic.

For Brouwer, finding a new mathematical proof should be like feeling your way through an unfamiliar landscape in the dark. You stumble here. You touch the squishy ground there. You find a place where the mud hides a sunken plank, like those mucky duckwalks in the trenches on the Western Front. You walk. Stay low. Have no idea where to go but keep going. See where the path leads you. Step by step. Squish by squish. Across the gooey mess.

In a dark World War I battlefield at night, so long as you know where you started, how far you have gone, and what turns you made, you can find your way through the muck one step at a time. Likewise, in mathematics, you can work your way through a proof using nothing more than the simple concept of natural numbers, adding 1 to 2 to 3 and so on.

It's a radical approach to mathematics and in some ways completely original. Still, because Brouwer borrows Kant's basic idea of intuiting and building on the senses, intuitionism is arguably a movement rather than a product of one man—a journey taken partly by Poincaré with a pinch of Kant and a good dose of *God-made-integers-and-all-else-is-the-work-of-man* Kronecker, whose approach blazes the path that Brouwer follows by denying existence to anything under a mathematical sun whose explicit construction cannot be demonstrated.

Brouwer is a radical, however, and advocates rethinking math along purely constructive lines. This results in a very controversial, even extremist worldview because rejecting anything you can't construct means rejecting the notion of countable versus uncountable, denumerable versus nondenumerable, or little versus big infinity. And that's a plague on set theory. Brouwer's constructivism also, of necessity, refuses any use of the law of the excluded middle. That's a pox on modern mathematical proof, a disease of sorts, which will explain a lot of what happens later in this book. People don't just dislike Brouwer's ideas or disagree with them. They, especially Hilbert, are actually incredibly threatened.

Understand this: While many see Brouwer as some sort of a second coming of Kronecker, he's really about much more than a simple reboot. He's more like *Kronecker II: The Reckoning*—that same zero-sum constructive philosophy with a lot more CGI explosions. Imagine Kronecker with a kick-ass crossbow. Or a howitzer. Or Kronecker shooting molten lava from his hands. That's what Brouwer is like.

◆

Thus, in the dawning days of Europe's awakening from the long, awful horror of World War I, the Great Math War quietly begins, with Hilbert taking one direction to secure the foundations of math and Brouwer going another way. Brouwer's impulse to reject anything you can't construct will soon put him deeply at odds with Hilbert. Nobody realizes it yet, but Brouwer has already fired the first shots by derisively labeling Hilbert's approach *formalism*, contrasting it in cruel mockery with his own cool-sounding, Kant-labeled *intuitionism*, his greatest ambition and hope.

That's the last thing I want to say on the subject of Brouwer v. Hilbert for now, except for this: It's a shame. They were friends. Sparkling with mutual admiration and respect. And they had in common one final, fundamental point of agreement: They shared a worldview marked by a singular concern for Queen Math—the conviction that the fundamental foundations of math needed fixing. But that common ground gave way because of their deeply divergent approaches. They were two mathematicians separated by a common cause.

The Great Math War doesn't exactly start in 1918. You can trace it back to 1900 or 1883, as I have done—or earlier still. But it's really not until the 1920s that the battle lines are drawn. In the years leading up to World War I, the foundations of math were more or less on hold as Russell and Whitehead were hard at work trying to fill the yawning hole in the

THE GREAT MATH WAR

foundations of math with logic. They brought out their incredibly complex three-volume *Principia Mathematica*, that triumph of logicism. But their work merely bandaged the wound. That sore scabbed over during World War I as the international exchange of ideas and travel came to a grinding halt. Machines. Monsters. The Western Front. But after the Armistice, the bandage is ripped off and the debate over math's foundations erupts as never before.

DAS KONTINUUM

Let's return for a second to the paradoxes. The paradoxes of set theory were the main motivation for wanting to shore up the foundations of math in the first place. For Russell and Frege at the turn of the century—and for Hilbert in the 1920s—the paradoxes revealed some sort of surface problem, however serious. Their existence showed that something was wrong with our formulation of logic or the language of mathematics we use but did *not* suggest some fundamental underlying problem with math or logic itself. They all saw the existence of these paradoxes as akin to pimples on the teenage face of mathematics—unsightly but not permanent. And easily popped, however painful that may be.

But Brouwer regards the paradoxes as fundamentally more scarring: No mere blemish but an indictment of the foundations of classical mathematics itself. They show that math as we know it is far from perfect, as Dutch American mathematician Ernst Snapper will say in 1979. The only solution Brouwer can see is to tear mathematics down, rip up the foundations, replace the rebar, repour the concrete, and rebuild it afresh using his step-by-step intuitionism and his constructivist *mathematics-as-a-mental-activity* approach.

———◆———

That's a major departure from Hilbert, but there's more—an even larger fundamental disagreement. Brouwer rejects one key aspect of Hilbert's worldview: the idea that any problem in math can be solved, either by finding an answer or showing that a solution is impossible. Hilbert has subscribed to this idea, which I call exuberant solutionism, for years. *If you can state it, then you can solve it*, he claimed in his famous 1900 talk in Paris.

But there's no reason to assume that, Brouwer thinks. Any proofs, and for that matter any mathematical object, must be constructed, in his view. Anything that *can* be constructed, even conceptually, is valid. Anything that *can't*, well...it doesn't exist! And it's easy for him to contemplate an object you can conceive but not construct, denote but not derive, or state but not solve. So he rejects Hilbert's exuberant solutionism just as much as he rejects approaches like pure existence proofs.

This isn't a small thing. It challenges decades of mathematical results, since mathematicians in Europe have been publishing paper after paper using pure existence proofs for years—the *exact* thing Brouwer's approach rejects. Because of this, intuitionism throws lots of treasured mathematical proofs into question. In Brouwer's view, what was once QED is now DOA—to which he basically says, *So what?* Hilbert says all problems are either provable or can be shown to be unproveable. Brouwer says, basically, *Balls!*

That's one of the most striking things about Brouwer. He doesn't just stake out an absolutist position that rejects a huge number of mathematical discoveries—he's completely fine doing so. He says, with complete self-confidence, *Throw it all out and start over.* Toss out any part of mathematics that can't be constructed via intuitionism—and by the way, that's no real loss, he claims, because such results are meaningless in the first place.

This position is revealing because it shows something else that's bizarre about Brouwer: a mild hypocrisy. He rose to fame in the years prior to World War I on the back of his groundbreaking work in topology. The single piece of work for which he was most famous was his so-called fixed-point theorem, which claimed that any function will have at least one fixed point. But his work on this theorem was by no stretch of the imagination an example of intuitionism—in fact, the opposite. The fixed-point theorem was constructed through a pure existence proof. Thus, the specific thing that launched Brouwer's career before World War I is the exact type of thing he rejects after the war is over, when he embraces his hard-line stance on constructivism. *The enemy of thy enemy is thyself!*

Hilbert is completely unaware of all this in the early days after the war, but that's soon to change after Brouwer makes an ally of one of Hilbert's all-time favorite students. Hermann Weyl is the smartest and most successful protégé Hilbert will ever have—and in many ways the son he never had. Weyl first arrived at Göttingen as a teenage grad student in 1904, finished his PhD under Hilbert's guidance around the same time Brouwer was finishing his in Amsterdam, and quickly became a rising star on the faculty at Göttingen in 1911–1912, when Brouwer visited the university. The two became fast friends.

Like Brouwer, Weyl was a mathematical whale. In 1913, he published a hugely popular book titled *The Idea of Riemann Surfaces*, which became an instant classic in math circles. He was a gifted writer—so much so that the late, great Constance Reid, perhaps the greatest math biographer of all time, once said in an interview that she had considered writing a book about Weyl but couldn't be bothered because he had basically said everything there was to say about himself already.

Such twin talents are rare among mathematicians, and Weyl's brilliance coupled with his eloquence made him a rising star. The Institute of Technology in Zürich snapped him up, hiring him as a professor in 1913. He accepted and became colleagues with Einstein for one year before Einstein left to take a position at the University of Berlin.

———◆———

Weyl had a keen sense of art, beauty, and aesthetics that was as fundamental to his outlook as his love of mathematics—something Brouwer also loved about him. Weyl was present in Zürich when the antilogical Dada movement started there during World War I, and he embraced it. In 1917, while the Great War was still going on, Weyl was working hard on logic, reading Russell and Whitehead's books.

Independently, Weyl came around to Brouwer's point of view that logic alone would not be able to establish the foundations of mathematics. His departure from Russell's worldview was based on the logical definition of real numbers, which Russell and Whitehead describe in *Principia*

ALL EYES ON ALBERT EINSTEIN

Mathematica. Weyl perceived that the definition of real numbers depended entirely on their nature. But in logicism, the definition itself was partly responsible for determining that very definition (thanks to our dear old friend the vicious circle). The nature of real numbers is as we define them, Russell and Whitehead say. That definition determines their nature. And from that nature, we derive our definition. And around and around the vicious circle we go.

Weyl rejects this absurdity, and he sees eye to eye with Brouwer in that regard. This is significant because Weyl is the best ally either Brouwer or Hilbert could ever want. Weyl has a natural gift for promotion. He's a force of nature as a communicator, well-spoken in more than one language, bold, and brilliant. He has a tasteful, even poetic writing style—one that, at least initially, is claimed by Brouwer as he sets out to rewrite the rules of mathematics along his own intuitionistic lines over the next decade. Thank God for Weyl, he thinks, because now he no longer has to do it single-handedly. And at least initially, he's not disappointed.

When Weyl comes back to Zürich after a brief stint fighting in the German army, he carefully studies Brouwer's work. He also writes a tiny book called *Das Kontinuum* that explores the ideas of people like Poincaré and Russell on the foundations of math. Some have accused him of committing an act of poetry in the booklet because of the beautiful way he describes the continuum of real numbers—that nondenumerably large aleph-one infinite set that Georg Cantor explored nearly half a century before—and the smoothness of the continuum as a revelation in constancy.

His assessment of the foundations of math is equally poetic, if far more grim. He says the foundations are not made of bedrock (or, in modern house building, concrete poured into molded shapes over reinforcing steel rebar). Mathematics is more like a wooden shack built on sandy soil, Weyl thinks.

Das Kontinuum is a short but difficult work—another one of those books that everyone gets but nobody groks. A modern scholar has said in recent years that it took sixty years for Weyl's work to be truly appreciated. Part of the reason is that the continuum is so hard to explain. Cantor says it's the power of the nondenumerable, uncountable "large infinity" set of real numbers, which he later tells the high Catholic church is equivalent to the absolute, unapproachable, unassailable, and incomprehensible absoluteness of God. A more modern, less biblical way of defining the continuum basically boils down to asking, how many points are on the real number line? Understanding this will show the difference between big infinity and small infinity.

That's an intimidating place to start an exposition, but here we go.

BURNING DOWN THE MATH

Think of small infinity first. Imagine the whole numbers as buckets of sand—one bucket, two buckets, three buckets, four. And so on and so forth, ad infinitum. The countable, denumerable small infinity of set theory is the infinite collection of all such buckets.

To contemplate how to approach imagining the larger, nondenumerable set of real numbers, think not just of each bucket of sand but of every grain of sand inside. Imagine buckets so large that there is an infinity of grains inside each one. Now imagine capturing a wealth of data on each grain of sand. Think of all the endless surface variations on a single piece. Think of the unique arrangements of the silicate crystals glommed together inside each grain. Imagine the countless atoms that make up those crystals. Consider all the valence states of all the electrons in every atom in each crystal. Think of the temporal quantum fluctuations inside each subatomic particle, not just at any one moment in time but from their stellar origins to their heat-death doom. Think of all the snapshots of every quantum state of every atom in every molecule in every crystal in every grain in every infinite bucket of sand over all time—further back than even the James Webb telescope can see.

Now consider a countless parade of people offering endless individual perspectives on all those single grains of sand, opining on every quantum state of every atom in each sand crystal. Think of an infinite number of different quantum universes where slightly different but more or less the same people in each separate universe have alternative hot takes on each single grain of sand. Imagine a galaxy in a single grain. Contemplate a never-ending, constantly evolving universe of single-grain galaxies. Think about an eternity of revelation in every single grain for every subsecond of existence. That's starting to get at the difference between large and small infinity—an infinite number of sand buckets versus the infinite variation inside every single bucket. Big infinity is *much* bigger. But even then, you're not close to defining the continuum.

You could never hope to map, name, count, cry, otherwise represent, or even fully contemplate every object in human existence at any one moment—let alone all those same objects from one moment to the next over a lifetime and beyond, on and on into the future. Nor could you conceivably account for, even in one instant in time, an infinitude of alternative quantum realities where each moment is slightly, almost imperceptibly different. That's the difference between denumerable and nondenumerable small and large infinities. One is an endless series of buckets, and the other is the hopeless inability to even contemplate what's inside a single one.

———◆———

Weyl and Brouwer meet several times after the Great War ends—each time with greater urgency. They vacation together in 1919, and Brouwer brings Weyl up to speed on everything. Weyl's attentive ear is welcome. So is his silver tongue.

Weyl's words are beautiful. Accessible. Inspiring. And—however hard the conceptual weight behind them—compelling. That's in strikingly stark contrast to the unforgivingly technically terse style of Brouwer's writing. At times, his style is hard. At other times, boring. Impenetrable. It helps little that his ideas are so novel as to seem strange.

That's why Weyl is a great help. In the early days of the Great Math War, he helps Brouwer launch his ideas onto the mathematical stage. In Weyl, Brouwer finds a mathematician of stellar reputation and one as willing as he is to challenge the status quo. They're cut from the same cloth in almost every respect except when it comes to communicating their ideas in writing. Brouwer may be a rare revolutionary in original thought, but he's a middling writer. Weyl's silver tongue and golden pen help Brouwer start a bona fide movement. Brouwer lights the spark. And Weyl blows the flames bright. He's like Trotsky to Brouwer's Lenin.

THE GREAT MATH WAR

"Brouwer is a man whom I love with all my heart," Weyl tells a colleague in November 1920, right in the midst of this bromance.

There is one glaring source of tension between the two, however: Their goals are not exactly the same. "Weyl is the cautious revisionist, who wanted to find the safe kernel of traditional mathematics," the modern historian and Brouwer biographer Dirk van Dalen will write decades later. He seeks tightening, incremental improvements rather than wholesale replacement, in other words. A controlled burn. To trim and brash the trees rather than cut them down and plant a new forest. He wants to find order in the chaos rather than what Brouwer wants: wholesale replacement—a complete teardown, not a renovation. He's a *throw-out-the-baby-with-the-bathwater* and *burn-it-all-down* revisionist who, "like Samson, wanted to bring the whole building of mathematics down before erecting it again," van Dalen writes.

A CRISIS IS BORN

Weyl is the one who gives a formal name to the struggle that ensues, calling it the *Grundlagenkrise* (the foundational crisis) in the title of a paper he writes in May 1920—*Über die neue Grundlagenkrise der Mathematik* (The new crisis in the foundations of mathematics). After he finishes the paper and before it's published, he immediately sends it off to Brouwer with relish. "It should not be viewed as a scientific publication," Weyl says to Brouwer in an attached letter, "but as a propaganda pamphlet." *Suited to rouse the sleepers!*

In this paper, Weyl compares the paradoxes that emerged from the works of Cantor, Burali-Forti, Russell, Zermelo, and others twenty years before to small border conflicts, regional strife ahead of global conflagration. "The troubles in the borderland of mathematics must be judged as symptoms, in which what lies hidden at the center of the superficially glittering and smooth activity comes to light—namely the inner instability of the foundations, upon which the structure of the empire rests," he writes. Trouble in Sarajevo is trouble in Europe. And trouble with the logic of set theory is trouble with mathematics.

The Baltics are not the only war or political metaphor he musters. He also says the whole of mathematics is a paper economy where proofs are bought and sold with no real intrinsic value. And he ends with a revolutionary flair. "For this order cannot in itself be maintained, as I have convinced myself," Weyl writes. "And Brouwer," he concludes, "that is the revolution" (*das ist die Revolution*).

Brouwer loves it. He showers heartfelt appreciation on his friend. "Your explanation, it seems to me, will also be clear and convincing for the public," he says. And as for the fit-to-be-roused sleepers Weyl mentions, Brouwer laughs. Weyl is merely flipping them as they slumber in their beds, Brouwer says. They'll soon roll back over and return to sleep.

Weyl doesn't name Hilbert in the paper, but after it's published in 1921, few have any doubt whom he refers to when he says the *old* order. It's Hilbert.

Hilbert has an awful reaction once he learns about all this. He will come to regard Brouwer's intuitionism as an existential threat. It's destructive since it demands discarding too much of modern mathematics, including some of the greatest treasures in the math chest, if they cannot be derived through the constrictive methods of constructivism. What would be left if you tossed out pure existence proofs would be lone, leathery scraps of math—the "wretched remains," Hilbert says. Like all those burned-out French buildings at Verdun. *Ridiculous!*

As for Weyl, Hilbert is dismissive. Troubles in the borderland of mathematics? A revolution? This is not a revolution but a dictatorship, Hilbert says. A mutilation of science.

———◆———

From Hilbert's perspective, there's far less to love in a revolution than in a war. Revolutions aren't glorious. They're destructive. Terror. Fire. Horror. Chaos. That's what Europe starts to see in Russia in the 1920s. And that's what Hilbert sees in Brouwer. And because he sees his former Dutch friend as representing an existential threat, he calls him out and calls him names—a false prophet. Some sort of flimflam man. A scandalous Mesmer. A corrupter of youth.

Hilbert knows how much Cantor was scarred by his interactions with his nemesis, Kronecker, and the latter's unfair criticism. After Hilbert becomes fully aware of Weyl and Brouwer's work, he's reminded of this injustice. Intuitionism is a threat to Cantor's legacy. Realizing that, he reacts with dismissiveness, decisiveness, and revulsion. He laments. He wails. He grows red in the face, tight in the fist, and white in the knuckles. Nobody loves an existential threat. Most painful of all, he sees Weyl, his best former student, promoting Brouwer's crazy ideas. And he can't stand it. His intellectual son!

The thing is, the Great Math War could have been avoided if Brouwer and Hilbert had wanted to avoid it. As old friends, they could have gotten together, discussed, and moved forward in parallel efforts and mutual respect, if not friendly competition, as Hilbert and Einstein had in the race for relativity half a decade earlier.

But there's none of that. The two certainly never sit down and have a one-on-one conversation about the foundations of math. Hilbert does attend a lecture Brouwer gives, but it devolves into an awkward public humiliation. They have dinner once, but it's a group thing and carefully orchestrated to be nonconfrontational, and the conversation is specifically steered to avoid the foundational crisis. With a little more olive-branching, who's to say how things could have gone?

On the other hand, maybe the Great Math War was unavoidable. Sitting down may have just hastened the inevitable conflict. Hilbert and Brouwer were both known for sticking to their guns. And once the wheels were turning, perhaps the clash became inevitable. Brouwer, as a person, was an unstoppable force—and Hilbert, by his very nature, an immovable mass.

CHAPTER THIRTEEN

THE PIED PIPER OF PARADISE

1921–1924

"
The poet must see what others do not...the mathematician must do the same.
"

—Sofya Kovalevskaya

Engraving of the Pied Piper from an 1871 book by John Greenleaf Whittier. Cropped from a public domain image via the New York Public Library.

Hilbert is good with math but bad with kids. That's a tiresome trait for an adult who has no children—but a tragic one for someone like Hilbert, who does have kids. It's a recipe for awkwardly distant, aloof, awful *Ordinary People* parenting and worse.

I sometimes wonder what's harder: to lose a loving father due to tragic loss, the way Ottoline and Bertie both did as young children. Or to lose your father to distance and essentially grow up without him, as I did (excellent stepparents notwithstanding).

Like Bertie, I lost my dad as a toddler—not to death but to divorce. During my childhood, I saw him on a few occasions. Vacations and holidays, usually for only a few days at a time and almost always with years of tears in between. The closest he ever lived to me when I was growing up was a thousand miles away. For half my childhood, he lived overseas. The first time I resided within two hundred miles of him was when I was in graduate school.

Looking back at myself as a kid, I see a lot of pseudopaternal psychological sublimation fantasy in my childhood choices of movies and novels. I read and reread sci-fi stories that featured wise and warm proxy father figures or distant or missing ones, like *The Chronicles of Narnia*, *Stranger in a Strange Land*, and *The Lord of the Rings*. As a teenager in the 1980s, I was drawn to movies depicting troubled heroes who struggled to survive in dystopian, rudderless, fatherless worlds. *Blade Runner*. *Road Warrior*. *The Terminator*. *Streets of Fire*. And in comic books—especially in comic books—I sought to live out all my faraway-father fantasies in the pages of Marvel comics.

I realized this a couple years ago after discovering what remains of my childhood comic-book collection. One strange theme I saw was that I seemed to collect any title I could find in the 1970s and 1980s that featured the woeful, titanic, planet-eating villain Galactus. He was no hero, to be sure. But I liked Galactus. I wanted to hang out with him. I felt like I should get to know him. To help him get to know me. That was a tall order in the Marvel universe Pantheon. Galactus never noticed anyone. He left that to a succession of humanoid helpers, his heralds.

THE PIED PIPER OF PARADISE

I fancied myself Galactus's herald. In the comics, if he chose you, he would shoot powerful rays of light out of his eyes and imbue you with cosmic energy. Earth-shaking might. Silver Surfer. Terrax the Tamer. The disco star Dazzler—she of groovy quad skates and cool ripped jeans. They all got noticed by Galactus. Heralds were lucky. How I used to wish Galactus would notice me too. Shoot me with those magic eye beams. Make me his herald. Should I have a ripped jean too?

───◆───

Hilbert is sort of like Galactus for his troubled son, Franz. But Franz is no fit herald for his father. The father's a genius. The boy's no good at math. "My son gets his mathematics from his mother," Hilbert would say—often and many times over to anyone who would listen. "All else is from me."

Just before World War I, at the age of twenty-one, Franz suffers some sort of mental breakdown and has to be involuntarily committed. That's a blessing in disguise in the end because it keeps him alive and out of the war. What exactly happens and why is a mystery.

As the story goes, Franz becomes deeply disturbed one day, walks off the job, jumps on a train, and disappears. Hilbert's wife Käthe panics. She prepares to take the next train herself and chase down her boy, but then Franz suddenly reappears. He's covered in mud and stark raving mad, having jumped off the train midroute and retraced his steps through rain-soaked fields. He's suffering anxiety. Dread. He's shouting about invisible evil spirits. They're coming for him. They're coming for them all!

"Oh, you stupid boy, there is nothing," Hilbert says. "There are no ghosts or devils."

The unfolding scene gets surreal. Hilbert's wife is distraught. His colleagues are concerned. His son is in crisis. A doctor arrives. And Hilbert is mad. He just keeps banging his fists on the table, over and over, repeating the same thing: *There are no ghosts!*

The doctor sedates Franz and takes him away to a clinic in a taxi. Later that morning, Hilbert says coldly, "From now on, I must consider myself as not having a son."

THE GREAT MATH WAR

Franz never gives up. He wants to see his dad. He wants his dad to see him. And he does—to a point. For the next few decades, Hilbert will find his son jobs here and there at the university. And later in life, a more mature Franz will take on a strikingly David-like appearance—one he will deliberately cultivate. He will dress the same. Groom his face the same way. And even mimic his father's famous speaking style, according to one biographer. As a performance, it's more disturbing than convincing, however. Most people will view him as they would an organ-grinder's monkey: as a sad, strange, aping pantomime. "The sound without the substance," people who knew the family will later say.

Hilbert loves his son. But he is not good with kids, and he knows from early on that Franz is not his mathematical heir—his son is no Dazzler in ripped jeans and roller skates dancing to the beat of his Galactus. That's why Hilbert feels a profound loss when he discovers Weyl's betrayal—his true *intellectual* son. Weyl's actions are too much for him to take. He decides he will take the fight to Brouwer and not back down, whatever comes. He is going to war!

VANITAS NOTAS IN PAGINA

Lots of people are intrigued by Brouwer's intuitionism, but not everyone. Hungarian mathematician George Pólya, Weyl's colleague in Zürich, calls Brouwer's program mathematics in short sleeves—in other words, not fully dressed. He and Weyl get into it. They argue. Finally, Weyl makes Pólya a friendly bet that Brouwer's ideas will ultimately win out. Within twenty years, he predicts, intuitionism will become the dominant theory.

After Weyl's foundational-crisis paper comes out, the lines are drawn. Mathematicians become divided into two camps of formalists, who follow Hilbert, and intuitionists, who follow Brouwer.

Hilbert's approach dominates. He is far more famous, after all, the beating heart at the beating heart of the mathematical universe. Besides that, there is a lot to love in what he says. His exuberant solutionism is intoxicating—the idea that if you can logically state a problem, given enough time, pure reason, and hard work, eventually you will solve it.

THE PIED PIPER OF PARADISE

At the same time, Hilbert's formalism as it evolves in the early 1920s is a strange beast for someone who has spent the bulk of his career working on problems in applied mathematics. Nothing could be more abstract, pure, and far from applied math adulation than the foundations of math. But the roots of this transition were there all along. Hilbert's 1899 book, *Foundations of Geometry*, basically takes what was always an applied subject and remade it into an abstract one. Now, in the 1920s, he wants to go further.

For him, trying to establish the foundations of math by relying on some other system, like logic, is like eating a cup of mushy peas when you have a cowboy steak on your plate. Instead, he takes mathematical objects, represents them with symbols, and combines statements using standard logical rules to form mathematical proofs—"metamathematics," he calls it, or "proof theory." His formalism is no more complicated than that: working on theorems made of strings of meaningless marks on the page, as Hilbert puts it. A game, really.

But that in no way diminishes the seriousness of formalism. Games have rules. So Hilbert advocates setting up his mathematical proofs along those lines: games with established rules where the focus isn't on the mathematical object playing pieces but rather on the rules of play. When rules are followed, meaning emerges. And results have meaning, even if the marks that get you there don't. That's the basis of his approach, proof theory, formalism, metamathematics, or whatever you want to call it. Making meaningful statements about meaningless symbols.

What does that mean exactly?

---◆---

Consider a standard game of chess. There are set pieces in chess—standard not in shape and appearance but in abstract representational meaning: A king is a king, a queen is a queen, and a pawn is a pawn. I have three chess sets in my house. One has fancy, hand-polished copper-and-coffee-colored pieces done by professional artists in the likeness of Winnie the Pooh characters. I bought the set in New York City years ago, thinking at the time that all those Piglet pawns and Eeyore knights would appeal to my small children. The pieces were too fragile to play with, however, and the set has been in its box since the day of purchase.

Another chessboard I have at home is a unique glazed and fired clay set my daughter made as part of her AP ceramics class in high school. It has plant pieces on one side and animal pieces on the other. The knights are trees and squirrels, and the pawns are bumblebees and tulips. And the third set I own is just a cheap commercial travel kit where the thirty-two plastic pieces are standard black or white Trojan-horse knights and minaret-spire pawns. They fit snugly inside the case, which folds out into a board.

The point is that the exact shape, material, and color of the pawns don't matter at all. Ceramic bees. Plastic horses. Hand-cast Piglets. They look nothing alike, and their appearance is meaningless. The only thing that matters is the rules—that you have eight pawns side by side on the second row at the beginning of play and that each can move only one or two squares forward on their first move and only one square forward thereafter, unless capturing.

There are other rules as well. Players take turns—that's a rule. There are rules governing the movement of all the other pieces. You can't move a king more than one space because to do otherwise is to play a different game. How the board is arranged is a standard rule, even in a variant layout like Fischer Random Chess, or Chess 960, as some call it, where the back row is randomized. Rules decide how the game proceeds and ultimately whether it's a win, loss, or draw. Nobody playing a fair game of chess would seek to bend the rules (the occasional grandmaster controversy aside).

And the question of whether chess is "true" is basically meaningless. The game is the game. And in the same way, math is math. As long as the rules are consistently defined and faithfully followed, that's enough, according to Hilbert. It doesn't matter if the pieces are plastic, wood, or diamond-encrusted hand-carved marble. They have no meaning unto themselves. Pieces are pieces. Rules are rules. And chess is chess.

Hilbert's approach in "gamifying" mathematics, so to speak, is broad. He treats all the playing pieces of math the same, whether real or imagined, complicated or obvious. They are simply objects written using the language of mathematics—not in the disappointing ambiguity of ordinary language but in the specific bereft-of-unintended-meaning symbolic language

THE PIED PIPER OF PARADISE

of math. Again, think of the ceramic-bee pawns my daughter crafted in high school. They have no meaning as objects until you arrange them on a chessboard.

◆

The game of math is harder than the game of chess, however. Why? First, the objects of mathematics aren't as meaningless as plastic pawns. What we talk about when we talk about the objects of math are things like numbers, lines, and geometric shapes. These things seem to have obvious, unambiguous, and objective external meaning. Those meanings are deeply ingrained in our understanding of math. Some of them, like whole numbers, may even be a neurologically hardwired part of our cognition.

Modern neuroscience is uncovering more and more how humans, crows, dolphins, clown fish, and many other species have innate quantitative abilities that directly concern numbers, geometric shapes, and other objects. Seen in the mind's eye, numbers, lines, and shapes are not petty plastic pawns but real-world things. They are the stuff of external reality, and to ignore that is to close your eyes to the world around you.

Mathematical objects often carry massive cultural significance as well. We define beauty and order in terms of geometric symmetry. We attach social and even moral meaning to numbers and shapes. *Just the two of us. Walk the line. Play it straight. A square deal. The circle is complete. Lucky number 7. Seat 11a. Unlucky 13!* But it's more than just that. At first glance, numbers, lines, and shapes are not just *part of* math. They *are* math. For thousands of years, in fact, they were the sole stuff of the subject—the only thing we talked about when we talked about math. For this reason, some experts call numbers, lines, and shapes "privileged" objects.

Starting in the nineteenth century, however, mathematics began to claw back that privilege and ask what happens if you remove it and begin to treat is as any old object, whether objectively real or arbitrarily meaningless. And when Hilbert starts to develop proof theory in the 1920s, he goes crazy with this idea. What he proposes is radical. He wants to remove the privilege from *all* privileged objects and turn them into meaningless marks on the page. It's all a bit shocking at first blush... math games, meaningless marks.

THE GREAT MATH WAR

◆

That's what's so astonishing about Hilbert's approach. Math isn't meant to be meaningless. One typically does math specifically because it *does have* meaning. Math solved practical problems throughout human history. Geometry is useful for everything in the ancient world. Surveying. Building. Estimating. Measuring. Navigating. We have 60 minutes in an hour and 60 seconds in a minute because we have two hands, each with 5 fingers, each finger with 3 phalanges. You can count to 12 with one hand, touching your thumb to one of 3 bones in each of your opposing 4 fingers. Holding up one finger on the other hand gives you 5 x 12 = 60 possible numbers—and an exceedingly useful numeric sign language for commerce. The division of minutes in the hour and seconds in the minute today is still a legacy of that Babylonian invention.

And Hilbert, of all people! He's specifically someone for whom meaning in math is sacrosanct—or so it seemed. He's basically been an applied mathematician almost his entire professional career. Why would someone who has spent a lifetime ensconced in applied meaning suddenly pivot to such pure meaninglessness? And yet here he is, late in life, "draining mathematics of any meaning whatever," as Nagel and Newman will write in 1956.

But boldness is Hilbert's ally. In the 1920s, when he starts developing proof theory, he faces a brutal uphill climb—harder than climbing from base camp to the K2 summit. Hilbert knows axiomatizing mathematics and fixing the flawed foundations of math will require those axioms to be consistent. So he needs to find proof of consistency. His whole motivation for developing proof theory is to do exactly that—introspectively study what math can and cannot achieve, according to Argentinian American mathematician and computer scientist Gregory Chaitin. That's Hilbert's summit. His destination. His stated goal. His North Star. His *I-climbed-it-because-it's-there* expedition. Climb that mountain. Secure those foundations. And save Queen Math.

Easier said than done—and not just because math is harder than chess but because, as any good mountaineer will tell you, the real danger is on the way down.

THE PIED PIPER OF PARADISE

TERTIUM NON DATUR

In creating proof theory, Hilbert doubles down on his commitment to a concept known as *Tertium non datur* (There is no third possibility), also known as the law of the excluded middle. It's a useful tool for mathematical proofs, which basically says that for a mathematical proposition, there can exist only two possible states. True or false. On or off. Yes or no. Go or stay. No ifs, ands, or buts. Black or white—never gray.

But that's not quite precise. A better way to state *Tertium non datur* is *True or un*true. That formulation is more accurate because the law of the excluded middle is not an A-or-B proposition but rather says you have either A or not A—no sort-of A. No A-like. No A-ish. No maybe-A.

Tertium non datur is an extremely handy tool for mathematical proof because it's expansive. It means you can establish the truth of a proposition by showing it's not false—the denial of a denial being equivalent to an assertion of truth. That means you don't have to exhaustively prove everything. You just have to show something is true in a single instance or that its opposite is untrue, often by demonstrating that the other possibility (¬A, or not A) implies a logical contradiction.

Think of the game of chess again. Who won the last game? Black or white? Zermelo's theory of chess holds that at game's end, one player or the other has won or the game has ended in a draw. But what if nobody can tell you whether the black side won? A *Tertium non datur* approach to discovery might ask, Did white win? No. Was there a draw? No. Then the only possible conclusion is that black won.

That sounds painfully simplistic, but it's a logical approach to reasoning that goes all the way back to ancient Greece. (Aristotle himself first formulated the law of the excluded middle.) But it wasn't until the late nineteenth and early twentieth centuries that *Tertium non datur* began to take on even more significance because mathematicians found they could use it along with set theory to treat infinite entities using finitary means. With the law of the excluded middle, you don't need to construct an infinite set in order to work with it. You can use *Tertium non datur*, for instance, to construct a pure existence proof—something Hilbert loves. "Existence proofs carried out with the help of the principle of excluded middle usually are especially attractive because of their surprising brevity and elegance," he says.

THE GREAT MATH WAR

Brouwer's approach of intuitionism and constructivist math says you can definitively answer that same question of whether black wins the chess game only if it can be demonstrated (perhaps by reconstructing the moves from the beginning and showing that on the final move, black knight takes white king). His position is revolutionary, if destructive. For thousands of years, nobody has really questioned the principle of the excluded middle. But in his view, anything that's not constructed need not be valid, including the law of the excluded middle itself. *Tertium non datur* is an unconstructed assertion in Brouwer's view. He doesn't just question it—he dismisses it as a meaningless combination of unproven words.

PARADISE LOST

Rejecting the law of the excluded middle is really upsetting to many mathematicians because it means...yikes! Taking the *Tertium non datur* out of math is like taking the cutting board out of the kitchen—stealing one of your most powerful tools and leaving you with nothing but dull knives.

Hilbert spits on the idea. "To prohibit existence statements and the principle of excluded middle is tantamount to relinquishing the science of mathematics altogether," he says. It would be like asking an astronomer to work without a telescope—or, as Hilbert says, instructing a boxer that he can't use his fists. Without throwing a cross, you're never going to score a knockout. Without your telescope, how can you ever hope to make an astronomical discovery? And without *Tertium non datur*, you would chuck out some of the best gems in the mathematical chest. That's too scorched-earth to even consider.

> *Rather than repeating Hamlet's*
> *Rhetorical prose:*
> *To be or not to be*
> $2(B) \vee \neg 2(B)$?
> *Intuitionism gives*
> *An absolutist pose*
> *To be only when* SHOWN *to be!*

THE PIED PIPER OF PARADISE

Brouwer realizes the significance of rejecting the law of the excluded middle, of course. He knows large parts of Cantor's so-called paradise will have to be thrown out if he does. But that doesn't bother him in the least. He doesn't care. In fact, he welcomes it. He is, after all, chasing revolutions, not rainbows.

From 1918 to 1923, Brouwer publishes a number of papers on intuitionistic set theory in Dutch and German journals specifically rejecting *Tertium non datur* as part of a mathematical proof, especially when dealing with infinite objects. He says that in an infinite set of possibilities, one cannot know for certain that A and not A are the only two states. And he rejects Hilbert's belief that all math problems are solvable on the same basis. Some theories may not be provable. Others may be false even if they present no apparent contradiction. A false theory not unstopped by contradiction is like a thief not yet apprehended by the law, he says.

That's only part of Brouwer's work. His big innovation is developing what he calls constructive set theory using something known as choice sequences, where a mathematical object is seen as a construction rather than an existing structure—a becoming rather than a being, to put it in purely mystical terms. With choice sequence construction, intuitionism seems to Brouwer to be rich and rigorous, if not righteous. He seeks a philosophical "reconditioning" of math away from mathematical objects as unreal entities to mathematical objects as real objects constructed by the human mind.

But it is specifically his rejection of *Tertium non datur* that makes intuitionism hard for many to love. Hilbert's formalism is manageable. Economic. Elegant, even. Radical and traditional at the same time. Who doesn't love a good game of chess? Constructivist proofs, on the other hand, are unwieldy, even ugly. Some are ten times longer than comparable traditional proofs, sometimes completely devoid of elegance, and sometimes seemingly pointless and destructive—like a gender-reveal party that starts a large wildfire. Strict intuitionism is so zero-sum that it outright rejects things you can't construct, including completely familiar concepts like irrational numbers and Cantor's transfinite numbers. To many people, that's ridiculous and worse.

Thus, when Hilbert throws his gauntlet and begins bombarding Brouwer with criticism, the crowded-room European math world mostly sides with him. Brouwer and Weyl are hopeful in the early 1920s that intuitionism will reimagine mathematics in a simpler, cleaner, more elegant way, but it never happens. And most people aren't willing to wait and see if it will. As the decade wears on, it becomes more and more clear that intuitionism is more cumbersome and less sexy than the simple elegance of formalism, which plays directly into Hilbert's hands. As he climbs his mountain and sets off for K2's peak, people clap for him along his journey all the way from base camp—while Brouwer somehow stagnates and turns into Tumbledown Dick.

Weyl later laments the "almost unbearable awkwardness" of intuitionism and the crumbling effect of excluding nonconstructivist proofs. "The mathematician watches with pain the larger part of his towering edifice, which he believed to be built of concrete blocks, dissolve into mist before his eyes."

For Hilbert, it's far worse. To accept Brouwer is to reject Cantor. Hilbert "could not bear this mutilation," Weyl says. "They would chop up and mangle the science," Hilbert says. "We would run the risk of losing a great part of our most valuable treasures."

That's far, far farther than Hilbert is willing to go. Nor does he buy Weyl's statement that Brouwer's intuitionism is revolutionary. You can't call it a revolution, Hilbert says. More like an attempted coup. "Doomed to failure from the beginning."

Shots fired!

"UNE AFFAIRE FRANÇAISE"

Brouwer and Weyl both appear at a conference in the German city of Nauheim in September 1920. They both give talks, and according to one account, Brouwer's verges on impossible to understand.

The Nauheim meeting is far more famous today for being the place where Albert Einstein first publicly defends his theory of relativity, then just exploding in popularity. But it's also interesting for its politics. Nauheim is organized intentionally to conflict with another meeting happening at the same time—the Sixth International Congress of Mathematicians in Strasbourg. The fact that Einstein, Hilbert, Weyl, and others are there instead of at Strasbourg is a sign of the times. All German nationals are forbidden to attend the sixth congress in Strasbourg.

Two years earlier, a group of Belgian, French, British, and other allied scientists convened at the Royal Society in London and issued a statement calling for boycotting German scientists as a general punishment and reprisal for unspecified yet horrific war crimes. *Atrocities that have shocked the civilized world!*

"The wanton destruction of property, the murders and outrages on land and sea, the sinking of hospital ships, the insults and tortures inflicted on prisoners of war, have left a stain on the history of the guilty nations which cannot be removed by mere compensation of the material damage inflicted," the statement read.

The architect of this exclusion was French mathematician Charles Émile Picard, a star of the Paris science scene who had trained at the Sorbonne under famed French mathematician Charles Hermite and later married his daughter. A picture of Picard taken before the war shows a bright, sneering milquetoast with soft eyes and a big black upward-facing handlebar mustache curled well past his lips. He looks like a modern hipster barista in the photo. Pour me a red-eye, Chuck!

In another picture of Picard taken years after the war, he looks monstrously different. He's older. Heavier. Balder. But it's more than just the fickle finger of Father Time. He's war-weary. Weather-beaten. Angry and with a piercing scowl. Something's changed. Those charming mustache handlebars are gone. He wears a shorter 'stache on his upper lip instead—an angry mustache—with a pince-nez, a permafrown, and hard, penetrating eyes. The worm turned for Picard during the war. His son was killed. And it becomes kill-the-kaiser all the time for him after that.

———◆———

He was not alone in his contempt and sorrow. Even though victorious, parts of France lay in ruins. Bombed out. A landscape of broken buildings and broken men. Trampled wrecks of trees. Stripped of bark. Choked of life. Splintered trunks. Shattered and burned. Tombstones to their former glory. Broken boughs filled with burned embers. Nothing grows in wartime. *Cut down like swaths of corn at harvesttime!*

Like many others, Picard blamed Germany for everything. And he saw the difference between the French and the Germans not just as competing national pride but as wholesale at-odds worldviews. French mathematicians stood for "liberating and ennobling mankind." German science was "an instrument for its enslaving."

Enough was enough, Picard proclaimed during the war. The old ways were dead. All ties with German mathematicians should be severed, he said. He advocated for German nationals to be forcibly withdrawn from international organizations, to boycott German scientists at international meetings, and to ban German as an official language of journals and meetings—even in German-speaking parts of Switzerland. He quoted Cardinal Désiré-Joseph Mercier, archbishop of Malines, a voice of Belgian resistance in the Great War: *To pardon certain crimes is to become accomplice with them!*

And Picard initially succeeded. German researchers were barred from two-thirds of the 195 international scientific conferences held from 1920 to 1924. The proportion of papers and books published in German shrank from 40 percent of international publications in 1909 to less than 30 percent in 1929. But the most visible of these exclusionary efforts came at the Sixth International Congress of Mathematicians in 1920, where Picard sat as queen bee. He saw to it that invitations were sent narrowly. Exclusively. To individuals, not groups. Speakers were decided in advance. Lectures were narrowly focused.

This all gave the meeting a "harsh atmosphere," the Spanish mathematician Guillermo Curbera will write in 2010. It was also a tiny meeting. Only two hundred mathematicians attended, mostly French. Some observers called it *une affaire française*.

In his closing speech, Picard told the audience that perhaps one day German mathematicians would be allowed back in—but then again, perhaps not. "Our successors will determine if a sufficiently long lapse of time and a sincere repentance could allow them to resume someday," he said.

There are two types of internationalism. One is warm, welcoming, and universal. The other is cold, clawing, and elitist. One seeks to expand the world. The other to constrict it. One seeks to unite. The other to divide. One is about contact—the other contraction.

DE INFINITO

As his battle with Brouwer heats up, Hilbert starts giving a number of lectures on the foundations of math and his new ideas. He begins splitting mathematics into its classical parts and the new part based on his formalist approach and proof theory, which in 1923 he dubs "metamathematics," since it treats math itself as an object. It's not just some pointless exercise. It will ultimately prove invaluable for the practical applications of mathematics. But Hilbert's main goal is to provide an unshakable foundation for mathematics. With his proof theory that will be attainable, he says.

Finally he develops what becomes known as "Hilbert's program," a complicated and ambitious set of goals that constitute eight stages of attack to free all existing branches of mathematics from contradiction: starting with basic logic and calculations, moving on to number theory and then set theory, and finally finishing with higher types of math—a brush-the-channel-with-your-elbow, road-to-victory plan that would make even the German grand strategic planner General Alfred von Schlieffen blush.

In 1925, Hilbert presents a mature form of his program in a paper. "The definitive clarification of the nature of the infinite has become necessary," he writes at the beginning of the paper, "not merely for the special interests of the individual sciences but for the honor of human understanding itself."

———◆———

Let's pause here to consider one of the most remarkable things about Hilbert—his views on infinity. Normally in philosophy, someone would be considered a "finitist" because they outright reject the notion of any actual infinite. In practical life, we are all finitists in our daily lives because we live only a small, temporal existence in a big yet still finite universe. We play out the hands of our ticking time-clock lifetime and face real, concrete limitations at each stage of it.

THE GREAT MATH WAR

A finitist mathematician thinks that only mathematical statements that can be proven in a finite number of steps are real. But Hilbert is a paradoxical finitist in a sense—he is only because he isn't. He believes that essentially anything can be proven in a purely finite, limited way. But that's where he hits the same major speed bump that has bedeviled mathematicians for millennia: how to deal with infinity. How can you manipulate, treat, and prove things involving infinite collections using purely finitary means?

"According to Hilbert, the most vulnerable point in the fortress of mathematics was the infinite," the American mathematician Steve Simpson will write in 1988. "In order to defend the foundations of mathematics, it was above all necessary to clarify and justify the mathematician's use of the infinite."

And why not? Physics is undergoing a revolution in the 1920s, so why can't mathematics have one as well? Quantum theory had shown that energy is not continuous at the subatomic level but rather discrete, specific, and quantized. Relativity theory had bent light and space all to hell. And the big-bang theory is soon to suggest that space, time, and the universe itself are not infinite at all but constructed according to that canonical narrative: beginning, middle, and end. According to Simpson, that leads Hilbert to conclude that "infinity does not correspond to anything in the physical world."

So what better way to capture infinity than with a small, meaningless mark on the page? The one problem, and the key to his success, Hilbert knows, is finding a "consistency proof" to show that proofs involving infinite sets can be provable by reducing them to meaningless marks on the page and manipulating them with finite methods alone. Provide proof of that, and BAM! He's done.

If he can do that, Simpson will write in 1988, it will show "that finitistically meaningful end-formulas of infinitistic proofs are true in the real world." That will mean he has paved the way toward proving anything, established a solid foundation for mathematics, and won the Great Math War.

Not everyone is down with this idea, though—least of all Brouwer. Unlike Hilbert, who is convinced that anything will eventually be proven given time, energy, and the successful establishment of consistency through his program, Brouwer has a pragmatic if not a sourpuss attitude reminiscent of the "triumphant pessimism" of French physiologist Emil du Bois-Reymond, whose philosophy Hilbert repudiates.

Brouwer firmly believes that some propositions cannot be proven. "In mathematics, it is not certain whether or not all logic is permissible," Brouwer says, "and it is not certain whether it can be decided, whether or not all logic is permissible." In other words, we don't know, and in some cases we can't know. And at the end of the day, we may not even know if we can or can't know. (Reminiscent of the "unknown unknowns" words of former US secretary of defense Donald Rumsfeld.)

TOSS THE MAYBE WITH THE MATH WATER

Brouwer is even more forthright in his criticism of Hilbert's whole concept of meaningless marks on a page, relentlessly mocking him for it. "Empty formalism," he says. Bereft of meaning. Devoid of interest. Nothing to see here.

Who cares about Hilbert's stupid approach if mathematics has no meaning? It's unfair criticism, to be sure, but Hilbert lobs bombs of his own. Hilbert calls his nemesis the worst form of devil's advocate—a destructive naysayer. Narrowness ad infinitum. He accuses Brouwer and his followers, particularly his former student Weyl, of willy-nilliness. They "seek to provide a foundation for mathematics by pitching overboard whatever discomforts them."

By now it's too late for détente. Almost nothing Hilbert says or does at any time in his life concerning anything or anybody compares remotely to his emerging attitude and nasty remarks toward Brouwer in the 1920s. "Never before or after in his life did Hilbert take such an activist, and outright personal, position in a scientific debate as he did with Brouwer," one expert has said in recent years. "His attitude in this sense led him sometimes to a frankly absurd behavior."

What behavior? you might wonder. That's coming soon. First, Hilbert faces another, personal and far more harrowing, climb. He's grown into an old man in the 1920s. A lifetime of work and a wartime of turnips have done no favors to his constitution. He has major health problems, and now he finds himself at death's door. People who interact with him daily see it on his face. He's sick. Pale. Gaunt. Weak. Diminishing physically.

Of course, Hilbert is still an icon. A beacon. He's the Pied Piper of Cantor's paradise. Students, peers, colleagues—everyone adores and reveres him. Even members of the public think highly of Hilbert. How many mathematicians find their face on a postage stamp, as Hilbert did in the early 1900s? He's famous. A living icon in a way almost no mathematician ever has been, is, or will be. He embodies superhuman greatness, up there in the cloud kingdom with Albert Einstein.

And he's a mathematician's mathematician.

◆

When Hilbert shows up at meetings, people hold their breath. When he stands up to speak, they audibly gasp. *Oh, my God!* When he gives lectures, they come early and stay late. He's the dean. The wizard. The captain. The coach. Indisputable genius. Indescribable greatness. Being in poor health doesn't dim his star at all. He's probably the most influential mathematician of his day.

He gets even more juice from the fact that many people love what he's doing with formalism. Brouwer's carping aside, most mathematicians look on Hilbert with awe and eye his work with envy. It's gorgeous. Economical. Elegant. They embrace its warm beauty. Bubblingly fresh. And even as they embrace Hilbert's formalism, they are souring on the cold, clunky, confusing constructivism Brouwer is peddling. "They were far more amenable to a plan devised by Hilbert not for purging mathematics from within but for justifying it from without," the mathematician Allan Calder will write in *Scientific American* fifty years later.

"No one will drive us out of this paradise that Cantor has created for us," Hilbert declares in the 1920s, just to put a fine point on it. And it's not for the faint of heart to question or stand against that.

Brouwer, however, is just crazy enough to do so. And in the 1920s, he's beginning to have breakthroughs of his own. He starts publishing them. From 1923 to 1928, he publishes twenty-seven original papers. In 1924, with his new intuitionist constructive methods, he pushes ahead, proving thorny theorems where classical methods have failed. He also develops parallel proofs of problems previously solved using classical methods, like the fundamental theorem of algebra, which Gauss first proved a century before. (The theorem states that for equations of degree "n"—which contain terms such as X^n—there are n solutions.)

But even though his success should help intuitionism take off, it doesn't. Partly it's because he's his own worst enemy as an atrocious writer. Even people who admire Brouwer the man hate Brouwer the writer. Brouwer's biographer van Dalen says his major papers in the 1920s are "dry as dust" and "scholarly to the point of dull."

Brouwer's output is spotty and disconnected besides. Some years he doesn't publish at all. Others he publishes multiple papers but fails to connect one to the other. He doesn't build his body of work, piling each paper upon the last, the way scholars do. On top of all that, the 1920s are a busy time for Brouwer. He's distracted with other commitments. Preoccupied at his pharmacy. Swamped with his duties at the University of Amsterdam. Active as a hardworking journal editor—at *Mathematische Annalen*, the same journal Hilbert oversees (a source of awkwardness for them both). As a result, instead of a steady progression of impactful publications, he produces work in fits and starts—and more fits than starts.

Some start to think of Brouwer's work, according to van Dalen, as a sort of fallow hope. "A land of promises, but with little fruit."

◆

Fruitless landscapes are an especially resonant metaphor for the more-than-just-barren-trees Germany of the early 1920s. The long whiplash of economic malaise following the Great War lashes Germany the hardest. Deprivation gnaws at the sides of the country. Old Europe is gone, lying in ruins somewhere in Belgium. And the Weimar Republic is obligated under the Treaty of Versailles to make good on reparation payments.

How much the war cost and how many payments Germany needed to make was beyond comprehension at the end of the war—even in purely monetary terms, setting aside the question of emotional distress inflicted

THE GREAT MATH WAR

by the grievous loss of lives. One estimate in the 1920s projects that the war cost $200 billion (more than $4 trillion in today's dollars). No bounty that large has ever been demanded in human history, let alone paid, and yet people in those days are demanding it, especially in France.

Reparations become an eye-for-an-eye, tooth-for-a-tooth fixation for the French—one aiming to be as punitive as it would be remunerative. Some 1.4 million French citizens have been killed in the war and another 4.3 million wounded. In penalty, France proposes that Germany should pay $80 billion. Germany counters with an offer of $12.5 billion. A compromise figure of just over $30 billion is embraced in the London Agreement of 1921, but it's a settlement destined to please nobody—far below what France wants and far more than Germany can pay. Nevertheless, the Weimar Republic is forced to accept the deal and make payments however it can. Cash. Coal. Ships. Stocks. Floating bonds. Printing money.

By 1923, German industry is in ruins. The country faces impossible inflation and crippling currency devaluation. In 1922, the exchange rate falls from its prewar level of 4.2 marks to 1 US dollar to 162 to 1. In January 1923, the mark falls to 7,000 to 1. By July, it has been devalued to 160,000 to 1. By October, it's 2.42 million to 1. And by December 1923, a single US dollar is worth 4.2 trillion marks.

The crushing cost of high inflation hits people hard. They sometimes pay rent in butter. A small sack of potatoes could cost a basketful of cash at the beginning of 1923 and several wheelbarrow loads of cash a few months later. Money loses value so quickly that if you are paid by check, it's worth far less by the time you cash it. Some mathematicians seek work as ditch diggers to supplement their income.

CHAPTER FOURTEEN
NOBODY LOVES A REVOLUTION

1924-1928

"
One cannot hurry, alone, into nowhere, into nothing.
"

—Frederic Manning

The Storming of the Bastille on July 14, 1789, a line engraving with etching after Henry Singleton. Cropped from a public domain image via Wellcome Trust.

The negative reaction against intuitionism that begins to form in the 1920s is partly informed by political strife and economic malaise. People begin to recoil from intuitionism not just because Brouwer is a bad writer or because Hilbert, his nemesis, is so popular and outspoken against it. It's more existential than that. Some start to see intuitionism as dangerous because Weyl and Brouwer framed it early on as revolutionary, tapping into the popular appeal of the Russian Revolution. *Das ist die revolution*, Weyl wrote in his famous paper on the foundational crisis.

But even as some looked longingly eastward to revolutionary triumph, many Europeans were coming to fear it above all else. If political revolution seemed intriguing, okay, or even ripe and juicy in the postwar days of the early 1920s, by mid-decade, it no longer was. The Russian Revolution cast a long shadow over Germany, and many feared that their country would fall to the same fever Russia had. And London seethed with the same stress by proxy—not that the average Brit feared that communist agitators would take over Piccadilly Circus anytime soon, but many did worry that the *German* government would fall. What would that do to trade? It was a concrete concern in London because in many ways, what was in the Weimar's interest was also in London's.

Even a socialist-sympathetic person like Bertrand Russell quickly came to fear the danger of runaway revolution after he traveled to Russia in the spring of 1920. Before he went, he thought he would find something different—something nice. "I imagined that I was going to see an interesting experiment in a new form of representative government," he wrote soon after. Instead, what he saw shocked him. He met Leon Trotsky at the Moscow opera, feasting high on the hog in the plush theater box that had once belonged to the czar. He had an hour-long interview with Vladimir Lenin. Strange and bizarre. And he witnessed widespread suffering in the countryside. "Bolshevism is internally aristocratic and externally militant," Russell wrote. He went to Russia a believer but returned a doubter.

The shifting views on revolutionary politics are toxic for people's views of Brouwer's math. Weyl's *das ist die revolution* seemed so cool right after the war, but by the mid-1920s, it begins to sound ominous and threatening.

NOBODY LOVES A REVOLUTION

And once that association is made, it cannot be undone. By then, nobody loves a revolution. Many begin to look askance at intuitionism and Brouwer as well. Who would trade buttermilk for bitter vodka? Common sense for political chaos? People hear Brouwer, but instead of thinking math and creativity, they start to think communism. Danger. Oppression. A "Bolshevik menace," one expert writes of intuitionism.

Hilbert, sensing this vibe, employs all sorts of references to politics and war in railing against Brouwer. Is Brouwer's work a revolution? Not really, he rages. It's more like a weak attempt at a failed coup. A pathetic putsch. It won't work, he says. Doomed. And why? If intuitionism succeeds, Hilbert says, "then there will remain of present-day mathematics, apart from tightly bordered impregnable areas... only immense ruins."

THE EXCLUDED DIDLE

Another International Congress of Mathematicians is held in Toronto in 1924. By now support for the selfish exclusionary position of Picard and the French is slipping. A growing chorus is calling for an end to the ban on German nationals. Mathematicians in Italy, Holland, Denmark, Sweden, Norway, Britain, and the United States are calling for the readmission of Germans—even if some, like Picard, are unwilling. The lobbying falls short, however, and the Germans are still not allowed to attend in 1924.

That's only part of the drama with this meeting. It was originally slated to take place in New York City, but when that fell through, Canadian mathematician John Charles Fields single-handedly saved the day by relocating it five hundred miles north by northwest to the University of Toronto. (He will also famously create the early-career award now known as the Fields Medal—and sometimes informally the Nobel in mathematics.)

Another meeting takes place that summer. The Great Math War has become personal by then, and when people hear that Brouwer is planning to speak at Hilbert's own university, they wonder what will happen. Will it be a showdown? Given Hilbert's prominence in the city—his own city—Brouwer's arrival there seems a bit like dancing barefoot into a viper's pit. Bitterness must follow.

THE GREAT MATH WAR

Brouwer has other things on his mind. He is still reeling from a terrible loss in the spring. His close friend and classmate, the poet Carel Steven Adama van Scheltema, has just died from a peculiar injury. A socialist poet who wrote "simple and understandable" poems for the working class, Scheltema had sold some sixty-five thousand books by the time he died, though he suffered from nervous attacks his entire life and was hounded by the peculiar affliction of living constantly in fear of developing a brain tumor—the same type of cancer that had killed his father when he was a young man. Scheltema was slain instead by a different sort of brain injury. He fell and seemed fine at first but then developed dizzy spells, deteriorated, and died a few weeks later.

Brouwer was devastated. Losing a friend is harder than losing yourself. Thus, in the summer of 1924, when he is invited to give a lecture on intuitionism at the Göttingen Mathematical Society, Brouwer comes mourning even as Hilbert arrives brooding. Brouwer's latest line of attack targets one of the central premises of Cantor's set theory: the idea that the nondenumerable, uncountable big infinity set of real numbers is so large that if you took just the real numbers in a single line segment between 0 and 1, it would be larger than the little infinity set of whole numbers in its entirety.

He directs this criticism at Hilbert as much as at Cantor. In his lecture, Brouwer says, "When the formalist creates the set of all real numbers between zero and one, these words are without meaning for the intuitionist," even though this is one of Cantor's most beautiful results, where he showed that the size of the nondenumerable infinite set of real numbers is invariant of dimension—the idea that there are as many points in a line as there are in a plane as there are in 3D space. This is exactly why the mathematician Kurt Gödel will later call Brouwer's intuitionism "utterly destructive in its results." Adopting intuitionism is hard if it means throwing out everything you love in modern mathematics.

——◆——

Hilbert is even more critical. During the lecture, he sits silently stewing. People look over at him from time to time. He bides his time. Then he chooses his moment carefully. Toward the end, he stands. The audience turns. Eyes on him. This is his turf, not Brouwer's. These are his people, not Brouwer's. They love him. This is his town. Who is this Brouwer guy anyway? A loner! A loser! This is Göttingen, goddamn it—it's *his* city. His home turf. So he gives the audience the mic-drop moment they long to see.

"With your methods, most of the results of modern mathematics would have to be abandoned," Hilbert says to Brouwer. "To me the important thing is not to get fewer results but to get more results." Mic drop! The message is clear to the audience: Choose Brouwer and limp into darkness. Choose Hilbert and bask in glory and hope.

Weyl, meanwhile, is clamoring for a way out. He wants to find some middle ground, some way of reconciling the two great mathematicians between whom he finds himself caught. He's also duplicitous—though not for devious reasons. He's long supported and sided with Brouwer, but he's beginning to appreciate the beauty of Hilbert's program. He never wanted to throw out the principle of the excluded middle—not really. He accepts it. More than that, he's beginning to have second thoughts about his mentor and appreciate the pure beauty of Hilbert's work.

The significance of this change in tune from Weyl is hard to understate. At the exact time when Brouwer needs friends the most, his chief ally begins to waver. Weyl has always been Brouwer's biggest supporter. He gets Brouwer's vibe, and early on, he beautifully communicates his friend's energy and the ideas of intuitionism to the masses. Now, with equal verbal eloquence, he captures its crumbling edifice.

He concedes the gorgeousness of Hilbert's formalism as well as the off-putting aspects of Brouwer's intuitionism. It's restrictive. Constrictive. Limiting. And maybe even, as Hilbert claims, a bit destructive. "With pain the mathematician sees that the larger part of his tower, which he thought was joined with strong blocks, dissolves into smoke," Weyl says.

Weyl revises his predictions of a coming revolution. Maybe it's just a little revolution. Perhaps a tiny one. An itty-bitty, incremental revolution— or maybe it's not one at all. How can he find some way of brokering peace? He fears the rip of Hilbert's jagged edge. "Hilbert turned in heated polemics against the stand taken by Brouwer and me," Weyl writes around the time of Brouwer's lecture. "I think with little justification." But Weyl has

THE GREAT MATH WAR

dug his own grave by acting Trotsky to Brouwer's Lenin and by wrapping intuitionism with the blah-blah bluster of revolution.

He knows he bears a lot of responsibility for this debate, even as he seeks to pull himself out of it. "We wanted to storm the heavens," he will write a few years later, "and we have only piled cloud on cloud that cannot support anybody who tries in earnest to stand on them."

THE MECCA OF MATHEMATICS

When Hilbert's mentor Felix Klein retires in the early 1920s, Hilbert feels the loss. It's a blow to him and to Göttingen. One of the most influential mathematicians of his day, Klein was something of a wunderkind in his youth, appointed full professor at the age of twenty-three. That's rare even today and was almost unheard of in nineteenth-century Germany, where it could take years, often decades, to win a professorship. A century before Klein, the great philosopher Immanuel Kant himself had had to toil away as a low-level teacher for twenty-five years before finally winning his professorship.

In 1883, Klein took a position at Göttingen University and a decade later recruited Hilbert. Together they built their department into an international juggernaut and succeeded spectacularly. The university becomes known as the "Mecca of mathematics" in the 1920s and the "center of the scientific world," attracting some of the most brilliant students in the world—mathematicians as well as physicists like Wolfgang Pauli, Werner Heisenberg, Robert Oppenheimer, Karl Compton, Paul Dirac, and Linus Pauling. The "very air of Göttingen" is full of excitement in the 1920s, the mathematician and faculty director Richard Courant will later reflect.

At Göttingen, Klein was an icon, but Hilbert was an even bigger draw. Students and faculty all over Europe and America knew of him. They wanted to see him. They wanted him to see them. Many a brilliant mind was drawn to the mind-bending, space-warping, gravitational pull of Hilbert's greatness.

Still, it's a strange backwater for many a cosmopolitan rising grad student. Why go to Göttingen when you could live in Berlin or Paris? Why move to a place that "was little more than a village between gentle hills crowned by the ruins of ancient watchtowers," according to math historian Constance Reid? Why move to a town where if you want to get a good meal, you eat at the train station—and if you want to get a *really* good meal, you hop on that train? Why go to a city whose chief claim to fame, besides a few academic giants, is a macabre footnote: It's where the Brothers Grimm wrote "Little Red Riding Hood." Why move to Göttingen at all? Why allow yourself to be swallowed whole by such a wolf-leaped-out-of-bed Podunk backwater?

Three reasons: *Hilbert, Hilbert, Hilbert.*

———◆———

After Klein retires in the 1920s, Hilbert's student Richard Courant takes Klein's place as the head of the mathematics institute (the university first offers the position to Weyl, who says no). Courant continues to build on Klein's administrative work and is hugely successful. He recruits great faculty and students. He helps build the reputation of his institute. If Klein is king and Hilbert the prince, Courant is something of the heir apparent.

He acquires new physical space as well, securing funds from the Rockefeller Foundation in New York City to construct two new buildings at the university—one for mathematics and one for physics. When these two new buildings open, they are both paradise and playground—brilliance around every corner. Three levels. Huge lobbies. Generous lecture halls. The auditoriums in the math building are dubbed "maximum" and "minimum." Courant is an empire builder.

Courant is also noteworthy in the history of math for being instrumental in creating those famous yellow textbooks produced by the publisher Springer-Verlag (now Springer Nature) starting after World War I. (Anyone who has studied math or physics for the last hundred years would be familiar with this publisher.)

At the same time, some say he's a strange sort of mathematician. "A mathematician who hates logic, who abhors abstraction, who was suspicious of 'truth,' if it's just bare truth," one of his former students and colleagues once said to Courant's biographer Constance Reid. He suffers no small amount of criticism for his work. Some even claim that his work is

THE GREAT MATH WAR

not truly his own—rather, it is the product of his assistants and others around him. For that reason, according to Reid, "He was sometimes referred to behind his back as 'Dirty Dick' or 'Tricky Dicky.'"

I don't know about all that. To the modern mind, US President Richard Nixon will claim complete ownership of the Tricky Dick franchise in the 1970s. Nothing Courant ever could have done in his entire life would seem to rise to the level of Nixon's Watergate malfeasance. But perhaps the moniker is owing to the fact that Courant was a ruthless climber in the early years of the twentieth century. From a family of small businessmen and what he himself said was a "not very intense intellectual life" as a child, Courant earned his reputation as a hard charger who was not always well-liked specifically because of his climbing. But he had to do it. His life wasn't easy. When he was a teenager, his family faced financial problems. His uncle shot himself because of business woes. The whole family business went bankrupt.

Hoping to help, Courant responded to a classified ad for a tutor in the local newspaper. He took the job and tried to teach the boy. But he quickly lost his cool. "Impatient, then angry, then abusive," the story, as told by Reid goes. "The boy's mother finally slapped him and threw him out of the house. It was a cold, wet day... [and] he found himself lying in the mud."

In college, Courant contrived to ask Hilbert to allow him to be a tutor for his son, Franz, who was just a schoolboy at the time—not yet broken, muddied, and intellectually diminished by his father. Courant the climber leveraged that close contact with Hilbert to grow closer to him still. And it worked. In 1908, Hilbert selected him as his assistant—to read journals, write summaries, help Hilbert prepare lecture notes, attend lectures, and give feedback. It basically rocket-launched Courant's career.

——◆——

When World War I started, Courant set out to be a hero, first in the trenches and then at the benches. A diary entry that he recorded on August 13, 1914, as he was mobilizing to the Western Front, was enthusiastic: "It seems to me that I am on a beautiful summer vacation." But another entry just a month later reflects the growing stagnation along the front: "I am afraid it will take much longer than anticipated."

He was later decorated for designing new ways of communicating in the trenches. Telephone lines often fell victim to shrapnel shells exploding above, throwing communications up and down the line into disarray. Courant designed a new form of communication that relied on devices that needed no physical connection. After a similar technology was found during a raid on French trenches in 1916, the German command clamored for the new device, spending millions of marks on the project and granting a third-class military distinguished service recognition to Courant—but not the first-class Iron Cross he had hoped for. He was only a reserve officer, however, and he was elbowed aside on the project by a higher-ranking regular army officer. All guts and no glory. Courant would not let the same thing happen again.

THE LIVER CURE

In June 1925, Felix Klein dies, and Hilbert's heart is heavy. It's a sad day for Göttingen, but it's about to get worse. A few months after Klein passes, Hilbert finds himself at death's door as well.

He is diagnosed with late-stage pernicious anemia (also called Addison's anemia), a nasty form of blood deficiency where the body lacks sufficient oxygen supply carried through the bloodstream. Lots of things can cause anemia. There's sickle-cell anemia, of course, where misshapen crescent cells in the bloodstream staunch the flow. Some cancers that damage bone marrow can cause it to underproduce red blood cells, the component of blood that ferries oxygen from the lungs around the body. Parasites, autoimmune diseases, radiation, chemotherapy, and even excessive bleeding can all variously deplete the body of red blood cells as well. And some causes of anemia are purely dietary—iron deficiencies or, in the case of pernicious anemia, a lack of sufficient vitamin B12 or folate.

Until the end of World War I, pernicious anemia was a harrowing diagnosis. The five-year survival rate was basically zero, and most people diagnosed with it died within a few years. Fortunately for Hilbert, literally just before his diagnosis, three American doctors discover an effective treatment for pernicious anemia. A pathologist at the University of Rochester shows that when dogs eat copious amounts of raw liver, their bodies respond by overproducing red blood cells. Two physicians at Harvard pick up on this result and conduct a human clinical trial of a liver diet based on that dog study. What they discover is a medical miracle: People with pernicious anemia who are fed a liver-rich diet for six months see their red-blood-cell counts triple. It's a lifesaving intervention, though somewhat an awful prescription, requiring a person to eat one-third of a pound of lightly cooked to raw liver every day. (That's one rare liver burger every day if you're counting—*a Mc-Barf-L-T.*)

Adding to the torture of taste is the pain of price. After reports leak of this wondrous new "liver cure," the news has the unintended consequence of driving up demand for liver among wannabe health nuts in the 1920s. Many people who are perfectly healthy seek liver, thinking it some sort of superfood. Demand, in turn, drives up prices. Butchers see a run on their nasty cuts of liver—culinarily categorized as "offal meat," which rhymes with *awful treat.*

Hilbert, however, benefits from an innovation developed by the same Harvard doctors, who take liver extracts and encapsulate them—liver pills, essentially, which are way easier to swallow than the real thing. The doctors prepare these extracts and fly them to Germany, and Hilbert's condition begins to improve immediately. The toll of the disease is obvious, however.

◆

According to the German historian and mathematician Reinhard Siegmund-Schultze, when physics professor and Princeton dean Augustus Trowbridge visits Göttingen in July 1926, he meets with all the mathematicians, including Hilbert, who appears strong of mind if weak of body. "Hilbert is said to be 64 years of age," he writes in his report on the visit, "though he looks much older." The most significant effect of the illness is not on Hilbert's outward appearance, however, but on his inner outlook. He thinks pernicious anemia will finish him. Prior to the liver cure, it was a death sentence, after all. He doesn't know how long the pills will work, and he has no idea how much time he has left. But one thing is sure: He needs to force a conclusion to his troubles with Brouwer. And his fear of imminent death is what drives the Great Math War to a head.

Brouwer makes plans to visit Göttingen again in 1926, but enthusiasm for intuitionism is now fully waning. This is partly because while many of the people who know Brouwer or know of him are willing to tip their hats to him, few by now feel that they can ever truly embrace intuitionism. They give it a forced smile, perhaps. One thumbs-up at best—or, failing that, a polite F-U. "I am very doubtful whether I could trust myself not to use the law of the excluded middle," Dutch American mathematician Arnold Dresden says in 1927. "But I am ready to admit that anyone who can, may do important work in that way."

People are also distracted by the emerging new field of quantum mechanics. Weyl's Zürich colleague Erwin Schrödinger has just published his famous eponymous equation describing the quantum mechanical wave function. (He later develops the concept of quantum entanglement, exemplified by Schrödinger's half-dead cat, who sits behind only Garfield, Felix, Sylvester, Jerry's Tom, Pooh's Tigger, Seuss's Cat in the Hat, and perhaps Simba in terms of feline fame.) And at Göttingen, the brilliant twentysomething postdoc Werner Heisenberg is about to shake the world by introducing the uncertainty principle.

Given that backdrop, even some of those who think the foundations of math are an important subject begin to lose interest. Many also begin to find the controversy distasteful, especially since Hilbert is so ill and appears to be dying. The basic feeling, by 1926, is live and let live. Let sleeping dogs lie. To each their own. *Mi summa, tu summa.*

———◆———

A few mathematicians in Göttingen, however, are motivated to resolve the conflict for Hilbert's sake if nothing else. They fear that the stress of the conflict between him and Brouwer will hasten his demise, and so they try to broker a peace. In 1926, they see an opportunity to do just that. Pegged to an upcoming Brouwer visit to Göttingen, they hatch a plan. Emmy Noether will host a dinner at her apartment for the two of them where this awkward kumbaya can play out.

THE GREAT MATH WAR

Noether's fortunes have changed somewhat. At the end of World War I, new laws in the Weimar Republic relaxed teaching bans against women. In 1919, she was finally allowed to take an *official* teaching position—more than a decade after many colleagues in her generation achieved such status. Now she no longer had to lecture under Hilbert's name or publish using a male pseudonym, as she had been forced to do during the war. Even so, she still had to work without pay. In 1922, she was promoted again—though again there was no salary. Finally, in 1923, she was given a small stipend.

As but one small measure of this glass-ceiling injustice, on June 4, 1919, when she gave her inaugural talk as a teacher (still unpaid at that point), she stood up and described what's now known as Noether's theorem, a follow-up work from her contributions to Hilbert's and Einstein's race for general relativity. Noether's theorem unites symmetry in nature with universal conservation laws, and some consider it a cornerstone not just of general relativity but also of a vast part of modern physics, including particle physics. Some consider it the backbone upon which all modern physics is built. On the basis of her work, Einstein himself will later call Noether the most creative female mathematician of all time—creative and yet unpaid for most of her career.

Brouwer knows Noether well. He's hosted her at his Blaricum hut—once for nearly a month while she was visiting Amsterdam. Now, as he visits her city in 1926, she returns the favor—and secretly hopes to reconcile him with Hilbert. Brouwer, Hilbert, and a half-dozen other mathematicians climb the stairs to her flat. The scheme is to steer the conversation toward common ground: cajole them into criticizing another mathematician they both despise. And it works. They dish. They nod. They eat. They toast. Tasty food. Niceties. Friendlier and friendlier, one person recalls. "The success of this undertaking surely exceeded our boldest expectations," one mathematician who is there will write. Hilbert and Brouwer have a high-spirited exchange.

Is it the night? The food? The alcohol? The company? The vibe? The setting? Their beloved colleague Noether, who is hosting? Whatever the case, the night ends with a glimpse of the colleagues they could've been rather than the enemies they have become. But what could have been is never found. The night doesn't end things at all. It becomes a blip of détente in the Great Math War.

———◆———

Why do I keep referring to this largely academic dispute as a "war"? It's simple: They all did. From the beginning, people use military metaphors to describe the war between Brouwer and Hilbert. Weyl describes Russell's Paradox as a border skirmish and talks of intuitionism as a revolution. Hilbert dismisses Brouwer's work and Weyl's effort to popularize it as a failed putsch, perhaps referring to Adolf Hitler's 1923 failed coup d'état, the infamous Beer Hall Putsch in Munich.

Nor do others at the time shrink from such language. In front of an audience at a joint meeting of the Mathematical Association of America and the American Association for the Advancement of Sciences in Nashville on December 29, 1927, one expert says that anyone in the audience should either fight or retrench. "It is necessary for mathematicians either to refute [Brouwer's] position decisively," he says, "or else to abandon the part of the field that is under attack and to retire to 'previously prepared positions.'"

Proximity to the Great War is part of what makes the Great Math War the entrenched attrition it is. And now, in 1927, the real shelling is about to start.

LEXINGTON, MAXWELL, BRISCOE, TEMPLAR

The last half of the 1920s is a time of firsts. Charles Lindbergh sets off on his transatlantic trip in 1927. Heavily weighed down with fuel and supplies, the plane barely clears the trees and telephone lines at the end of the runway as it climbs. "The soft glow came above the clouds, the first of the sun breaking through. Far off above the trees the silver wing dipped and was gone," an article in the *New York Times* says.

The 1920s did not see the invention of mass production, mass media, fast credit and easy financing, wild speculation in investments, and an amazing new must-have consumer culture built on fashion and tech—but it brought all those things to bursting fruition. New cars tooled, tooted, and belched. Those lost, forgotten models—Lexington, Maxwell, Briscoe, Templar—as listed in Frederick Lewis Allen's 1964 book. In the 1920s, the past was gone, the present was the future, and the future was...well...less sweet than bitter.

By early 1927, Weyl and Hilbert have at least reconciled their differences and both say so publicly. Weyl still tries to defend Brouwer, clarifying and pointing out the good things in intuitionism. But it's too little, too late. Hilbert has already forgotten Noether's dinner and sees his Dutch colleague as ever more a threat.

In 1927, Brouwer goes to Berlin and gives a series of lectures on the foundations of math that has his audiences buzzing with excitement. The talks cause "a considerable stir," according to van Dalen. Then in early 1928, he gives two lectures in Vienna. The lectures themselves are unremarkable, but the extraordinary thing is that Wittgenstein is in the audience. After the splashy publication of his famous book *Tractatus Logico-Philosophicus* in 1920, Wittgenstein declared that he had accomplished all he wanted to in the field. The problems he once concerned himself with "were solved now and forever" in his view, the Dutch mathematician Dennis Edwin Hesseling will write in 2003.

But after Brouwer's lecture, Wittgenstein rediscovers his taste for philosophy. He decides to return to studying philosophy after a long hiatus and begins sketching out new ideas with revived interest that same night. "I believe that evening marked [his] return to strong philosophical interests and activities," says one person who was there, according to Hesseling.

Brouwer, meanwhile, is still basking in the détente from Noether's dinner the summer before. Feeling friendly, he decides he wants to bridge the gap with Hilbert and throw him a bone. He writes Hilbert a letter claiming that the differences between formalism and intuitionism are but a matter of taste and not hulking substantive issues. Formalism is not wrong, Brouwer concedes, and their disagreement will vanish. In the future, he predicts, mathematicians will simply choose one or the other approach according to their taste.

But Hilbert slaps that olive branch away. He travels to Hamburg in the summer of 1927 to give a talk and vows to finally remove "any question as to the soundness of the foundations of mathematics." He attacks intuitionism, saying that the principle of the excluded middle is both logically sound and road tested. Clear. Comprehensible. *Misuse is precluded!*

As he speaks, people in the audience are disturbed—not so much by what he says as by how he says it. He shows visible signs of illness. His pernicious anemia has been under control, but now it's coming back. His lecture is painful to watch, like US President Joe Biden in the summer of 2024. His voice is soft. People struggle to hear him. They feel sadness. Remorse. Does he need more liver pills? Another batch is quickly ordered from Harvard.

In a strange twist of fate, Brouwer's wife, Lize, also develops pernicious anemia in 1927. On top of that crisis, the Brouwers face another challenge when the city of Amsterdam exerts eminent domain over their pharmacy, claiming the property for some municipal use. The Brouwers push back, making an emotional pitch on behalf of the two thousand patients they serve. Deliberations take place throughout 1928, and in the end, they lose. The next year, the building is demolished and they have to relocate. It won't be the only loss for Brouwer that year.

"MATHEMATICS KNOWS NO RACES"

In 1928, another International Congress of Mathematicians takes place—this time in Bologna, Italy—presided over by none other than "his excellency" Benito Mussolini. Okay, that's weird, but the organization does finally allow its decade-old boycott to lapse. German mathematicians are finally allowed back into the meeting after a sixteen-year absence, and Hilbert triumphantly leads a delegation of sixty-seven German colleagues to Italy. He gives one of the keynote talks, making a resounding statement about his beloved field: "All limits, especially natural ones, are contrary to the nature of mathematics." A written copy of his speech includes the phrase "Mathematics knows no races."

Not everyone is so positive. The anti-Germanic French mathematician Picard, for one, continues to bitterly push for the exclusion of Germans from the meeting—to no avail. And there is another source of protest.

THE GREAT MATH WAR

Some German mathematicians decide to boycott the meeting as well, hoping to torpedo it the way the 1920 meeting was damaged when they held a separate meeting in Nauheim.

The beef of these would-be boycotters is obscure, if legit. They are objecting to the fact that while German mathematicians are finally again allowed to attend, participate, and give talks at the meeting, they are nevertheless forced to register under the category of "observers" and not allowed to attend as "members." It's sort of like getting invited to a wedding and drinks but not the banquet dinner that follows.

It's a distinction that would be functionally meaningless if all you want to do is attend the meeting, but it still smacks of some sort of exclusion—especially since organizationally, the second-tier status may prevent some in Germany from fully participating in governance or administrative leadership. Several activist mathematicians are appalled at this perceived partial exclusion.

The most prominent protester is a German mathematician named Ludwig Georg Elias Moses Bieberbach. He rallies Brouwer to his cause, which is not a hard sell because Brouwer has a strong sense of justice and is a keen champion of what he perceives as righteous causes. He's also an "expert at nurturing grudges," according to the modern mathematician Hesseling. He supports the boycott and agitates to make it even larger and more impactful, writing a pamphlet intended to shame German mathematicians if they attend. Why should they? "Each mathematician should ponder for himself whether participation in the planned congress is possible without mocking the memory of Gauss and Riemann, the cultural meaning of the science of mathematics, and the independence of the human spirit," Brouwer writes.

Hilbert is aghast—horrified by the boycott and the pamphlet, calling it "political blackmail of the worst sort." He sees the congress reopening as a triumph, not a tragedy. And he believes, as many do, that the love of mathematics is universal, cross-cultural, and transnational. Even a superficial glance at the field suggests that many people from multiple nations take part in making discoveries, he says. *Mathematics knows no races!*

◆

NOBODY LOVES A REVOLUTION

In the end, the boycott fails miserably. Nobody cares. Everybody goes. The congress succeeds. And many are left scratching their heads over why Brouwer even involved himself in the first place. Running into Albert Einstein a few months later, physicist Max Born is puzzled about why Brouwer would make such a noise, since he's not even German and was never excluded in the first place. "Hilbert considered, as we all did," Born tells Einstein, that Brouwer's behavior was foolish—and bizarre. He was even more nationalistic than the Germans themselves.

Einstein holds his tongue. Brouwer can be as combustible as he is volatile, and Einstein knows this. He's done poking bears. Hilbert, on the other hand, is pissed. Once again, he sees himself staring across no-man's-land at none other than Brouwer. While in Bologna, he bends the ear of one of his fellow editors of the journal *Mathematische Annalen* about their Dutch colleague. Could they remove him from the journal's editorial board? he wonders.

Then, in the fall of 1928, returning home from the congress, Hilbert has another health setback and life-threatening scare. A bad relapse of pernicious anemia throws him back into the sickbed, hospitalizes him for five weeks, and places him again at death's door. When he finally recovers and checks out of the hospital, he's ready for war.

CHAPTER FIFTEEN

1928: THE BATTLE OF THE FROGS AND MICE

> "
> Life is a magic garden. With wondrous, softly shining flowers. But among the flowers the gnomes are walking & I am afraid of them.
> "
>
> —L. E. J. Brouwer

An etching from 1688 illustrating the battle of the frogs and mice—the *Batrachomyomachia*. A kite descends and kills the combatants. Cropped from a public domain image via Wellcome Trust.

Hilbert writes an angry letter to Einstein on October 15 asking for his permission to act as puppet master in an odd power play to oust his rival. Brouwer has got to go! Hilbert tells Einstein he plans to send a second letter to Brouwer the following week, thanking him for his service and dismissing him from *Mathematische Annalen*—a critical and necessary decision that is entirely in his power to make. "I would like to point out," Hilbert adds, "that my decision is firm and unalterable."

Under no circumstances will Hilbert continue to serve on the same editorial board as Brouwer, he says. He can't. This isn't a personal beef, he assures Einstein. It's a matter of principle. There's no personal pride or pernicious ego at play, Hilbert claims. This is preservation. Salvation. Righteous indignation. *It's him or me*, Hilbert basically says.

The journal was fine before Brouwer came, and they'll be better off with him gone. Hilbert and Einstein are two of four chief editors, along with the Greek mathematician Constantin Carathéodory and Hilbert's protégé Ludwig Otto Blumenthal. Dismissing Brouwer unilaterally is within their power, Hilbert says. And why not?

———◆———

History doesn't record Einstein's face when he reads the letter, but most likely he groans, furrows his brow, and rolls his eyes. Provoking Brouwer is not what Einstein wants. And firing him is no small provocation. It's not just like poking the bear—it's like slapping the hungry beast awake out of hibernation, shaking *Ursus amsterdamus* from his long winter slumber, eye-goring the monster, threatening his young, smearing your neck with honey, and sticking it in the bear's mouth. Why beg the beast to snap his jaws with your neck in his mouth? Firing someone, anybody, is no small thing. And Brouwer is *not* just anyone. He's not one to trifle with, and Einstein knows this.

But they need to fire Brouwer, Hilbert insists. He is an embarrassment to himself and to them. He humiliated Queen Math at the Bologna meeting by helping organize a German boycott of the meeting. Brouwer went to

1928: THE BATTLE OF THE FROGS AND MICE

the extra trouble and added expense of making, printing, and mailing his silly little subversive *Don't go to Bologna!* pamphlets. Sending them "insulted me and, as I believe, the majority of German mathematicians," Hilbert says.

To further support his case for Brouwer's removal, Hilbert claims that he spoke to Felix Klein about this issue three years before, just before Klein died. Klein agreed with him then, Hilbert says. So it's Caesar up, thumbs down, let in the lions, and *Avē Imperātor, moritūrī tē salūtant* (Hail, Emperor, those who are about to die salute you). Brouwer's gotta go!

Hilbert sends two more letters to the other two top-tier editors who serve with him and Einstein and then waits a week. One, the German mathematician Otto Blumenthal, is compliant in his response. The other, Greek mathematician Constantin Carathéodory, is silent, and Einstein is...well...*ai-yi-yi!*

THE SLEEPING GIANT

Einstein says he can't. He won't. He'll not sign off on the decision. No, no thank you, and hell no! He doesn't want any part of it. He knows Brouwer the sleeping bear well. They spent time together in 1920 when Einstein was visiting the Dutch Academy in Amsterdam. Like many others before him—Weyl, Noether, Carathéodory, and even Hilbert—Einstein was the honored guest of a gracious Brouwer at his Blaricum hut. He knows the man *too well*. A voracious mind. A great thinker. True and tireless. But also a bit of a maniac. From Einstein's perspective, some other, far less heavy-handed way should be found to deal with the problem. You should never shoot an angry bear with a paint gun!

Hilbert's insistence on reminding Einstein of how, just a few months before, Brouwer fought for the stupid boycott of the Bologna congress backfires. It only serves to warn Einstein further. Einstein knew that nothing was at stake personally for Brouwer in that lost cause. It was something that nobody really cared about, and yet Brouwer went out of his way to fight for it. How will he respond to something like Hilbert's attempt to fire him? This is a substantive issue—something that threatens him *directly*. Have you ever seen a vegan get really, really angry? It's not a pretty sight! Einstein's in no mood—not now, not ever. No way, nohow!

"I consider him, with all due respect for his mind, a psychopath," Einstein writes Hilbert a few days later. "And it is my opinion that it is neither objectively justified nor appropriate to undertake anything against him." In other words, *Please leave Brouwer alone, and for God's sake, leave me out of it!*

But the skids are too greased to grind the brakes at that point. Brouwer receives a telegram on October 27, 1928, begging him not to do anything until he hears from his friend Constantin Carathéodory, who should be coming by to see him in Amsterdam the following Monday. He wants to see you, the message says. He will "inform you" about something. "An unknown fact of the greatest consequence." *Don't do anything until you hear from Carathéodory*, the letter implores Brouwer.

"The matter is totally different from what you might believe," the telegram adds—if you have read the letters.

What letters?

―――◆―――

Brouwer has no idea what any of this is about, as he is enmeshed at that moment in his legal wranglings with the city of Amsterdam over its plans to tear down his pharmacy. He has little time to consider it. Still, he is curious when he goes to the post office and finds two letters addressed to him from Germany. But he's too busy to deal with them. Carathéodory will be there in a day or so to tell him what it's all about. He's happy to set the letters aside until then, and so he does. Heeding the warning, he puts the letters away unopened. He'll wait to hear from Carathéodory. They're good friends, after all. Carathéodory will clear it up. *Don't do anything until you hear from Constantin!*

When Carathéodory shows up the next day, they sit down and look at the letters together. One of them is the dismissal notice from Hilbert, thanking Brouwer for his service, curt and cold. "Because it is not possible for me to cooperate with you, given the incompatibility of our views on fundamental matters," the letter says, "henceforth we will forgo your cooperation in the editing of the *Annalen* and thus delete your name from the title page." Brouwer will be sacked, it essentially says in its own stilted, formal, awkward human resources way.

1928: THE BATTLE OF THE FROGS AND MICE

Brouwer is deeply shaken. He feels sick for days afterward. He runs a fever. This is far worse than anything he could have imagined the letter would contain. He had no idea anything like this was even being discussed, let alone in the works. He's not prepared to accept or even consider this outcome. As Einstein predicted, he's pissed. It's unthinkable! Brouwer expected to be an editor at the journal for years to come—decades. And why not? He turns on his friend Carathéodory in a venomous shoot-the-messenger reaction.

Carathéodory is caught off guard and deeply shaken by Brouwer's lashing out. There was some kind of communication mix-up. Carathéodory was expecting Brouwer to have already read the letter and to have had time to calm down before they spoke. Carathéodory was not expecting Brouwer to be dropped into the first Kübler-Ross stage of grief before his very eyes, exploding in rage at the world in general and at him in particular. But that's always the way it is with rage. Anger is a point source—heat that burns hotter the closer you stand to the flame.

That's how the meeting ends. Badly. Brouwer is hot and bitter, and Carathéodory is sour and scorched. "I consider this visit as a farewell," Brouwer says as they depart. And according to his biographer van Dalen, he adds, "I am sorry for you." Carathéodory soon resigns his position on the journal's editorial board.

WHAT WOULD EINSTEIN DO?

Einstein is not the only innocent bystander to fear goring the bear. Brouwer's stubborn hardheadedness is no secret to anyone. As the *New York Times* will report years later, in the same obstinate way that Wittgenstein once refused to admit there was no rhinoceros in Russell's office, Brouwer would always refuse to admit a thing existed unless an "explicit mental construction admitted to its existence." That was the whole basis of his constructivist mathematics.

And Brouwer's indignant intransigence is fed by ego as well. His best friend, the Dutch poet Carel Scheltema, who died from the head injury in 1924, saw this side of him. When Scheltema wrote a book of literary criticism in 1908 about an avant-garde early-twentieth-century literary movement known as "the Eight," Brouwer reviewed his friend's book for the academic outlet *Tijdschrift voor Wijsbegeerte* (Journal of philosophy). The journal rejected his review, however—much to Brouwer's chagrin. He was outraged. Brouwer wrote to Scheltema that the editors had dared to make notes and suggestions on his writing in the margins of his submission and sent those back to him. "Pencil scratches on statues of Donatello," Brouwer mocked. He later burned the marked-up copy and said, "I did not want it to survive the insult."

That bullishness of Brouwer the man is on full display in Brouwer the editor. His career has been a long hopscotch across a muddy peat marsh of scholarly skirmish—a splashy jump from one stepping-stone *foolishness-of-great-men* academic squabble to the next. In 1911, for instance, Brouwer fought with a mathematician named Henri Lebesgue over which of them deserved credit for a proof they had both developed separately. Lebesgue thought he deserved the credit, since he was first to drink from that discovery well.

Lebesgue came off as vain and snobby, looking to put a younger man in his place. Brouwer was defiant and indignant—if not righteous and outraged. The proof was in the putting, he basically said. First or not first didn't matter, Brouwer claimed. Lebesgue's proof was faulty. Bad. *False and incomplete!*

Blumenthal was just starting out as an editor at the journal *Mathematische Annalen*, when the dispute was playing out in 1911 (the same journal Brouwer would later join and Hilbert is conspiring to fire him from in 1928). Blumenthal was immediately thrust into the middle of this fight. He became embroiled in it, even though he tried to stay officially objective.

In a letter to Hilbert at the time, Blumenthal confided that he found Lebesgue more compelling. Lebesgue's work was more accessible, for one thing. Topology was an emerging field—one of the very reasons the journal would hire Brouwer as an editor a few years later. But in 1911, there was no such subject matter expert on staff, and Blumenthal admitted to Hilbert that he couldn't even really understand Brouwer's work because of his atrocious writing style. He struggled as best he could to mediate, but one thing he didn't like was Brouwer's tone. Who did he think he was?

1928: THE BATTLE OF THE FROGS AND MICE

His correspondence, Blumenthal said to Hilbert, had "in my opinion an unfriendly and unpleasant ring."

Blumenthal decided to write to Brouwer as mediator. Your note to Lebesgue was very rude, he said, and easily open to misinterpretation. Your accusations that his proofs were false and incomplete were too much, he added. They smacked of hyperbole and were illogical besides. False and incomplete are not the same thing, after all, and they seem at odds with one another. (Like that old joke: "The food in this restaurant is awful! / Yes, and the portions are so small.") Nobody contests your priority for your own fundamental proof, Blumenthal said, so why not reconsider your letter?

Brouwer rejected the advice. He wrote to Blumenthal several times that summer to complain about Lebesgue—of his lack of politeness and of how his proofs were so bad that they "hardly, in my opinion, merit that name [proof]," he wrote. Here is my interpretation of Brouwer's perspective in a spike of a few lines:

> *Lebesgue, oh, Lebesgue—*
> *(Rhymes with Tsk! Tsk!)*
> *While he's in motion,*
> *his brain's at rest.*
> *You can barely*
> *even call his*
> *stuff proof.*
> *More like*
> *a dumb*
> *goof!*

THE TROUBLES WITH BROUWER

The Lebesgue affair was just one of several squabbles that Brouwer fell into before World War I. In 1909, he had an angry, argumentative correspondence with Arthur Moritz Schoenflies, which Hilbert had to mediate. Before that, he had a separate run-in with the mathematician Paul Rudolf Eugen Jahnke, who saw one of Brouwer's earliest papers.

The Jahnke affair was very eye-opening because it fully revealed Brouwer's nature, even as a graduate student. Jahnke was hardly rude. He noted that while Brouwer's methods were new and original, his results were the same as work he himself had previously published. Brouwer must not be aware of his work, Jahnke said. Maybe he could add a "short note in the same journal and if possible in the next issue to acknowledge my priority for the mentioned results," Jahnke suggested. But Brouwer was unrepentant. "I see my treatise interests you," he responded to Jahnke. But then he dropped a bombshell, claiming not just priority but supremacy. "Your final result is a byproduct of my principal," he said.

In his own notes at the time, Brouwer wrote, "When [Isaac] Newton found the laws of attraction and deduced the laws of Kepler from it, Kepler would not have wanted to didle the credit from him." (A didle is a long pole used for scooping up muck in canal dredging. It's a noun as well as a verb—and a uniquely Dutch insult.)

In early 1912, Brouwer exchanged three tense letters with Friedrich Engel, another elder statesman in the field, and the troubles with Brouwer continued even after the war. Some became convinced that Hilbert's mentor and colleague Klein had resigned from the *Mathematische Annalen* because he was so exhausted by Brouwer's incessant squabbles.

All this is to say it's not unreasonable to claim that there is more going on in 1928 than just Hilbert's feelings about intuitionism boiling over. Brouwer has made a number of other enemies. He has caused trouble more than once—all of which becomes Hilbert's implicit justification for dismissing him. At the same time, each of those episodes was relatively minor. Even together, they are a paper-thin cover Hilbert simply uses to justify ejecting Brouwer from the editorial board. But use them he does. The journal will be the better without Brouwer, Hilbert says. Demand is high for spots on that prestigious board. Becoming an editor of the journal is a high honor—one of the highest. The brightest minds in the field. He and Einstein are both editors. Who wouldn't want to be one as well?

Still, could anyone wonder at the truth of the matter? What really motivates Hilbert in 1928 are his 1900 dreams of redeeming Cantor and saving Queen Math. "No one will drive us out of this paradise that Cantor has created for us," Hilbert says. Formalism will secure it, and he seeks to remove any barrier that threatens that paradise. And if that means ejecting Brouwer from the journal, he will.

"OF UNSOUND MIND"

Brouwer, striking a typically indignant reactionary posture, takes it all as an unjust insult. He's an extraordinary editor. Everybody *must* know this. And he is being wronged by Hilbert. He literally wrote the book in his subject-matter area of topology. He works hard to understand all the papers he edits. He corresponds extensively with the authors. He discusses changes at length. Okay, he's slow and difficult—but in a good way. Demanding but not troublesome. Thorough without the least hint of obstinacy. Comprehensive though not unreasonable. Penetrating but never petty. He's not just a gifted editor. He's God's gift to editing.

And now he's outraged. This isn't just an egregious insult—it's actually illegal. The *Annalen* belongs to its editors, including him. It's *their* collective responsibility together, not the property of one man. He is on the masthead along with everybody else. No one editor can "embezzle" the authority of the entire editorial board, he thinks.

What is Hilbert thinking? He's someone Brouwer has long admired, after all. Brouwer calls him "the first mathematician of the world." He considers Göttingen his "second scientific home," according to van Dalen. He considers the other two chief editors, Carathéodory and Blumenthal, close friends. And Einstein as well, of course.

But there's something else. This firing hurts Brouwer to his core in ways no one could have expected. One of Brouwer's friends later notes that he leads a simple, beautiful, and pure life. Sunshine. Exercise. Mathematics. Hard work. Light supper. Sleep. His position at the journal is a big part of this normal daily routine. The threat of losing his position on top of losing his pharmacy is too much for him to bear, as if someone had knocked down his hut at Blaricum.

So when he sends a short note by way of formal response to Blumenthal and Carathéodory on November 2, 1928, it drips with poison. "After close consideration and extensive consultation," he says, he can only conclude that Hilbert is "of unsound mind." He insists on seeing a doctor's note and being provided with a written statement from Hilbert's wife regarding his mental state before proceeding further.

Next he sends letters to all the other editors and to the journal's publisher, spinning the story and appealing to their honor if not their sense of outrage. The real reason why Hilbert wants to dismiss him is nefarious, he says—an angry wish "to harm and damage me in some way." It has nothing to do with Brouwer's editing, of course. In Brouwer's version, the spat goes back eight years to when Hilbert tried to recruit him to Göttingen in 1920. Brouwer said no. He spurned Hilbert, and now the old man is trying to make him pay. Sour grapes and all. "Hilbert had developed a continuously increasing anger against me," Brouwer writes.

Hilbert's actions are those of a sick man, he writes in letters he sends to the other editors. He claims that Carathéodory admitted as much to him when they met a few days before in Amsterdam. And he suspects the only reason nobody has spoken up yet is because of Hilbert's failing health. Who wants to kick a dying man who has one foot in the grave?

"Carathéodory begged me," Brouwer says, "out of compassion with Hilbert, who was in such a state that one could not hold him responsible for his behavior, to accept this shocking injury in resignation and without resistance."

Einstein probably faints when he reads this last letter.

And yet Brouwer is not done. He follows up by sending yet another letter—this one directly to Käthe Hilbert. "I beg you, use your influence on your husband, so that he does not pursue what he has undertaken against me," Brouwer writes. "Not because it is going to hurt him and me, but in the first place because it is wrong." *In his heart, he is too good for this!*

Brouwer sends a copy of the letter to Richard Courant, asking him to follow up with Mrs. Hilbert. Courant does. And it's awkward. He writes back to Brouwer to say he's met with her. She has nothing to do with any of this. Nor does she think she can influence her husband. She said he makes his own decisions, Courant reports.

Managing Editor Blumenthal is friends with Courant as well, so he hears everything. Blumenthal is outraged. He calls Brouwer's letter to the editorial board "frightful and repulsive." He loathes the letter to Hilbert's wife. He has his own issues with Brouwer, dating back almost twenty years

1928: THE BATTLE OF THE FROGS AND MICE

to when he was in the middle of Brouwer's nasty 1911 dispute with Lebesgue. Now, all these years later, he sees Brouwer up to his old tricks—only nastier, more unpleasant, overtly unfriendly, and even cruel. Just awful. Blumenthal tries to stay officious. An objective mediator. But he finds the whole thing intolerable. How dare Brouwer! Pity poor Hilbert. So sick for so many months, and now this.

◆

Blumenthal knows exactly what to do. He jumps into action. Hearing that Carathéodory plans to resign, Blumenthal asks him to keep his resignation quiet so it won't appear as if he's resigning to signal support for Brouwer's case. And he immediately writes letters to all the other editors and the publisher asking them not to engage with Brouwer until he has had the chance to prepare a proper response. He drafts his response over the next week and sends it to Courant to review.

By that time, however, Brouwer is already on the move. He takes the train to Germany, steps off onto the platform in Berlin, and meets up with his friend Bieberbach, his main coconspirator in boycotting the International Congress of Mathematicians meeting in Bologna earlier that year. Bieberbach, like Brouwer, is an editor at the journal, and he agrees to return the favor Brouwer did him in supporting his Bologna protest by becoming Brouwer's key ally in his firing. Together they show up unannounced at the journal's business office in Berlin on November 13, 1928.

Publisher Ferdinand Springer considers ducking out the back door but thinks better of it. He's well aware of what's happening, of course, and that gives him pause. Meet with Brouwer and Bieberbach—*ai-yi-yi!* But maybe he should. Snubbing them might provide Brouwer with further ammunition for a public smear campaign. Besides that, Springer is a businessman. *Prepared to place everything on a single card!*

So Springer invites them in. He sits down, and he hears them out. But he makes no secret of the fact that he's siding with Hilbert "out of friendship and admiration." He also holds firm to a convenient public posture of editorial independence. He says that in his position as business head of the journal, he can't get involved. He shouldn't even be weighing in on such matters.

Brouwer and Bieberbach push back, insisting that Springer clarify Hilbert's legal standing. Can Hilbert as chief editor actually make such unilateral decisions? Springer says he can't comment without reading Hilbert's contract, and he's not prepared to do that this afternoon. He probably needs to talk to his lawyers first as well.

Okay, okay, Brouwer and Bieberbach say. But what about arbitration? Could he appoint a mediator? Maybe, Springer says. Truth be told, he adds, his company owns the journal as a legal entity, but intellectually, it has never really belonged to them. It's always been Hilbert's—just as it was always Klein's before him.

They leave, and Springer is left stunned.

BITTER, MEAN, AND VINDICTIVE

Brouwer is "an embittered and malicious adversary," Springer writes in an internal memo that he dashes off that afternoon. Brouwer and Bieberbach threatened him and threatened the journal, he writes. They are threatening to "damage the *Annalen* and my business interests," Springer says, insisting that they'll resign, take other editors with them, and set up their own journal in competition. Maybe that's not such a bad idea, Springer considers. It could end the Great Math War with the least amount of spilled blood. He won't tell them that, though. Ever the businessman, Springer plays it conservatively. Cautiously. Coolly. Closely. Hackles up. Cards down. Radar's on. Still, he admits, "founding of a new journal, wholly under Brouwer's supervision, would be the best solution to all difficulties."

Springer comes away from this ambush meeting girding for a major fight, which is where things appear to be heading. That's what Brouwer seems to want. "It seems," Springer predicts in his memo, "that he will carry the fight to the bitter end."

The lawyers become involved at this point. The publisher's lawyers tell Springer exactly what he wants to hear: He has a contractual relationship with the four chief editors *only*—Hilbert, Einstein, Carathéodory, and Blumenthal. Not with Brouwer! These four main editors, the lawyers agree, are fully empowered to fire Brouwer. (Though this is made more difficult if Einstein wants to stay on the sidelines through the whole affair and decline to make the vote of the chief editors unanimous.)

An outside legal adviser, who is a colleague of Carathéodory at the University of Munich, offers other advice. The lawyer says Brouwer actually *does* have a contractual relationship directly with the journal, at least implicitly. There's a good case to be made for that because he is compensated for his work. The lawyer concludes that dismissing him without appropriate grounds could be illegal under German law.

The only other possibility, the outside lawyer says, is what we would call today a "nuclear" option: Dissolve the editorial board entirely. That would avoid the dangerous litigious tangle of a potentially extralegal firing. There's nothing unfair in firing everybody. Springer's house lawyers agree.

So plans are made behind the scenes to dissolve the journal's entire editorial board. But first, Blumenthal wants to make the case to the editors, hoping to resolve the issue through a more bloodless coup. He prepares an indictment of sorts. At this point, he is driving the whole affair. Hilbert is hiding. Einstein is fleeing. And Carathéodory is wilting. It's left to Blumenthal to make it happen.

———◆———

Blumenthal is emotionally prepared to prosecute the case, since he's genuinely offended by Brouwer's nasty comments. "I have thoroughly misjudged Brouwer's character," he says in a letter to Carathéodory, recalling Brouwer's letter to Hilbert's wife a few weeks before. Repulsive. Only Hilbert has clearly seen all the flaws in Brouwer's character these many years, he concludes. "Hilbert has known and judged him better than we have." *Headstrong! Unpredictable! Domineering!* (Nothing helps fuel anger so much as righteous indignation.)

Still, when Blumenthal sends a letter to the entire editorial board explaining Hilbert's position, it comes off less unconvincing. Hilbert fears that Brouwer will bend the journal "to his will," Blumenthal writes. "And he has judged this such a great danger for the *Annalen* that he wanted to stand in his way as long as he could do so."

THE GREAT MATH WAR

Blumenthal describes the troubles with Brouwer. His arguing with other contributors before World War I. His slow pace at reviewing papers. That nasty letter he sent Hilbert's wife a few weeks before. All this reveals the sort of man Brouwer truly is. "I must confess," Blumenthal concludes, "that I have been thoroughly deceived in Brouwer's character." Just like Hilbert, he says he wants to be free of him. "I too am no longer in a position to cooperate further with [Brouwer]."

The letter lands with a dull thud instead of a wet splash. Almost no other editor responds. And why would they? Who even knows what to make of this tempest in a teapot? Three of them do respond—Bieberbach, of course, as Brouwer's main defender. He argues strongly against removing Brouwer, reiterating that it's not just wrong but illegal. The other two responding editors don't really say much. One indicates that he doesn't agree with the removal but asks to be left out of the whole affair. The last one says he also cannot approve of Brouwer's removal, specifically because he doesn't believe being a slow editor and contributing to delays in publication is sufficient grounds for dismissal. (Long delays have traditionally been par for the course in scientific publishing, and in fact, some journals today premise their business models on rapid review and shorter time to publication—or have special express publishing tracks.)

Everyone, meanwhile, wants to know what Albert Einstein thinks. Behind the scenes, several people lobby Einstein, hoping he'll take Hilbert's side. They see in Einstein a key figure—not just in the dispute or for the journal overall but in life, if not human history. Einstein wears his godlike fame like a second skin, whether he intends to or not.

But Einstein doesn't take the bait. He wants nothing to do with any of it. Even less now than in the beginning. And why would he want to get involved? The whole thing is aimless and petty. He calls the affair the *Frosch Mäusekrieg* (war of the frogs and the mice). Surely some other way could have been found to deal with the problem—ideally one that would demand no input or attention from Einstein.

"If Hilbert's illness did not lend a tragic feature," Einstein writes, "this ink war would for me be one of the most funny and successful farces performed by people who take themselves deadly seriously."

1928: THE BATTLE OF THE FROGS AND MICE

It's not just that. Recall that Einstein knows Brouwer to be something of a madman. They've already seen evidence of that in the letters Brouwer has written and his impromptu visit to Springer's office. Einstein can sense the sleeping bear is stung. There's a long list of mathematicians Brouwer has beefed with over the years, Einstein knows. He will not add his own name to that list. He sees Brouwer as "an involuntary advocate of Lombroso's theory of the close relationship between genius and madness."

PETTINESS IS A PAPER LANCE

"Allow me to stick to my role as astounded contemporary," Einstein says, articulating his neutral-to-the-point-of-being-almost-invisible position. In other words, *Leave me out of it!* He has no taste for it, writing to a friend near the end of 1928, "I do not intend to plunge as a champion into this frog-mice battle with another paper lance."

Instead of choosing sides, Einstein decides to resign from the journal. But by then, the wheels are already spinning. The Rubicon has been crossed. Icarus has fallen. Blumenthal and Springer have already decided that there is no choice but to pull the trigger, initiate the nuclear option, and dissolve the editorial board. Blumenthal begins putting that decision into place, crafting his exact moves with the help of lawyers. On the pretense of claiming to want to reorganize the journal, they dissolve the board, rehire Hilbert on a new contract, and leave it to him to choose a new panel of editors.

Blumenthal convinces Einstein to at least play along with the dissolution and keep quiet on the whole matter, since he is now being fired along with everyone else. Einstein does as he is asked. He wants nothing to do with it anyway. And with the whole editorial board dissolving, including him, it doesn't really matter anyway.

As 1928 rolls into 1929, Blumenthal writes another letter to the journal's editorial board, this one cosigned by Hilbert and Springer, informing the editors that the board has been dissolved, effective immediately, and will henceforth be reconstituted de novo. *The board is dead! Long live the board!*

The triggering event for this shake-up, Blumenthal's final letter claims, is the fact that the journal is finishing its 100th volume. (Because everybody knows meaningless milestones always demand wholesale reorganization.) Nobody buys it, of course, but it doesn't matter anymore. In early 1929, the deed is done. Brouwer is out—as is everyone except Hilbert.

Few are happy with the way things turn out, especially Brouwer. "I would consider a possible dismissal from the editorial board not only a revolting injustice, but…in the face of public opinion as an offending insult," he writes to Carathéodory.

Others are shaken. Carathéodory considers moving to California, where he has been offered a position at Stanford University—a place to run and hide. Einstein never returns to the journal, though some hope he will. He just moves on. Blumenthal begins to reconstitute the editorial board early in 1929—sans Einstein and sans Brouwer.

———◆———

One of Brouwer's biographers says the whole affair deeply affects him. It leaves him bitter and disillusioned and forces him into a sort of brief, sulky retirement. "He did not give up his mathematics," van Dalen writes, "but he simply became invisible." Brouwer stops going to meetings. He stops publishing. And he withdraws into his other activities.

To add insult to injury, Brouwer is standing on the platform of a Brussels train station one day in 1929 when a thief slips by, lifts his briefcase, and walks away with his bag and all its contents. Tucked away inside this briefcase is one of Brouwer's notebooks—a scientific diary of sorts filled with three years' worth of work, drafts, thoughts, and innermost ideas. Many of these he has never written down anywhere else.

"This event means for my scientific personality a serious personal mutilation in a way that is like the 'decapitation' for a pine tree," he writes to a friend.

The police say there's almost no hope of recovering the stolen case. Still, Brouwer does what he can. He hires a private detective. He takes out ads in local papers. He canvasses the area. He offers rewards. He even goes to see a clairvoyant. "But so far," he writes to a friend, "no success."

1928: THE BATTLE OF THE FROGS AND MICE

Some in the mathematics community gossip endlessly about Brouwer and about the affair. One physics professor at Göttingen calls the whole thing a scandal bomb—the closer you are to its epicenter, the more destructive it seems, but the farther away you are, the more humorous it appears.

The one exception is probably Hilbert. Even though he absented himself to ensure that his fingerprints were not on the executioner's axe, he must be happy about the whole thing. "Everything was glorious," he says to Blumenthal about his colleague's handling of the affair.

If Hilbert is flying high, however, he is too close to the sun, and he is about to fall full Icarus. He's just won a battle, but he's about to lose the war.

CHAPTER SIXTEEN

GOODBYE TO ALL MATH

1929–1932

> "
> Regard the present with satisfaction, and anticipate the future with optimism.
> "
>
> —Calvin Coolidge, 1928

Unemployed men queued outside a Depression-era soup kitchen opened in Chicago by Al Capone. Cropped from a public domain image via US National Archives [306-NT-165319c].

The 1920s are the decade when incomprehensible horror gives way to paper-thin hope for the future, thickens into irrational exuberance, and then slips into overwhelming dread. The decade begins with the Spanish flu and the squeamishly unsound Treaty of Versailles and ends with Black Tuesday carnage and the liquidation of the stock market.

In his last State of the Union address, one month after the 1928 election, lame-duck US president Calvin Coolidge reflects on the strength and clarity, if not the beauty, of the US economy. Herbert Hoover has just been elected thirty-first president of the United States, but Coolidge is not going to miss one last occasion to toot his own horn. His strength, America's strength, is his policy. Contentment! Tranquility! Harmonious relations! Freedom from industrial strife!

Coolidge gives himself a world-class verbal self–fist bump for his foreign policy. Peace. Goodwill. Mutual understanding. Manifest friendship. "The requirements of existence have passed beyond the standard of necessity into the region of luxury," he says. World peace and financial prosperity, those twin pillars of democracy. A firm foundation for the American dream. "No Congress of the United States ever assembled, on surveying the state of the Union, has met with a more pleasing prospect than that which appears at the present time," he says. Never in the history of American politics were more infamous famous last words ever spoken.

Nine months later, the liquidation of the US stock market on Black Tuesday, October 29, 1929, becomes one of the most devastating days in human history. It's worse than anything anybody's ever seen. Terrifying to behold. Indescribable. Most blue-chip companies close the day at only one-third of their morning value—if they're lucky. The defining feature of the crash, according to Canadian American economist John Kenneth Galbraith, is that it just keeps getting worse. Every day and every hour, people become convinced that the plummeting market has hit rock bottom, only to see it fall further.

The effect on the US economy is devastating. Long-lasting. US gross domestic product shrinks by a third. Unemployment balloons. Even ten years later, 20 percent of Americans are still out of work.

GOODBYE TO ALL MATH

CONTENTMENT, TRANQUILITY, AND FREEDOM

The US stock market crash is a global crisis as well, and it forms a sobering backdrop to a jubilant celebration in Hilbert's Germany a few months later, when Hilbert turns sixty-eight. Having reached his institution's mandatory retirement age, he gives a final farewell lecture at the University of Göttingen on January 23, 1930. Crowds cram the halls. The city of Göttingen names a street after him. The Great Math War appears to be over. The old East German city of Königsberg, where he spent his childhood and youth, piles on even more honors, inviting him there the following summer to give his final farewell public address.

Königsberg in summertime. He spent his youngest years in this rotten little river town (later renamed Kaliningrad in 1946 after the old Bolshevik Mikhail Ivanovich Kalinin). When Hilbert was born, the city was of particular significance to mathematicians since it was the site of one of the most famous math problems of all time, the solution to which was one of the towering achievements of the eighteenth century.

In the 1700s, a number of bridges spanned the Pregel River, the main body of water upon which the booming city was first built in the thirteenth century. In the middle of the river was an island, and a little way downstream from the island, the river split into two separate waterways—the Alter Pregel and the Neuer Pregel—forming a triangle of mainland that pointed back to the island. A total of seven bridges spanned the river. Four of them connected the riverbanks to the small island, two on each side. Two more downstream crossings connected the outer banks of the Alter and Neuer Pregel with the triangle of land in the middle, where the river splits. Finally, one more bridge connected that tranquil, triangular spit of land with the small island. Seven bridges in all.

The mathematical problem is simple to state but hard to solve: Is there a way to walk across all seven bridges once and only once? Can it be done? Most people say no. In your mind, if you start tracing paths, none of them seem to work, and it doesn't take long to prove to yourself that you can't. You always have to cross at least one bridge twice—unless maybe you jump in the Alter Pregel!

But mathematics is special among human endeavors in that it bestows the ability to transform wet, cold supposition into warm-blanket proof.

THE GREAT MATH WAR

Almost a hundred years before Hilbert was born, Swiss mathematician Leonhard Euler solved the seven bridges problem by proving it can't be done. There's no way to cross every bridge once and only once—not without strapping on Icarus wings or getting your socks soaking wet.

Euler is widely regarded as the most prolific mathematician in history and considered by many to be, in fact, history's greatest (although that's another one of those Julius Caesar versus Queen Victoria wild-card matchups). In any case, he was almost unapproachable in his genius. He extended the work of Isaac Newton and Gottfried Wilhelm Leibniz and used calculus to explore physical phenomena like vibrating strings. He coined the Greek letter pi as shorthand for the transcendental number 3.14159.... He was the grandfather of topology, the inventor of complex analysis, an early innovator of number theory, and the mathematical chef who set the table for centuries of quantitative feasting to follow. Far from the least of his accomplishments, he solved the seven bridges riddle and won worldwide fame in his day for it. What was so remarkable about his solution was not just *that* he solved it but *how* he did it.

He swept aside the obvious, stupid iterative approach that everyone contemplates when they look at the problem: sketching imaginary banks in their minds and sampling combinations among the more than five thousand possibilities. Euler took a different approach, something nobody in the history of math had ever thought to do. He simplified the problem by representing it as an abstract graph, and in the process, he laid the foundations for two entirely new fields that continue to thrive to this day: graph theory and topology. Topology, "the geometry of distortion," which explores fundamental geometric properties that don't change when we stretch, twist, or squeeze objects, is the same subject that would be made even more famous by Brouwer 150 years later.

———◆———

For some, Euler's legacy was as cultural as it was mathematical. That rich local math history legend was still alive when young Hilbert was growing up in Königsberg. Euler's example was profound for the boy, inspiring him the same way a guitarist growing up in the 1970s in East Liverpool would be inspired. He could see the boyish faces of barefoot Beatles in every crosswalk.

As young Hilbert walked around town, stubble-faced Euler must have appeared everywhere. Greatness around every corner. What Euler thought. What Euler did. How he tackled tough problems head-on. Do it. Solve it. The seven bridges problem is the classic example. Euler jumped in feetfirst and invented a math raft, floating new ideas. Those bridges were still there when Hilbert was a child. The rivers gushed Euler's name if you listened. That dude walked on water.

Inventing new math was exactly what Hilbert did best for his entire career, and he succeeded wildly. But the problems Hilbert wrestled with were far harder than even Euler could have imagined. During the Great Math War in the 1920s, and while he was sketching out his program, he was trying to do a lot more than just find a one-off selfish path across a bunch of goofy bridges. There was *a lot* more at stake than that. (Hilbert was never great at building bridges anyway.)

Now, in 1930, at the end of a long and accomplished career that would make even Euler proud, Hilbert is in the city of his birth, being honored for his career achievement—pomp for the great Queen Math and her sunset prince.

"Our entire Modern culture, insofar as it rests on the penetration and utilization of nature, has its foundation in mathematics," Hilbert tells the jubilant crowd at his celebration. Königsberg feels good. It's a victory lap. A final bow. The sun is shining on the city that summer, and it's shining on Hilbert as well. The mayor presents him with the key to the city. He drinks in the crowd. Glasses are raised. Toasts are made. Anything is possible. He knows this.

Hilbert ends his final speech exactly where he landed with his most famous one, an echo of the exuberant solutionism contained in his 1900 Paris talk. Königsberg is an echo of that talk and a strong reiteration of his lifelong conviction. There shall be no *ignorabimus*, no not knowing, he declares. That was his main message thirty years ago, his chief life lesson ever since, and it's his final thought now. What he said in Paris thirty years before, he now says again, and he and steps away from public life: *Wir müssen wissen, wir werden wissen* (We must know, we shall know).

THE GREAT MATH WAR

Hilbert might as well have said, *We won*. Brouwer is defeated, deposed, and depressed. And here Hilbert is, sun in his face, applause in his ears, fancy key-to-the-city ribbon pins on his chest, and Queen Math in his heart. Okay, then maybe it's *He won*. After his speech, he's whisked away for a live radio interview. He steps into a cab and away from math.

That could have been the end of the story. But at almost the exact same moment, another mathematician stands up in a stuffy lecture hall across town in Königsberg during a scientific conference taking place that week where three members of a panel of young experts gives a talk on logicism, intuitionism, and formalism.

Following that is a Q&A session, and that's when it happens. Something nobody expects. A bombshell. *We can't know, we shan't know!*

GRUNDZÜGE DER THEORETISCHEN LOGIK

Step back a few years to the summer of 1927, when a young mathematician from Vienna named Kurt Gödel became interested in logic in general and Russell and Whitehead's book in particular.

A year later, he wrote a letter to a friend saying he'd finished the book but was somewhat disappointed with it. Oh, sure! He had heard everything about the book. Its reputation—their reputations. A Herculean task to write—a miracle, bound and printed. All that was true. But the stature of the work was also puffed up by *bought-it-but-never-read-it* enthusiasm. The massive weight of perceived importance is always somehow heavier in the mind, and never was this more true than with Russell and Whitehead's famous work.

"I was less enthusiastic than I expected," Gödel wrote. The reason for his disappointment, according to Finnish philosopher Jan von Plato, is the same one Hilbert and Brouwer identified earlier: that the *Principia Mathematica* ultimately failed to establish the foundations of math through formal proofs.

So Gödel moves on. He is soon reading Hilbert's work as well, burying his head in the pages of a book titled *Grundzüge der theoretischen Logik* (*Principles of Mathematical Logic*), which was just published early in 1928 based in part on David Hilbert's 1918 lectures on logicism in the closing days of the Great War.

GOODBYE TO ALL MATH

One of the things the book does is to describe interesting mathematical questions that need to be answered, one of which is proving the completeness of first-order logic, or "predicate logic." This area of logic is the most familiar to most of us—it basically involves assertions, logical *connectives* like "and" and "or" and quantifiers like "there is" and "for every." Proving the *completeness* of predicate logic means showing that anything you can prove using those first-order connectives and quantifiers is in fact true. The reason Hilbert described the problem in his 1928 book is that it would go to support one of his crucial assumptions for formalism: that all axiomatic systems are complete in the sense that they will be powerful enough to allow any result to be derived.

This, of course, gets back to and is informed by Hilbert's fundamental *everything-can-be-solved* conviction—his exuberant solutionism, which he first described in Paris in 1900 and which becomes the theme of his farewell lecture in Königsberg. It's a compelling conviction: that if you can logically formulate a problem and apply enough time and energy to it, you will eventually solve it. And Hilbert is a true believer. The whole point of his formalism program in the 1920s is striving to reach the state of solvability. And the important first step along the way is to prove the completeness of first-order logic.

It's an odd challenge for young Gödel to take on, however, because everybody already assumes that first-order logic is complete in 1928. Hilbert knows it still needs to be proven. He needs that solid-bedrock-beneath certainty along the way to proving anything. Still, few people would bother with the Herculean task of exhaustively proving something they already believe to be true—and fewer still would even know how to do so. But Gödel takes the baton and runs with it. When his PhD thesis appears a year later, in 1929, he has done just what Hilbert asks, proving the completeness of predicate logic.

Hilbert is thrilled. It's a tremendously positive sign for his program. Accomplished in just one year—and by a graduate student, no less. He's impressed. Now he can retire, satisfied that as he fades, his program will thrive.

But a few months later, the same week that he gives his farewell *we must know, we shall know* speech in Königsberg, the Second Conference on

THE GREAT MATH WAR

Epistemology of the Exact Sciences features a Q&A session after a panel combining three presentations by three leading scholars of the three major foundational camps in the Great Math War: formalism, logicism, and intuitionism. Gödel stands up during that Q&A session and steps into history. He articulates a result nobody sees coming: Classic *logic* is complete, he says, but classical *mathematics* is incomplete.

He's only twenty-four years old, but he's about to get a whole lot older.

◆

Ten days before the conference, Gödel meets his friend Rudolf Carnap at the Cafe Reichsrat in Vienna. They are both on their way to Königsberg, and in discussing their upcoming trip, Gödel tells Carnap what he's been thinking. Carnap has trouble understanding him. So too, it seems, do most people at the conference when Gödel stands up ten days later and drops his bombshell: There are true statements you can make using the logical tools Russell and Whitehead established in *Principia Mathematica*, Gödel tells the crowd. But some of them are neither provable nor disprovable using those same tools.

Gödel says it's possible to create a true formula that claims of itself that it cannot be proven—something like *This statement cannot be proven*. Such a statement can be proven only if it is, in fact, itself a lie. But since we start with the premise that it's a true statement, we know that's not the case. Then, since the statement is true by definition and it asserts that it can't be proven, this shows that classical mathematics is incomplete. Why? Because it contains at least one statement P for which neither P nor ¬P (not P) can be formally proven. Basically, this is the idea of what becomes known as Gödel's incompleteness theorem—which will win him everlasting fame. He also develops a second incompleteness theorem, which says that if a system is inconsistent, it cannot be proven consistent using its own inconsistent means.

Gödel's results come as a complete shock. Nobody has ever suspected that you could actually prove that there are unsolvable problems. And his articulation of incompleteness is a moment of singular importance in the three-thousand-year history of logic. Some call him the most important logician since Aristotle (though the same is sometimes said of Russell earlier and Frege before him).

In the days and weeks that follow, rumors of the groundbreaking discovery spread quickly. "What is going on with Mr. Gödel?" German philosopher Heinrich Scholz writes to Gödel's friend Carnap in early 1931. "I hear all kinds of exciting things, but I cannot make out what this is all about."

He's not alone. As far-reaching and ingenious as Gödel's work is, it's also "very fiddling and difficult to grasp," in the words of British philosopher J. R. Lucas in 1961. A formal written record of the Königsberg meeting glosses over what Gödel says, no doubt because its significance is unappreciated in the moment. But as word spreads, and as the months roll on, people will begin to ponder, study, and reluctantly, if confusedly, accept his work. It's highly technical, employing bizarre methods and coming to startling, sweeping, and unexpected conclusions—so much so that some greet incompleteness with suspicion and horror. Others just want to vomit.

"ASTONISHING AND MELANCHOLY"

Gödel's two theorems basically end Hilbert's argument *We can know, we will know*. They deflate formalism, and they destroy logicism in its claim (or rather hope) to be able to secure the solid foundations of mathematics. Incompleteness, in short, kills the foundational crisis and draws the debate to a close in confusion and ignominy. The Great Math War is not won, really, so much as it's lost. Loss of the ability to prove anything under a mathematical sun. Loss of the promise of logicism to find the logical roots of math. And above all else, loss of interest in those foundations.

Interest in even the foundations of math as a topic—let alone the continued debate underlying the Great Math War—fades in the wake of Gödel's work. The foundational crisis completely fades. It's not simply that people forget about the subject, it's more like they simply no longer care. Many sour on the whole subject after Gödel's paper appears in 1931. "Mathematicians on the whole threw up their hands in frustration and turned away from the philosophy of mathematics," the prominent Dutch American mathematician Ernst Snapper will say fifty years later.

THE GREAT MATH WAR

Gödel's results are not simply the final nails in the coffin of the debate over the foundations of math—they are the nail, the box, the body, the hearse, the driver, the wake, the grave, the tombstone, the digger, the dirt, the tear-soaked hanky dabbing a wet, mournful eye, the officiant's eulogy, and the obit that appears in the paper the next day.

"Astonishing and melancholy," Nagel and Newman will say, describing the end of the foundational debate twenty-five years later in their definitive little book *Gödel's Proof*. The philosophical weight of incompleteness is smothering. "No final systemization of many important areas of mathematics is attainable, and no absolutely impeccable guarantee can be given that many significant branches of mathematical thought are entirely free from internal contradiction," they will write.

Those melancholy results greatly brighten Gödel's future, of course. Princeton University will recruit him in a few years. After World War II, he will win the very first Albert Einstein Prize. Harvard University will award him an honorary doctorate for "discovering the most significant mathematical result of this century." And decades after his death, at the end of the twentieth century, *Time* magazine will include Gödel on its list of the twenty greatest thinkers and scientists of the previous one hundred years. Some have argued that the underlying concept of the digital computer is owed to Gödel. "A vast industry has arisen founded on logical algorithms," British mathematician Andrew Hodges will write in 2008. And on and on.

───◆───

Gödel's genius will be universally recognized and lauded until the day he dies—though his weirdness is often noted as well. (For instance, he is a raging hypochondriac and suffers from the striking, strange, and debilitating delusion of feeling certain that someone is trying to poison him.)

All his future *someone-is-surely-trying-to-kill-me* paranoia notwithstanding, Gödel's results in 1930 are the blindingly bright light of his own future—though they will darken David Hilbert's retirement permanently. Nobody feels the frustration more severely than Hilbert does. He is angry when he hears of Gödel's results. Crushed. And he's not the only one.

GOODBYE TO ALL MATH

"It became an almost universally held opinion that Hilbert's program was dead and buried," Finnish philosopher Panu Raatikainen will write seventy-five years later.

Ever the optimist, Hilbert continues to toil through illness and retirement, even after Gödel's results are published, in a fruitless effort to prove all of mathematics consistent. He thinks perhaps his program could be modified to avoid the consequences of Gödel's theorem. He changes his thinking in hopes of establishing the foundations of math in a modified way. But he is retired then. And more than that, he is tired.

Still, his anger is understandable. The older I get, the easier it is for me to understand the mentality of an angry old man. Friends, even close ones, disappear or drift away. All the better, perhaps, since to continue seeing them would only serve to remind you of their emerging shortcomings. Even best friends fail. Some die tragically. Some dissolve like old chrome photos, leaving you with sepia-streaked prints or dark developer splotches on your archival memories: glory days long gone. Your own ideas and all your ambitions fade as well. Your health fades. Maybe you develop a limp. A wheeze. A spinal tweak. A heart condition. Gout! Whatever it is, your body becomes an angry sack of lies as you age—a bitter bag that's supposed to contain the health span of your better days but now carries only chronic pain. Aging is a grab bag of cruel, cold physical insults in an increasingly foreign and friendless psychologically damaged world. It's no wonder old men feel anger. It's amazing they feel anything at all!

Hilbert's anger is exceedingly raw. Thanks to Gödel, he sees the flotilla of his life's ambitions not just beached but utterly smashed to pieces on the rocks. He sees burning crew members jumping into the water, better to drown than burn alive. He sees the ocean's unstoppable surf pound his ships to shreds. He sees bulkheads splinter. Hulls split. Sails rip. Decks flood. Mizzen masts heave. Passengers spill into the drink. And all his ships sink.

Hilbert watches little Jonah swallow the mighty whale whole.

CHAPTER SEVENTEEN
THE GREAT MIGRATION

1933–1935

"
The art of asking questions is more valuable than solving problems.
"

—Georg Cantor

Moses draws water from the rock, and the Israelites quench their thirst. Etching by Jean Le Pautre. Cropped from a public domain image via Wellcome Collection.

Again, the Great Math War could have ended there—twisting in ignominious defeat at the feet of Gödel's incompleteness theorems. But there is a strange, political, Earth-shattering epilogue to this tale: a long, withering tail of brutality and human horror. Crimes against humanity. The dismantling of Göttingen. Another dawn of destruction like nothing the world has ever seen.

In January 1933, one of the most inhuman regimes the world has ever known installs the stupid buffoon and racist murderer Adolf Hitler as chancellor of Germany. The violent evil that follows begins very publicly with a brutal national boycott of Jewish businesses. Windows smashed. Doors blocked. Storefronts defaced. Hate-crime graffiti. Innocent people beaten in the street. Brown shirts. Black moods.

The same repression runs unchecked through German universities. In April 1933, the government passes a new law called the Restoration of the Professional Civil Service Act, which formally bars anyone of Jewish descent from holding a civil service position. Since all German universities are state institutions, that means any university faculty member of Jewish descent can be fired.

Albert Einstein probably would have been the first to be fired had he been in Germany at the time. But he is in the United States when the law takes effect in 1933, and he refuses to return. The Nazis respond by seizing his property. Einstein protests in kind by resigning his prestigious elected membership in the Prussian Academy of Sciences—an honor that few receive and none have ever refused or renounced. In response, the Nazi minister of education demands that the Prussian Academy censure Einstein. The academy does as it's told and sends out a press release denouncing their most famous member of all time and the very face of modern science itself. "No reason to regret Einstein's withdrawal," the idiotic press release reads.

"MORE WILL FOLLOW"

In Göttingen, things are bad and worse. The city and the university have long been a hotbed of right-wing student activism. Now Göttingen seethes

THE GREAT MIGRATION

with loathsome anti-Semitism. College students wearing swastikas become a common sight in classrooms. The German Student Association begins to spread fake news propaganda. Anti-Jewish slogans are shouted at rallies outside academic buildings. Radical students even hold a massive book burning. (Nothing says "Nazi" like a good book burning.)

Göttingen professor James Franck takes the bold and dangerous step of standing up to the Nazis. He publicly resigns his post in protest, a singular act of defiance and courage.

Franck is a remarkable person. He's the director of experimental physics at the university, a Nobel laureate, a decorated veteran, and a bona fide war hero. He was wounded on the front lines and was awarded no fewer than two iron crosses. He's Jewish, but the new law can't oust him because it includes exemptions for anyone like Franck who was decorated in active-duty service in World War I—or anyone whose father or son was killed in the war. Those exemptions become the exact basis for Franck's protest. "I refuse to make use of this privilege," he says in his resignation letter.

The whole thing shakes Franck's family and friends. They fear for his life. It's a fiery political climate, and they try to talk him into resigning quietly. But he ignores their advice and quits loudly, in protest, while making as much public noise as he can. He sends letters of resignation to university officials and to the Nazi minister of education—a copy of which he's already leaked to a sympathetic newspaper reporter. "The whole affair was carefully timed," one member of his family recalls later, "so that the reception of the letters coincided with the publication in the newspapers."

The reaction in Göttingen is predictably hostile, even among his fellow faculty. Some are silent, but many are outraged by Franck's actions. "A completely perverted version of this affair was spread by rumors and garbled nonsense," says a sympathetic Richard Courant. Some forty-two professors, including Franck's young colleague the German mathematician Werner Weber, sign a letter condemning him. They call his resignation a form of active sabotage. Anti-German propaganda. And they call for a purge of the university ranks in response. "We hope that the government will therefore accelerate the realization of the necessary cleansing measures," their letter reads.

THE GREAT MATH WAR

———◆———

Two days later, the hammer falls. A local newspaper called the *Göttingen Tageblatt* announces the dismissal of six prominent Jewish faculty members under the ominous shouting headline "More Will Follow." Courant, who for ten years has been head of the university's world-famous mathematical institute, is one of the first six fired. The irony is that for two days, he and another top faculty member were avidly discussing Franck's protest and considering whether they should resign in protest as well. They resolved instead to stay. Better to resist from within, they think.

But the Nazis take that chance away. Having decided to stay, Courant is sacked the next day—despite his war decorations and testimonials, petitions, and personal appeals from many of his colleagues. "I see the future of our institute as very dark," Courant says in a letter a few days later. "It is a pity to think what treasures are going to be destroyed." *What senseless damage!*

Pause and consider Courant's case for a moment: Here was someone who, like Franck, had fought in World War I, served as an officer in the trenches, had been wounded in battle, worked ardently on research for the war effort even as he was recovering in the hospital, and had been decorated for that service. After the war, he literally wrote the definitive book in his field—the hugely influential 1924 *Methods of Mathematical Physics*. He pioneered public-private partnerships with publishers and industrial companies. And he was a talented organizer and fundraiser. At Göttingen, he led one of the greatest math departments in the world, which he built despite the crushing inflation of the 1920s. A gifted fundraiser, Courant brought in a huge pile of money from the Rockefeller Foundation, used it to leverage matching funds from the German government, and built a state-of-the-art facility.

But in the 1930s, those patriotic achievements mean nothing, and the very money he has raised only serves to fuel rumors against him. It was foreign money, after all, and that adds poison to an atmosphere already toxic with rumors. Some students accuse him of being a communist. One story says he used to carry a red flag around everywhere. They say he has been a card-carrying socialist. They accuse him of being an ardent Zionist engaged in reverse discrimination—that he unfairly promotes Jewish scholars.

All this shock-and-awe propaganda overwhelms his ability to defend himself. He is fired from Göttingen without a second thought. "This 'elimination' hits me with an almost unbearable force," Courant writes to a friend. "I am very much afraid, quite aside for my own person—that something irreversible is happening."

He lingers in Germany with his family for a few months but flees the country that summer, shortly after his youngest son blurts out an interest in joining the Hitler Youth. "There was nothing sensible left to do but emigrate," Courant will recall in a radio interview years later.

———◆———

The loss of Courant plunges the Göttingen math department into chaos. Several remaining professors discuss resigning en masse, but the proposal never progresses beyond the idea phase. Mathematician and later historian Otto Neugebauer is appointed as Courant's successor. He's also a war hero and former prisoner of war, having served as an officer in the Austrian army in Italy in 1917–1918, where he was captured and imprisoned at Monte Cassino for a year.

But Neugebauer has no taste for the job and lasts only one day. He refuses to declare an oath of fealty to the new Nazi bosses and is sacked. Neugebauer flees Germany for Denmark a year later under a cloud of anti-Semitic suspicion. He famously declares he's not Jewish by saying, "I did not have the *honor* of having a Jewish grandmother."

Hermann Weyl, who is now a professor at Göttingen, having been recruited in 1930 to replace the retiring Hilbert, is hastily named its new mathematics director. That doesn't last long either. Weyl initially accepts the position, thinking he can resist the Nazis from within. But he departs within a few months, perceiving that his family will soon be targeted because his wife is Jewish. They too emigrate to the United States.

Emmy Noether is also forced out. She is one of the first six professors fired, along with Courant. She receives an official memorandum from the education minister saying simply, "I hereby withdraw from you the right to teach at the University of Göttingen."

Stripped of her teaching role, she initially decides she can still host students at her apartment informally. At the first meeting, however, one of her favorite students shows up in a Nazi uniform. Things are changing fast.

After Noether is fired, Göttingen professor Edmund Georg Hermann Landau tries to give a lecture. When he arrives at the auditorium, he is met by a mob of seventy student protesters, many of them wearing Nazi SS uniforms. They are led by none other than Landau's own former assistant, the same former student of Noether's. The mob blocks anyone from entering the hall, and his former assistant tells Landau that Aryan students want Aryan mathematics—not Jewish mathematics. "The speaker for the students is a very young, scientifically gifted man," says Courant, who witnesses the episode, "but completely muddled and notoriously crazy."

THE EMERGENCY COMMITTEE

Hundreds of scholars flee Germany in the coming months in a historic academic exodus that becomes known as the Great Migration. Some quit in protest. Some leave quietly. Some fight to keep their position but are forced out anyway. Most are fired. "At the time we really thought there was still hope left that the worst could be warded off," Weyl will write a few years later. "It was in vain."

In the United States, the Rockefeller Foundation sees the dismissal of these scholars as an opportunity in crisis. The foundation has already spent huge sums on constructing new buildings in Göttingen and elsewhere in Europe. And it has been supporting the work of researchers by funding scholarships for years. Here was its chance to redirect some of those funds to support some of these exodus professors in the United States.

Just a week after Franck's protest, a Rockefeller program officer writes a report highlighting the urgency of the situation. Could the foundation find fellowships for some of these professors who will soon be out of work? Many of them are exceptional, the memo points out. It calls for salvaging "the traditions of learning that have taken so long to evolve and that are now so seriously threatened."

THE GREAT MIGRATION

The memo works. Within a month of Franck's resignation, the Rockefeller Foundation and other nonprofits establish a special emergency fund in Manhattan called the Emergency Committee in Aid of Displaced German Scholars. (They soon rebrand to replace the designation "German" with "Foreign.") Similar groups follow in London, France, and the Netherlands, and hundreds of scholars fleeing Germany are soon sponsored. It seems like a promising program at first glance.

"Universities are at once the storehouses and the manufactories of the culture of a society," writes Edward R. Murrow, the executive secretary of the committee and later a famed CBS broadcast journalist in London who wins acclaim for his radio broadcasts during the London blitz and for standing up to US Senator Joseph McCarthy during the 1950s. "An attack upon them is an attack upon the very symbols by which a State lives."

The committee finds spots for the exodus professors by prodding America's top institutions to clear space for them in their ranks. They sweeten the deal by offering private grants to defray the cost of their salaries. The awards are relatively small, typically $1,000 or $2,000 a year. (A full professor's annual salary in the mid-1930s is usually closer to $4,000.) But they are immensely attractive offers to exodus academics fleeing Europe because each position also offers a hard-baked path to US citizenship. American immigration laws passed in the early 1920s had established strict country-by-country caps on migration. But the laws did carve out exceptions for scholars who resided in the United States for at least two years while holding an academic appointment. Thus, in the 1930s, the aid for displaced scholars program could reliably offer not just a temporary job but permanent US citizenship.

But some glaring problems still confront the program. First and foremost, anti-Semitism is strong in the United States in the 1930s, both at universities and in the government. Ninety percent of the mathematicians who flee Germany are Jewish, and there is backlash because of that. While some universities welcome fleeing professors with open arms, many do not. Some American schools have in place long-standing formal limits on the number of Jewish students they admit each year. By one count, fewer than one hundred Jewish professors taught at all American universities combined prior to the Great Migration.

THE GREAT MATH WAR

There is also rampant xenophobia in the United States—pervasive mistrust of foreigners in general and specific hatred of Germans in particular. *Kill-the-kaiser* resentment still lingers from the war, and panicky anticommunist red scares are what led to the stringent 1920s anti-immigration immigration laws in the first place. America in the 1920s and 1930s is in many ways the land of the free and home of the rage.

On top of all that, the United States of 1933 is absolutely reeling from four long years of the terrible Great Depression. Things are at their utter worst, even when they're at their very best. Universities have been forced to cut jobs. Cut costs. Cut everything. Faculties have been trimmed to the bone. Across 240 colleges and universities in the United States at the start of the Depression, almost 10 percent of faculty jobs have already been eliminated due to belt-tightening.

The faculty members who remain often have to make do with lower salaries. Endowments have shrunk. Gifts have been withdrawn. Hundreds of mathematicians are out of work in 1933, and things are tough all over. One major US mathematical society passes a resolution waiving fees for its members who are unable to pay. Young American scholars who cannot get jobs are "forced to become hewers of wood and drawers of water," American science historian Nathan Reingold will write fifty years later.

That financial pain causes blowback. Many American professors complain that with the belt-tightening, they must carry heavy teaching workloads, and when the Great Migration starts and hundreds of European academics flow into US universities in the years prior to the outbreak of World War II, some existing faculty members at these institutions complain vigorously that the universities have reached saturation. They claim that the German scholars will force US academics out of work. "To say that there are too many foreigners in American universities is not chauvinism," the newly appointed chair of mathematics at the University of California, Berkeley writes in 1933. "Merely that the careers of promising students in America are being cut off at the top."

Anticipating blowback, the Rockefeller-funded committee vows no American will lose their job during the Depression only to have their place taken by a foreigner. The committee decides to guarantee that vow by structuring its offers such that the Great Migration professors will be

THE GREAT MIGRATION

awarded primarily research positions as opposed to teaching appointments. That will minimize their classroom exposure and keep them somewhat out of the public eye.

The committee also decides to make the awards heavily biased toward established merit instead of promising potential. The exceptional. The extraordinary. The unrivaled. The top thinkers in their respective fields. That's whom the committee of displaced scholars will help. Think Einstein. Think Weyl. Think Noether. It would seek homes for established rock-star scholars who already have giant reputations on the world stage, hoping that these extraordinary aliens will duck scrutiny for potentially depriving a low-level faculty member of a job. To achieve this, the committee sets a *minimum* age requirement of thirty for any professor they help.

With those rules and that intention in place, the committee begins recruiting. The list of professors who move to the United States is impressive. James Franck finds himself at MIT. Hungarian physicist Leo Szilard, who later conceives of the nuclear chain reaction and plays a key role in the Manhattan Project, settles at Columbia University. Edward Teller, who will also work on the Manhattan Project and later become known as the "father of the hydrogen bomb," finds a position at George Washington University. Physicist Hans Bethe spends the rest of his career at Cornell University. Swiss physicist and later Nobel laureate Felix Bloch moves to Stanford. Courant goes to New York University. Courant's student Hans Lewy spends two years at Brown University and then goes to the University of California. Hungarian mathematician Paul Erdős moves to the University of Pennsylvania and becomes the most prolific mathematician of all time in terms of number of publications.

But no amount of starlight can ever brighten a winter night's sky.

And therein lies the tragedy. Because most of these professors are so prominent in their fields at the time of the Great Migration, many of them probably could have found new positions when they fled anyway—or those positions would have found them.

When famed physicist Max Born leaves Göttingen with his family, for instance, he's inundated almost immediately with recruiting calls from universities eager to court him. Einstein, who enjoys the same privilege a thousand times over, laments that there will never be enough money, time,

THE GREAT MATH WAR

or open positions to get all scholars out of Germany—and those under thirty don't even qualify.

"My heart aches when I think of the young ones," Einstein writes.

◆

Göttingen becomes a weird place after 1933. After the first round of firings, the university is diminished. "By the following semester," one student at the time writes, "the great days of Göttingen were fatally interrupted." By spring of 1934, most of the faculty who do remain are searching for jobs elsewhere, and their students are "hurrying to get theses done," the student recalls.

You can't fire your top talent and hope for continued success. It takes less than a year to destroy what had been one of the twentieth century's greatest communities of math and physics scholars. What happens next is predictable—perhaps inevitable. Its faculty depleted, Göttingen goes in search of new professors. Again, it woos Brouwer, but again, he declines. Next a Nazi apparatchik becomes head of the math institute at the university after Weyl leaves. Hilbert lectures one last time at Göttingen over the winter of 1933–1934. Once, toward the end of this final run, Hilbert is sitting next to the new minister of culture at a tasteless meal. The minister turns to him and asks, "How is mathematics in Göttingen now that it has been freed of the Jewish influence?"

Has it suffered?

Suffered?

"It hasn't suffered, Herr Minister," Hilbert says. *"It just doesn't exist anymore."*

THERAPEUTIC IMPERATIVES AND HORROR

Now, one word on those Nazi bastards. What begins with faculty firings at Göttingen and other German universities is only the start of a campaign of academic restrictions. Soon it's not just a question of who teaches whom but what is taught. And how. Whole fields give themselves up to Nazi reform. Some willingly. Some even enthusiastically.

THE GREAT MIGRATION

Salaries. Permission to publish. Everything falls under state control, and over time, the restrictions get worse. A 1937 law expands the justifications for firing a professor to things like shopping at a store owned by Jewish proprietors. Making an offhand comparison between Nazis and communists will get you sacked. Too much public complaining becomes a crime. Failing to clap loudly enough draws scrutiny. Unenthusiastic saluting invites investigation. And not objecting immediately and violently to anybody else's criticism of the party will get you quickly canned in 1938.

There are biological directives as well. A national forced sterilization code, called the "law for the prevention of genetically diseased offspring" is passed in 1933 at the same time as the law governing Jewish civil servants. This law is informed by eugenics, a vile pseudoscience and racist creed explored and nurtured in British thought, nursed in the fertile soils of US state law, and blessed by the US Supreme Court in the case *Buck v. Bell*. Justice Oliver Wendell Holmes, writing for the majority, infamously said sterilizing poor Carrie Buck against her will was in the best interest of society because "three generations of imbeciles are enough."

In Germany, forced sterilization reaches its full toxic flower and awful expression. "No one really knows how many people were actually sterilized" in the 1930s, American psychiatrist Robert Jay Lifton will write fifty years later. "Reliable estimates are generally between 200,000 and 350,000."

Numerous conditions are listed in the 1933 law as grounds for forced sterilization: having schizophrenia, bipolar disorder, epilepsy, Huntington's disease, blindness, alcoholism, and "feeble mindedness." Castration of homosexual men is offered as an alternative to prison in the 1930s. Thousands of men accept this punishment rather than being sent to jail. Countless sterilizations are performed on people considered to be of low intelligence—not based on any quantitative measure, modern scholars point out, but simply by virtue of a doctor's order deeming them so. That means doctors can essentially sterilize anybody for any reason at the stroke of a pen.

The Nazis also corrupt medicine from within. They rework medical school curricula, emphasizing dehumanizing theories of the biology of race. They push the German medical establishment to accept state-sanctioned murder on the basis of a simple stupid eugenic concept of "life unworthy of life," a phrase coined by an infamous German psychiatrist in the 1920s who also referred to people as "human ballast" and "empty shells of human beings." Killing them is not just okay, he advocated, but useful. These people are already dead, he wrote.

———◆———

Modern scholarship suggests that the German medical profession was more than willing to go along and in fact fully culpable in these gross horrors. As a professional class, more doctors than members of any other occupation join the Nazi Party. Fully 45 percent of all doctors are card-carrying party members—more than teachers, lawyers, and business executives.

Euthanasia is soon taught in medical schools, propagandized as a merciful treatment for the incurably ill. It's also cynically sold as a boon to society. Some make economic arguments: Institutionalization is expensive, so why not relieve the state of its burdens? they say. Anyone who cannot be cured must be killed. One doctor at the University of Berlin argues that society has "an obligation" to kill the unworthy. As the 1930s march on, it only gets worse. Life itself begins to be considered a disease in German medical circles. Some doctors argue that medical murder is both compassionate and completely consistent with medical ethics. Doctors justify murder as a corrupt, utilitarian, *ends-justify-the-means* salvation. It's a "therapeutic imperative," some German experts say.

Modern experts call all this murderous equivocating by doctors the "incomprehensible perversion" of the Hippocratic oath. Under their corrupt and morally bankrupt utilitarian philosophy, *Primum non nocere* (First do no harm) is mutilated to *Primum societati nihil nocet* (First do no harm *to society*). And it's only a short step from there to mass murder. The cold sweep of Nazi horror starts at state hospitals, which become places where people are killed rather than healed.

———◆———

By 1938, medically ordered sterilization programs have basically become shells for state-sanctioned murder. Psychiatric patients. The poor. The weak. The politically marginalized. Medical ethics bends and twists itself into an effective set of arguments to excuse and rationalize this murder. What comes next is history and horror. Hospital-based killings in the 1930s pave the way for the creation of concentration camps, death camps, and the Holocaust.

One should always be skeptical of utilitarian philosophies and of any argument that excuses illegal or unethical behavior by claiming that it serves some greater good. Corrupt utilitarianism with its *ends-justify-the-means* nonsense fueled forced sterilization, legitimized state-sanctioned murder, and introduced human horror—all in service of a cold, cruel, and corrupt creed and the paper fantasy promise of world purity. Murdering innocent people has never made the world a better place. And it's important to remember how it happened leading up to World War II: the cruel laws. The awful propaganda. The co-opting of science. The perversion of medicine. The bending of ethics. The bitter taste of human horror.

All these things need to be remembered. Germany in 1938 is a lesson not just in cruel butchers and crimes against humanity but also in how ordinary people were turned into monsters and became the purveyors of those horrors, especially health workers. If doctors can become detached from ethical codes and complicit in murder, then any professional can.

CHAPTER EIGHTEEN

∀ THINGS ∃ ND

1935–1938

> "
> *Entia non sunt multiplicanda praeter necessitatem.*
> (Things should not be multiplied beyond what is required.)
> "
>
> —William of Ockham

Scientific instruments: gyroscopes, telescopes, pestles and mortars, cosmological manuals, natural history specimens, etc. Cropped from an 1835 lithograph by Jean-Baptiste-Joseph Jorand. Public domain image via Wellcome Collection.

In 1938, Hungarian mathematician George Pólya meets Hermann Weyl at Stanford University to remind him of a bet they made two decades before, toward the end of World War I, when they were colleagues in Zürich. Weyl had insisted that within twenty years, intuitionism would come to dominate math. That was in 1918. Now, Pólya says, it's 1938—the bet has expired, and so has intuitionism. Weyl considers this. He admits he's lost the bet, but he is eager to avoid any public statement—and he's more eager still not to think about the ashy taste.

Some fifteen years before, in the early 1920s, he was uncomfortably caught in the middle of the nasty dispute between Brouwer and Hilbert at the heart of the Great Math War. For years, he was trapped in the mix. He was beaten up for framing intuitionism as *die revolution*. He was criticized for calling the foundational crisis a "crisis" in the first place—a word seen by some as hyperbole, a contrivance cooked up to promote intuitionism. Now, in 1938, the past is the past. Sleeping dogs are lying. All is forgotten, if not forgiven. Interest in the foundations of math has faded to none.

Hilbert retired in 1930, and he has done almost nothing since then. He tragically watched his university and his country fall to the Nazis, and after World War II starts, he himself will fall, slipping and breaking his arm in 1942. He will die a year later—largely due to inactivity.

———◆———

Russell has also drifted away from the subject. He more or less gave up work on the foundations of math after serving time in Brixton Prison in 1918. He went on to raise a family and start a school after that. As a final footnote on his role in the foundational crisis, when a new edition of his *Principia Mathematica* appears in 1938, Russell insists in an updated preface that the main thesis is still valid—that mathematics and logic are the same thing. And he again makes arguments against both formalism and intuitionism. But by then, the field has moved on.

$\forall_{\text{THINGS}} \; \exists_{\text{ND}}$

He is shaken by the tragic death of his good friend Ottoline Morrell that year. She has spent the last two years of her life in and out of hospitals, and in 1938 winds up under the care of a doctor who, according to one of her biographers, basically kills her by overprescribing an off-label drug. The same doctor faces criticism and scrutiny from British medical authorities, and with the walls closing in on him, he commits suicide. But before he dies, he leaves orders for his staff to continue Ottoline's treatment. For the next two days, the disgraced doc's assistant administers the drug. Ottoline dies on the third day.

Bertie is not quite done with the Great Math War, and he will go on to describe much of its history in the first volume of his expanded autobiography. That will come decades later, late in his life when, running low on cash, he will pick up a personal memoir he finished some twenty years before. He will lard it with letters from famous people and publish it in three fat, hugely successful volumes. The first one will come out alongside *Sgt. Pepper's Lonely Hearts Club Band* in 1967, the second alongside the *White Album* in 1968, and the third alongside *Abbey Road* in 1969. Russell will die the following year, five days after I am born.

———◆———

Back in 1938, Brouwer also appears to be done with the foundations of mathematics—though not with math itself. After his ouster from the editorial board of *Mathematische Annalen* in 1928, he is true to his word and his threat to Ferdinand Springer. In 1935, Brouwer launches a new, competing journal (called the *Compositio Mathematica*), which he will publish for five years until the onset of World War II halts production. After the war, Brouwer will fall into a power struggle with his new publisher and will lose once again. The publisher will demote him to "special projects" and boot him from the masthead.

Seeing him emotionally drained, Lize will seek to comfort him. You are not a failure, she will say. "Don't be sad about mathematics. I know that you can do it all like before, if only tranquil returns to your mind." And, she will add, "even if you could not do it, you have already done enough."

THE GREAT MATH WAR

"I don't like mathematics," Brouwer will go on to say later in life. "It basically bores me." He will be killed in the 1960s when he is struck by a car while crossing the road in front of his house. He is knocked into oncoming traffic, is hit by two more cars, and dies on the scene.

His stepdaughter, Anna Louise Elisabeth de Holl, had felt sidelined by the attention Brouwer lavished on her school friend Cor, who moved in with them and became his assistant. Anna Louise left home during World War I intending to become a nun, and she is said by Brouwer's biographer to have retained a "deep and bitter hatred" for him her whole life. In the 1980s, however, years after Brouwer's death, she will see an article in a Dutch newspaper about a local symposium held in his honor on the occasion of the centenary of his birth. People will be celebrating his achievements—especially his work in topology. Louise will be so impressed in reading this that she will rethink her stepfather. Believing in divination and the occult, she will try to contact Brouwer's ghost, and she will later claim that she was successful. His spirit was pleased with the symposium, she will say, and his soul has finally found peace.

◆

Weyl in 1938 has moved on from the foundations of math as well. He spends the last decades of his life at the Princeton Institute for Advanced Study, arriving there in October 1933 in the first wave of the Great Migration and going on to become an iconic figure for a whole generation of mathematicians and physicists. He will work on things like quantum theory, relativity, and the geometry of space-time. He will invent the concept of wormholes and develop gauge theory, an area within physics that will become so important that the great physicist John Archibald Wheeler will reflect in 1986 that more people are working on it that year than all the physicists alive in the world when Weyl created it.

"Weyl was—is—for many of us, and for me, a friend, a teacher, and a hero," Wheeler will say. "His style is that of a smiling figure on horseback, cutting a clean way through, on a beautiful path, with a swift bright sword."

Weyl will die suddenly in 1955 while in Switzerland. He overexerts himself walking back and forth to a post office to check for letters from the many people writing to congratulate him on his seventieth birthday. On his way home from one of these trips, he gasps, collapses, and dies.

THE FALLACY OF SEEING

The last concept I want to introduce, to wrap up this book, is something I like to call the "fallacy of seeing," a simple notion that the harder we cling to something, the more true it can seem—and, by virtuous circularity, the more true a thing seems, the harder it is to let it go when it turns out to be false.

We are often fooled by the rational, mathematical nature of our minds as well as the deliberate, quantitative methods of our measure. We view the world, form beliefs about it, and often arrive at what we deem true or evident based on carefully weighing the data. We observe, collect, consider, conclude, and BAM!—achieve certainty. But often we're wrong, even when we carefully weigh the facts and data-drive ourselves the whole way there. Why? Because of basic psychology: Ego is our copilot.

Underlying the fallacy of seeing is a psychopathology whereby our egos, obstinacy, cautious natures, and unconscious biases largely prevent us from letting go of a certainty once it has wormed its way into our brains. The fallacy of seeing says the more we invest in a concept mentally and the more we come to consider it true and meaningful, the harder it is to let it go. It's hard to admit being wrong when you've convinced yourself that you're right. Nowhere is this more relevant than in math and science. The history of science is full of wrong ideas that take a generation to die because their chief adherents cannot let them go.

But time is a long road. And what seems scientifically certain one day rarely is a century later. Data accumulates. Models change. Knowledge penetrates. Something that seems mysterious to one generation is completely understood by the next. Some things that seem insignificant at one point in time, like junk DNA, turn out to have hidden richness. Other things that shout significance lose importance as time marches on—like crumbled Ozymandias and his *look-on-my-works-ye-mighty-and-despair* lonely heel.

People often narrowly and foolishly ask *Why?* when it comes to history (and especially when it comes to the history of science). In general, this is a good question. What were the societal forces at play? Who was involved? What were their personal motives? Those are the meaningful *Why?* questions. Those are a historian's *Why?* questions. But often people don't ask them. Many instead ask a more meaningless *Why?* A selfish *Why?*—what I like to call the transactional *Why?* People often eschew the broad question of *Why did this happen?* and embrace instead the far more narrow, selfish one of *Why should I care?* For many people today, all knowledge in general and science or math in particular is not so much a question of *Why does it matter for history?* as *Why does it matter for me?*

That's a simple-minded, linear way of thinking about the events of the past—and an ironic one given how much we concern ourselves with futurism today. History matters whether we like it or not—whether we learn it or not—and it shapes our future whether we care about it or not. The purely mercenary *what's-in-it-for-me-ism* misses what really matters in history: history itself. There's a reason why half of Shakespeare's plays were histories: History is *interesting*. And that's the point many people miss when asking why it matters. History doesn't matter because it's relative. It matters because it's absolute. Narratives. Journeys. Characters. Minds. Behavior both bad and saintly. Lifetimes of energy. Mysteries solved. Unintended outcomes. Unexpected consequences. History matters because it's interesting, that's all. It matters because it's alive, not dead. Only the fallacy of seeming says otherwise.

———◆———

The history of the foundational crisis has almost been forgotten. It ended with a whimper, thanks to Gödel. His incompleteness theorems smothered a Great Math War debate already starved of oxygen in 1930. By 1938, the debate over the foundations of math—if not the foundations themselves—seemed unimportant. Interest had fallen off sharply by then, and that would continue after World War II. Mathematical logic and the foundations of math are no longer taught in math departments—perhaps sometimes in computer science or philosophy classes, but "in a purely technical way" and with "no concern for history or philosophy," the American mathematician David A. Edwards writes in 2011.

And the Great Math War itself—the *Grundlagenstreit* (foundational debate) or *Grundlagenkrise* (foundational crisis), as Weyl called it, is generally not taught at all. And if it is, then just barely and in passing. I read one claim that this entire story (which took me more than five years and 110,000 words to describe) can be explained to first-year graduate students in a few minutes—a single lecture essentially.

In the eighty years since World War II, the foundations of math have become a highly specialized, "esoteric subject," according to the European logician Mirna Džamonja. "Considered by mathematicians as an esoteric subject of little interest to their daily life," she wrote in 2019. I almost gave up the subject after reading that because it *seemed* unimportant.

Some today also say the debate was a bad thing. Overblown. Silly. Useless. Unnecessary. A contrivance without consequence. People assume it could have been settled amicably and easily, and they aren't wrong. Like World War I, the Great Math War could have been avoided. But instead, its mighty flames consumed some of the best minds in Europe for years until they fizzled out by 1938. Despite the fact that it's considered one of the great intellectual arguments of the twentieth century (and the most important one you probably never heard of), many people today ask only that selfish *Why—Why should I care?*

Part of the reason why the Great Math War has been forgotten is that it seems unimportant despite its odd outcomes. It's almost unique in the history of math and science as one of the only cases where a fundamental split between competing worldviews is decided not by the emergence of winners and losers but by a chess-like Zermelo draw and the dissolution of the competition. No one worldview wins out, and logicism, formalism, and intuitionism all remain as valid as approaches to math as they are equally invalid at solving the question of its underlying foundations. "They complement each other and can coexist peacefully," Spanish mathematician José Ferreirós writes. It's Vesuvius quenched. Colosseum shuttered. Marathon canceled.

But there is triumph in that failure to find a winner. The most major consequence of the Great Math War was its liberating effect on the field. After 1938, mathematicians were set free to use any of the elements of logicism, formalism, proof theory, or intuitionism every day, all day without

THE GREAT MATH WAR

ever once having to worry about the foundations. People can be a formalist if they want. A logicist if they wish. And an intuitionist if they must. The three separate camps, though distinct, have never really been at odds or competed with each other again since 1938.

"We are less certain than ever about the ultimate foundations of (logic and) mathematics," Weyl will write in 1946. And yet, he says, "outwardly it does not seem to hamper our daily work."

Some might say, simply, that there's much to admire in the foundational crisis for its underlying rigor. The mathematical purity of the ideas. The fact that logicists, formalists, and intuitionists all climb separate faces of Mt. Everest. But how they summit isn't really the point. The value of the *Grundlagenkrise* lay not in journey but in destination. The unintended fruits of the debate were golden outcomes. Results. New discoveries.

BECAUSE IT'S WHERE?

There's a farcical quality to science history because even when we're wrong, we often make great advances for stupid reasons. A lowly cleric in England named Edward Stone isolated aspirin from the bark of the willow tree because he foolishly thought that plant had some hidden essence that allowed it to thrive in wet, moist substrate exactly the same way as a lung infection (though he would never have thought of it as an infection since people had no concept of microbiology in his day). Still, it's an amazing discovery and a transformative medical revolution based on that idiotic, incorrect understanding of the underlying mechanism.

Arguments in favor of basic science funding often focus on such unintended consequences. The microprocessor and GPS were originally developed for the Apollo space program. RNA vaccines designed for Ebola outbreaks were retooled for COVID-19. Adaptive optics designed to spy on Soviet satellites found use in correcting the optical distortions of ground-based telescopes to search for faraway exoplanets. Tim Berners-Lee invented the World Wide Web at CERN in Switzerland for the narrow application of allowing scientists to share massive amounts of data collected in particle physics experiments. The same sort of off-target outcomes emerged from formalism and logicism.

"Just like Columbus had set sail to reach the Indies and found other unexpected—not less rich and exciting—lands," Paolo Mancosu writes in recent years, "in the same way the adventure of reason and logic and the foundations of mathematics led to surprising new finds that dramatically reshaped the original goals."

◆

Logicism, channeled by Gödel into his incompleteness theorems, brings unexpected dividends through computer science, model theory, computer algorithms, theories of recursion, and computability. It opens the gates to the golden age of logic, from 1930 to 1970, when economics, political science, applied mathematics, clinical medicine, and computer engineering rapidly advance algorithmically.

Formalism, despite its own failure to secure the foundations of mathematics, is also wildly successful in the end. Proof theory is now not just an important tool but an essential premise of modern math. Hilbert's basic approach, "in spite of its failure," Ferreirós wrote several years ago, "established itself in practice as the avowed ideology of 20th century mathematics."

So what do those stunning *phoenix-from-the-ash-heap* successes of formalism and logicism mean for intuitionism? What should we make of the abysmal failure of intuitionism—so gleefully expressed by Pólya as he came to collect the winnings of his bet with Weyl? Is Brouwer the great big loser? Not really.

There's no question Pólya won his bet with Weyl. Even before Gödel and his incompleteness theorems, intuitionism fell out of favor. Mathematicians by the late 1920s had largely come to dismiss it and to abandon Brouwer's thesis that only restrictive constructivist methods should be relevant to math. "The mathematical community," Ernst Snapper writes in 1979, "almost universally rejected intuitionism."

But Brouwer's ideas of constructivist math may be due for a revisiting renaissance today. Mathematics may be a basic sense, like the ability to discern salty from sweet. A "human primitive," I like to call it. The ability to count numbers, analyze patterns, and manipulate data for decision-making purposes may be an innate, if emergent, part of human cognition—a piece of our basic cognitive tool kit that evolved in our species over time. Math may have actually come before language in human evolution, and it may have been our mathematical abilities that allowed us to develop language in the first place. Basic quantitative skills and abstract representation may have given our human ancestors huge survival advantages, affording them the ability to count "winter dinners, one a hill," as Robert Frost once said.

Brouwer believed this. And Hilbert was also convinced that was the case. He always saw math as fundamental to our basic makeup, writing in 1913, "The axiomatic method is not new, but is deeply rooted in human thinking." (It's not that the world is mathematical, as one of my college professors used to say—it's that *we are*.)

FISH GOTTA SWIM

Possible proof that our human math ability is fundamental can be seen in a 2023 study of infants in Barcelona, where neuroscientists found evidence of logical reasoning emerging as early as nineteen months. The babies were able to develop competing hypotheses, determine probabilistic outcomes, and weigh them accordingly. Spontaneous logical thinking, those researchers speculated, plays a crucial role in young humans by filling gaps, compensating for a lack of learned evidence, and reducing uncertainty.

More evidence can be seen in the fact that math ability may not be unique to our species. For many years, we thought math and logic were uniquely human—to the point we thought that the ability to do math was specifically what makes us human. But it's not. Other animals can count, estimate, and analytically reason as well. The ancient Greek philosopher Chrysippus of Soli may have been the first person to realize this. Some 2,500 years ago, he saw a dog chasing another animal. The dog loses it. Reaches a crossroads. Looks right. Looks left. And then tears off down the middle road. Animal ecologists have seen a similar thing in recent years in Chicago. Ten years ago, I saw a presentation by one who showed videos of urban coyotes waiting for the walk/wait signal to change at crosswalks on busy roads and looking both ways before stepping into the street—showing their capacity for abstract representation.

Some animals reason superbly. Researchers at the University of Tübingen in Germany recently showed that crows can associate signs and symbols with numbers and interpret them complexly. They trained the crows to peck out numbers and found they could learn to peck out recursive sequences—performing quite well, the researchers write: "On par with children and even outperforming macaques." Clever birds!

Abstract mental recursion, the ability to have one mental representation embedded within another mental representation, has also long been considered a key feature of *human* intelligence. For years, some have even gone so far as to make the caw-caw claim that it defines what it means to be human. (Though those crows may have something to say about that.)

As I was finishing this book, researchers at the Okinawa Institute of Science and Technology in Japan discovered that the anemonefish *Amphiprion ocellaris*—more sweetly known as the clown fish—recognizes other fish by counting their stripes. Unlike the fun Pixar clown fish Nemo, anemonefish are not Ulysses-like adventurers but mostly stay-at-home, stand-your-ground defenders who ruthlessly bite and attack other fish they perceive to be threats. Apparently they can tell friend from foe by counting the number of white stripes on their body.

That's the bizarre lesson here. One math expert once said fish swim, but they don't solve hydrodynamic equations, and in the same way, "good mathematicians are not much good at the philosophy of mathematics." But maybe this misses the point. Maybe mathematics is nothing more than a system of evolutionarily unlocked, neurologically developed cognitive processes that allow us to navigate and negotiate our way through life.

Intuitionism, then, may have unrealized potential today. Maybe it can help unlock our primitive math sense, help people see the world around us, and—like Nemo—attack life accordingly. And maybe all this evidence in neuroscience supports Brouwer's basic premise that we intuit the world through constructive mental exercises.

As such, Brouwer in particular and intuitionism in general may grow in significance in the future—perhaps not by replacing Hilbert's proof theory as the avowed ideology of twenty-first-century mathematics but by becoming more relevant nonetheless. Brouwer's work could be due for a renaissance from a neuroscience point of view. The human brain may indeed represent mathematics somewhere in its selfish folds, manipulating numbers and doing calculations through cell-cell signaling and electrochemical depolarization impulses. Math ability may come to be regarded as just another form of muscle memory in the future—like shooting clutch free throws in a basketball game.

And because Brouwer based his idea of math on a mental activity proceeding by the performance of mental constructions, I wonder whether there might be something useful in his constructivist approaches that could help shed light on our human ability to do math. I wonder if intuitionism might be able to provide a useful clinical correlate of the underlying neurocognitive processes that take place as we recall numbers and interpret data. Maybe understanding that will improve our grasp of the human mind, help us develop a better taste for mathematics and enhanced quantitative abilities, lead to new approaches to math education, or all three.

Only the future will tell. And that's exactly why it's useful to study history.

—Bethesda, Maryland

ACKNOWLEDGMENTS

As I sit finishing this book, the final embers of the last fire of the winter are dying in my fireplace, and my mind is burning with gratitude for all the people I need to thank. I have now been working on this book for more than half a decade, and I never would have finished it without the help, input, guidance, grace, and encouragement of many people along the way. Here are just a few.

The original idea for *The Great Math War* landed on my lap quite randomly and unexpectedly during my father's retirement party from Penn State University in the fall of 2017. I was telling his colleague Kira Hamman about my book *The Calculus Wars*, and she said it reminded her of the foundational crisis in mathematics. Months later, she sent me a nice email detailing a rough summary of Russell, Brouwer, Hilbert, and the whole affair. That began my research.

Two years later, in the summer of 2019, I was buried in investigating historical topics. Set theory. The Boer War. British politics in the 1910s. World War I. The Weimar Republic. Versailles. The Treaty of Locarno, which I spent untold hours researching only to ultimately cut all mention of it from the book for the sake of space. The 1920s. Bertrand Russell and his complicated relationship with Ottoline Morrell. Anemia. Insomnia. Incompleteness. *Despair!*

That's when an unexpected breakthrough occurred. I was interviewing for a managing editor position at the Council on Foreign Relations in Washington, DC, and the think tank's vice president James M. Lindsay opened my final interview by grilling me about my book. This was actually the first time I had ever discussed the concept at length with anybody. I didn't even have a finished proposal at that point. And even though Jim and I never discussed my book again, the conversation had a lasting impact on

ACKNOWLEDGMENTS

me. For the next five years, and long after I left the Council on Foreign Relations, I constantly returned to this conversation. Throughout the process of writing the proposal, selling it, researching the book, writing my first draft, and editing it, I constantly imagined myself back in that corner-office conversation, wondering what would Jim say? It's funny to say this now because I haven't spoken to him for years. I doubt he even remembers our conversation, but I have thought of it many times.

My boss at the Council on Foreign Relations, Tom Bollyky, was also encouraging, having just written a book of his own. I like to think I returned the favor, though, because in the early days of my research, I stumbled upon a quote from Immanuel Kant: *All human knowledge begins with intuitions, proceeds from thence to concepts, and ends with ideas.* We were just launching the online magazine *Think Global Health* at that time, and I adapted and shortened Kant's quote into the magazine's present tagline: *Better health begins with ideas.*

When I began working for Jane Metcalfe at *Proto.Life* magazine a year later, I found another key supporter—a generous patron, a true mentor, and a great boss. The four years I spent working for her were wonderful professionally, and she lavished me with encouragement for my book. After the magazine ended publication, Jane and I continued to discuss *The Great Math War*, and she read an early draft of my manuscript. I can't imagine a more supportive friend and colleague, and I thank her many times over.

Several institutions and one individual I must thank for their awesome image resources are: the Wellcome Collection, the Library of Congress, the US National Archives, the New York Public Library, Nordiska Museet, and the artist Anna Gorban, whose amazing CC-BY sketch of David Hilbert was done as a commission for the Royal Society. Also thanks to the McKeldin Library at the University of Maryland for its community borrowers program, without which I could not have done my research.

Many thanks as well to the Anderson Literary Agency. My agent, Giles, has been a tireless champion and friend for twenty years. He gave me valuable feedback in the early stages, when I was framing the book, and handled the business aspects of negotiating my contract. I love my publisher Basic Books, part of the Hachette Book Group, and I owe a huge debt of gratitude to my editor Thomas Kelleher for his wisdom, guidance,

ACKNOWLEDGMENTS

and incredibly useful edits. He helped shape the book into its current final form in a thousand substantive and subtle ways. His editorial assistant Gillian Sutliff was also extremely helpful throughout, and I owe her many thanks as well. Overall, the whole Basic Books/Hachette team was fantastic. I specifically want to acknowledge and thank my amazing production editor, Michelle Welsh-Horst, my incredible copy editor, Connie Oehring, my hardworking publicist, Meghan Roberts, and Angela Messina, the director of publicity and marketing associate William Hearn at Basic Books. This was the best group of professionals I could have imagined.

I also want to thank my many friends who offered specific encouragement, discussed my ideas, or read early drafts of this book and provided feedback. Most substantively, my old friend John Comeau carefully read an early draft of my book and offered lots of nice feedback. Aneil Mallavarapu discussed the book with me at length, and Rasmus Grønfeldt Winther read an early draft and provided valuable comments and welcome encouragement.

◆

Finally, I want to acknowledge the support of my family, including my parents, Lucy and Al, and my father, JB. But most of all, I want to thank my wife, Jennifer, whose support was rock-solid throughout the whole process. She discussed ideas, listened to me read passages, commented on specific wording, and gave me suggestions many times over several years along the way. Her help ranged from the specific, like copyediting an early chapter for my proposal, to the expansive, such as discussing how I should frame the concept of uncountable infinity for the general reader. She also put up with me taking over our bookshelves, using the kitchen table for months at a time, and leaving piles of papers all over the house. For all that, I owe her a lifetime of spring cleanings at the very least!

My son, Zack, was also very encouraging, and his teenage years paralleled my production schedule. When I started the book, he was just starting middle school, and when I finished, he was a senior in high school and interested in history in his own right. I fondly recall taking him with me on research runs to the University of Maryland.

ACKNOWLEDGMENTS

I also want to thank my daughter, Georgia, for her amazing help. She became a dedicated assistant at times in the crucial final stages of this project. She chased down references, found papers I was looking for at the UMD library, and spent countless hours helping me format my references. It was also fun to share ideas for chapter titles with her—some profoundly stupid (e.g., "Throw Out the Maybe with the Math Water").

ANNOTATED BIBLIOGRAPHY

This book covers a variety of topics, from the legendary to the obscure, from famous figures to notorious world events, from enduring philosophies to forgotten scientific ideas, and from the bitter poison of politics and warfare to the overwhelming beauty of the human intellect. Annotated descriptions of its various topics follow.

MATHEMATICAL BACKGROUND

First, if I had to give a highly curated short list of additional reading, I would say that readers interested in more of the math should start with articles by Calder (1979); Bishop (1975); Putnam (1967); Courant (1964); Snapper (1979a); Ferreirós (2009b); and Cohen and Hersh (1967), which is perhaps the best *Scientific American* piece ever written about mathematics. Finally, Halmos (1990) is a brilliantly well-written state-of-math-type paper that I highly recommend.

For a deeper dive into the technical mathematical history of the Great Math War, see van Heijenoort (1967), Mancosu (1998), van Dalen (1999 and 2013), and Dauben (1979). For the general reader interested in colorful depictions of the era, see autobiographies by Kovalevskaya (1978), Russell (1967a, 1968, 1969), and Morrell—listed under Gathorne-Hardy (1964 and 1975) as well as the excellent memoir by Graves (1929), which also served as the inspiration for the pink...black...slimy description of the relationship between Russell and Lawrence during the war. Other great books are Wilson (2011) and Dangerfield (1980).

BERTRAND RUSSELL

The once gifted and completely forgotten-in-the-attic copy of *The Basic Writings of Bertrand Russell* I mention in the preface is Russell (1967b). See

ANNOTATED BIBLIOGRAPHY

Clark (2012) and Monk (1996) for good biographies of Russell. Shorter treatments of his life can be found in the New York Times obituary by Whitman (1970), Warnock (1966), Whittaker (1945), the long obituary by Kreisel (1973), and my own earlier article timed to the fiftieth anniversary of his death—Bardi (2020). An interesting description of his relationship with his first wife, Alyssa "Alys" Whitall Pearsall Smith, appears in Krieg (2000). Russell's brief stint at Harvard resulted in some of the most hilarious skewering in the history of university life. A great description of this is Willis (1989). A good description of Russell late in life can be found in Grattan-Guinness (2009).

Logical atomism is described by Russell himself in a lecture he gave on the eve of being thrown into jail during World War I. See Russell (1919a, 1919b, and 1919c). He is called the Pablo Picasso of modern philosophy in Ayer (1941). Russell's own autobiography is an excellent read, if somewhat hyperbolic at times—see comments in Bell (1985). Russell himself deserves the last word, of course, having won the Nobel Prize in Literature. His selected letters are excellent and now published in books—Russell (1992b and 2001). His late-life, larded-with-famous-letters, three-volume autobiography (1967a, 1968, 1969) is, despite flaws, a great read.

DAVID HILBERT

Hilbert's speech in 1900 and the significance of his twenty-three challenge problems are discussed in Nunemacher (2003). Two papers that stunningly demonstrate the inspirational legacy of Hilbert's twenty-three problems are Morgenstern (1972) and Smale (1998). Another great paper that discusses his legacy more generally is Davies (2005). Hilbert's life and legacy are nicely detailed in the obituaries by Weyl (1944) and Taussky (1943). The book by Reid (1996) is excellent and the best overall biography of Hilbert. Hilbert's boyhood town of Königsberg and its eponymous seven bridges problem are described in Brinkle and Griesacker (1990) and Mallion (2008).

Hilbert as an applied mathematician is described in Corry (1999). His book *Grundlagen der Geometrie* (*The Foundations of Geometry*) is still in print—see Hilbert (2021). Its significance is discussed in Peckhaus (2003). Finally, pernicious anemia and the liver cure are discussed in *The British*

ANNOTATED BIBLIOGRAPHY

Medical Journal in the articles "Dietetic Treatment of Pernicious Anaemia" (1926) and "Nature and Treatment of Pernicious Anaemia" (1934), the article "Pernicious Anaemia" in the *Canadian Public Health Journal* (1936), and the article "Conquest of Anemia One of Medicine's Great Epics" in *The Science News-Letter* (1934).

L. E. J. BROUWER

Everything under the Brouwer sun, from Dutch science in the Netherlands to L. E. J. Brouwer's childhood, his schooling, his work in topology, his impact on the field, his faculty job at the University of Amsterdam, his marriage to Lize, his relationship with stepdaughter Anna Louise Elisabeth, her classmate Cor Jogejan, and his book *Life, Art, and Mysticism* can be found in the excellent books by Dirk van Dalen (1999, 2011, and 2013). These are invaluable, amazing resources for anyone who wants to take a deep dive into Brouwer's life and work.

A great depiction of Brouwer as a difficult colleague can be found in van Dalen (2000). Calder (1979) gives a fantastic, easy-to-read introduction to constructivist mathematics. Brouwer's inaugural address at the University of Amsterdam, titled "Intuitionism and Formalism," is found in Brouwer (1913). Hilbert's biography by Reid (1970) also has good information on Brouwer, including both his editorship of the *Mathematische Annalen* and his meeting with Brouwer while vacationing along the Dutch coast in the 1910s.

LADY OTTOLINE MORRELL

The best sources for Lady Ottoline Morrell's childhood are her own memoirs, listed under Gathorne-Hardy (1964 and 1975), as well as the books by Darroch (1975) and Seymour (1992). If you want something shorter, Brink (1977) is a good review of Darroch's book, and Griffin (1993) is a review of Seymour's. Both have summaries of Ottoline's life. Another interesting source is Vytniorgu (2018). The Bloomsbury Set is described in Edel (1977). Examples of what I term the mean-girl exclusion of the Morrells from history can be found in Hepburn (1976), Lyon (1994), and Schwerin (1999). Sources on her affair with Russell include her biographies as well as Brink (1976 and 1977) and Moran (1991). Garsington is described in Gill (1974).

ANNOTATED BIBLIOGRAPHY

LOGICISM

Russell and his early work in mathematical logic, particularly his interest in Giuseppe Peano, is described in Jourdain (1911, 1912, and 1916b) and Kennedy (1960). A nice early essay on Russell's 1903 book *Principles of Mathematics* appears in Jourdain (1910). The product of his collaboration with Whitehead is still in print—*Principia Mathematica*, vols. I, II, and III, Whitehead and Russell (2011). Early insightful reviews of it are in Jourdain (1912) and Bernstein (1926).

Russell's "ramified theory of types" is described in Kamareddine (2002). Russell's own description is Russell (1908). Russell's "Logical Atomism" lectures (1919a, 1919b, and 1919c) are fascinating reading—especially given the fact that it's toward the end of a brutal four years of war and horror and he's about to be jailed for sedition when he delivers them. Musgrave (1977) gives a good general introduction to logicism. A modern edition of *Introduction to Mathematical Philosophy*, the book Russell wrote while in prison, is Russell (2017). The legacies of logicism in the birth of computer science, computer algorithms, theories of recursion, computability, and the golden age of logic in the 1930s are described in Chaitin (2002).

FORMALISM

Hilbert's brief flirtation with logicism, which motivated him to develop proof theory, is described in Ferreirós (2009a) and Mancosu (1999a and 2003). See also Hilbert's own 1928 book, Hilbert and Ackerman (1950). His axiomatic methods and formalism are discussed in Irvine and Simons (2009) and Webb (1997). The development of metamathematics and proof theory is described nicely in Zach (2003) and Avigad and Reck (2001). An older but quite excellent basic explanation is given in DeSua (1954). A longer treatment of Hilbert's program is covered in Simpson (1988) and Raatikainen (2003). The disposition of mathematicians to be "far more amenable to a plan devised by Hilbert" comes from Calder (1979). Finally, legacies of formalism and Hilbert's program are discussed in Peckhaus (2003), Sieg (1999, 2019, and 1988), and Halmos (1990).)

ANNOTATED BIBLIOGRAPHY

INTUITIONISM

Good, in-depth discussions of intuitionism appear in Dresden (1924) and Bauer (2017). Intuitionistic mathematics and its origins appear in van Dalen (1995). A good summary of Weyl's *Das Kontinuum* can be read in Mancosu (1998), which also reprints Weyl's famous *Grundlagenkrise* (foundational crisis) paper. The source book by van Heijenoort (1967) also contains a number of original papers describing intuitionism. A discussion of *Tertium non datur* (law of the excluded middle) can be found in Church (1928). Reactions in Europe to intuitionism are discussed in Hesseling (2003) and Avigad (2006). The "three crises" paper by Snapper (1979a) has a nice short treatment of intuitionism and is a great paper besides.

THE FOUNDATIONAL CRISIS

Good technical discussions about the foundational crisis can be found in Wavre (1934), Snapper (1979a and 1979b), Robič (2015), Poirier (2024), and especially Ferreirós (2009b), which is an excellent and short summary of the crisis and its players. Good general works on mathematical logic are DeLong (1970) and Mancosu (2010a). A great older paper on mathematical philosophy is Dresden (1928), and a more modern source is Linnebo (2017). Another good source is Bishop (1975). A long discussion of the frog-mouse battle appears in van Dalen (2013). The outcome of the foundational crisis—that it soured many people on the whole subject—can be found in Mancosu (1999b) and Dawson (1984). A good paper reflecting on the philosophy of mathematics decades after the debate is Mehlberg (1960). An extensive discussion of what mathematics became after the early 1930s can be found in Gray (2008) and Rowe (2013)—the latter being a short review of the former. A well-written and provocative take on the foundations of mathematics can be found in Putnam (1967). A great paper on mathematics in the "modern" world—modern being the early 1960s—is Courant (1964).

MATHEMATICS AND MATH SOCIETIES IN 1900

Good sources on mathematics in the late nineteenth century include Hille (1953), Miller (1900), and Pierpont (1999). The creation of the

ANNOTATED BIBLIOGRAPHY

international math community is described in Cassels (1999) as well as in Dauben (1979). The formation of the international congress in the late nineteenth century is covered in Curbera (2010). The European math scene in the late nineteenth century is surveyed in Ewald (2003) and Graham and Kantor (2006).

Nice descriptions of the Second International Congress of Mathematicians in Paris in 1900 are by Scott (1900) and the article "The International Congress of Mathematicians" in *Nature* (1900). The separate Paris expo is described in Thompson (1901). The Third International Congress of Mathematicians is described in Tyler (1904). The Fifth International Congress is covered in Snyder (1912) and Young (1912)—as well as the biography by Monk (1996). The Nauheim meeting and the Sixth International Congress in Strasbourg—*une affaire française*—are discussed in van Dalen (1999). The move to exclude German scientists from the international community after World War I is discussed in several sources, including the article "International Scientific Organization" in *Science* (1918). Charles Fields and the Seventh International Congress in Toronto are discussed in Barnes (2005). The eighth congress in Bologna is detailed in Tonelli (1929) and Siegmund-Schultze (2016).

OTHER MATHEMATICIANS AND SCIENTISTS

The best source on **Georg Cantor**, his childhood, family, religion, and mathematics, is Dauben's book (1979) as well as his papers (Dauben 1978 and 1983). Elsewhere, Dauben (1977) discusses Cantor's interactions with Catholic authorities and Cardinal Johannes Franzelin of the Vatican Council. Other good discussions of Cantor's work include papers by Tahta (2007), Ferreirós (1996), Johnson (1970), and Oberschelp (1982), which is really just a review of Dauben's book.

Good sources about **Albert Einstein** on relativity and his race with Hilbert for general relativity are Einstein (1923) and Corry et al. (1997). The claim that Einstein achieved his godlike social status when he "awoke in Berlin on the morning of November 7, 1919, to find himself famous" is in Elton (1986). General stereotypes of math in pop culture are in Wilson and Latterell (2001).

ANNOTATED BIBLIOGRAPHY

Discussions of **Gottlob Frege** and his *Begriffsschrift* (concept notation), his *Grundlagen der Arithmetik* (Foundations of arithmetic), and his influence appear in Boolos (1995 and 1998), Frege (1915), and Heck (2003). Frege's influence on Russell is described in Klement (2004).

Kurt Gödel and the Königsberg conference Q&A session in 1930 where he drops the first announcement of incompleteness are discussed in Davis (2006). The symposium itself is in Carnap et al. (1984). Gödel's 1929 thesis on predicate logic is described in von Plato (2018). Kurt Gödel's strange death and legacy are described in Dawson (1999) and Sigmund (1997).

Felix Klein and his influence on mathematics are described in Reid (1995) and Rowe (1986).

An excellent short biography of **Sofya Kovalevskaya** can be found in Rappaport (1981) and Rygiel (1987). Her own autobiography, *A Russian Childhood* (1978), is an excellent read. The significance of Kovalevskaya's mentor Weierstrass is discussed in A. R. F. (1897). Politics and women in Russia in the 1860s and 1870s are discussed in Koblitz (1988). The papers by Kelley (1996) and Fabricant (1990) contain excellent short bios of many of the pioneering women in mathematics, from Hypatia to Kovalevskaya.

Hermann Minkowski is described in Galison (1979) and Pyenson (1977). A short Weyl biography is Newman (1958). Weyl's legacy is discussed in the essay by Wheeler (1986).

One work on **Henri Poincaré** that's worth reading is Davis and Mumford (2008)—it presents his mathematical worldview just a few years before he dies. Also see Gray (1998).

Photos of **Charles Émile Picard** appear in "[Photograph]: Emile Picard" in the *American Journal of Mathematics* (1895) and Miller (1926). A great obituary can be found in Hadamard (1942).

Emmy Noether has one main biography, quite an old one by Dick (1981). As a note to any aspiring writers out there, she is due for a major modern treatment. Shorter pieces on her life and work appear in Kimberling (1972 and 1982) and Weyl (1981).

Alfred North Whitehead and his mathematical background are described in MacColl (1899), Macfarlane (1899), Lowe (1975), and Lutskanov (2011). His collaboration with Russell is described in the obituary

ANNOTATED BIBLIOGRAPHY

by Whittaker (1948). As a dramatic backdrop to that collaboration, Doxiadis and Papadimitriou's graphic novel *LOGICOMIX* (2009) has a juicy though mostly historically unsupported depiction of Russell's infatuation with Evelyn Whitehead. (Strangely, that comic imagines this relationship as full-bore remains-of-the-day asexual tension while completely ignoring Russell's real physical and emotional relationship with Ottoline Morrell.)

Ludwig Wittgenstein's childhood is described in McGuinness (1988) and Monk (1990). Snyder (1912) and Young (1912) cover the troubles with Wittgenstein and Ottoline. The significance of *Tractatus Logico-Philosophicus* can be found in Russell's introduction to Wittgenstein (1969) and in Kreisel (1978). Wittgenstein's return to philosophy in 1928 is described in Marion (2003).

For more information on **Ernst Zermelo**, see Moore (1978) and Kanamori (2004). I wish I could have said more about him in this book, but space would not allow. (Sorry again, Zermelo!)

INFINITY, SET THEORY, AND THE PARADOXES

Troublesome infinity, its history, and how it bedeviled Archimedes, Gauss, Locke, Descartes, Spinoza, Galileo, and Aristotle can be found in Moore (1995). Zeno of Elea and his relevance to the foundational crisis are described beautifully in Lasley (1942). For a more modern discussion of set theory, see the absolutely outstanding paper by Cohen and Hersh (1967) as well as Kanamori (1996). Good general discussions of the continuum hypothesis can be found in Stillwell (2002), Gödel (1947), and Moore (2011). The Burali-Forti Paradox is discussed in Copi (1958), Rosser (1942), and Menzel (1984). Cantor's Paradox is discussed in Ferreirós (2009b). Russell's Paradox is detailed in Klement (2007).

SOUTH AFRICA AND THE BOER WAR

First-person accounts of the Boer War appear in Makins and Ashton (1900), the article "The War in South Africa: The Battle of Colenso" (1900), Beadnell (1900), the article "The War in South Africa: The Medical Aspects of the War" (1900), Cheyne (1900), and Thomson and Hartley (1900). A very British view of the Boer Army appears in the article "The Boer Army" (1900). Guerrilla hit-and-run tactics are discussed in Farwell

ANNOTATED BIBLIOGRAPHY

(1976). British "land clearance" policies, blockhouses, and not-so-pretty counterinsurgency approach can be found in Spiers (2004), Cosgrove (1980), and Krebs (1992). Non-British perspectives on the concentration camps are covered in de Reuck (1999), Heyningen (2008), and Moore (2010). The lingering effect of the war in the British mind is presented in Donaldson (2013). The propaganda nature of the war appears in Kuitenbrouwer (2012a and 2012b).

BRITISH POLITICS IN THE PRE-WORLD WAR I ERA

Ferguson (2002) gives a nice summary of Great Britain at the start of the twentieth century. Meyer (2000) talks about British influence around the world in 1900. C. P. Scott and the British revolution in journalism are described in Hampton (2001). The Liberal revolution in England and the ultimate, almost ironic downfall of the Liberal Party are described in Moorhouse (1973), Clarke (1984), Cregier (1970), Koss (1968), Richards (1975), Wilson (2011), and Dangerfield (1980). The general nasty, vitriolic, and violent elements of English elections are described in Gregory (2003). The 1906 election is described in Irwin (1994), Klug (2001), and Hazlehurst (1970). Notestein (1916) covers the life and career of Prime Minister Herbert Henry Asquith. The House of Lords veto is described in Stephens (1982) and Weston (1968). The naval arms race is described in Massie (1991), Rahn (2017), Ross (2018), Grainger (2014), and St. John (1971). David Lloyd-George is described in Brooks (1981). His *who-killed-Cock-Robin* coup of Asquith in 1916 is described in Fry (1988) and McEwen (1978a and 1978b).

Irish home rule is covered in McCready (1963), Foy (1996), and Buckland (1967). The crisis it creates for the Asquith government is covered in Jalland and Stubbs (1981) and McEwen (1972). The Larne gunrunning and its impact are described in Jackson (1993). The armed Easter uprising of Irish republican nationalists in Dublin is described in Buckley (1956).

WORLD WAR I

There are many references for World War I, the Western Front, and the human cost of the war. Several interesting general sources are Martel et al. (2013), Sheffield (2007), Mallard and White (2019), Fromkin

ANNOTATED BIBLIOGRAPHY

(2004), Kiester (2007), Doyle (2014), Stokesbury (1981), and Coetzee and Shevin-Coetzee (2002). The books by the incomparable Barbara Tuchman (1962 and 1966) are above all else highly recommended for their wonderful style and penetrating insight. The first-person account of Graves (1929) makes for one of the best autobiographies ever written in the English language. Another account is Manning (1930). Spy fever is described in French (1978). US army spending is explored in Ayres (1920), the source of the statistic that the US alone bought ninety-six million pairs of socks in 1918. The fates of Kitchener and Churchill are detailed in Puleston (1931).

ANTIWAR ACTIVISM

Russell's peace "conversion" is covered in Kennedy (1984) and Hare (2002)—as well as in the graphic novel LOGICOMIX by Doxiadis and Papadimitriou (2009) which is reviewed in Mancosu (2010b). Russell's commitment to peace is reflected in the American Mathematical Society naming one of its top prizes the "Bertrand Russell Prize of the AMS" (2025). Antiwar protests in England are described in Rempel (1975). The women's movement involvement is detailed in Hirshfield (1982). Conscription, Britain's Military Service Act, and the exceptions for conscientious objectors are described in Adams (1986) and the article "The Military Service Act" in *The British Medical Journal* (1916).

GERMANY IN THE 1920S

The characterization of Germany after World War I as a "joyless republic" and a "republic without republicans" is from Hertzman (1960). The characterization of the Weimar Republic as one that was "born in defeat, lived in turmoil, and died in disaster" comes from Roth (2004). Germany's war debt is discussed in Lutz (1930) and Fischer (2005). The causes and effects of Black Tuesday in 1929 are explored in Galbraith (1961). Mowry (1965) discusses the culture of the 1920s more generally—especially in the United States.

THE GREAT MIGRATION

Adolf Hitler's rise to German chancellor is covered in many places. Frankel (2003), Clemens (1999), and Gaab (2011) describe his early days in Munich,

ANNOTATED BIBLIOGRAPHY

for instance. Marx (1937) discusses the impact of the restoration of the Professional Civil Service Act, and Segal (1980) discusses academic activism in Göttingen in its wake. Courant's perspective is described at length in Alexanderson (1980) and the excellent book by his biographer Constance Reid (1976). Otto Neugebauer's brief tenure is described in Pingree (1991) and in Neugebauer (1936). Backlash at American universities against exodus professors is detailed in the excellent paper by Reingold (1981) and Lamberti (2006). Long lists of exodus professors can be found in Dresden (1942).

EUGENICS

This is a topic I would have liked to cover in more depth, but space would not allow. I touch on it briefly in the context of 1938, but there is much more to say—especially of American and British involvement. The German medical profession's culpability in eugenics comes from Czech et al. (2023), which describes how euthanasic murder was being taught in German medical schools by 1938. Sterilization in Germany and America is discussed in Tanner (2012), Hart (2012), Braslow (1996), and Reilly (1987). The impact of the eugenics movement in Germany is discussed in Weikart (2003). A good concise overall summary of eugenics is Allen (2001). A nice paper on the overlap between eugenics and statistics is Louçã (2009). Other good sources on eugenics in America and Britain are Burke and Castaneda (2007), Jones (1992), Stillwell (2012), and Field (1911). A good paper on the overlap of eugenics and genetics can be found in Ludmerer (1969). Finally, there are lots of soul-searching, genuinely mournful apologies for past eugenics, one example of which is Trent (2023), which came almost twenty-five years after Wikler (1999) wrote a thoughtful paper on the social justice legacies of eugenics within the field of genetics.

INCOMPLETENESS

There are lots of sources on incompleteness. Nagel and Newman's short book and long paper (1958 and 1956) are good easy points of entry. Rebecca Goldstein's outstanding book (2005) is an excellent long-form treatment with one of the most beautiful personal essay openings I have ever read. Short appreciations of Gödel's proof can be found in Hodges (2008) and Davis (2005).

ANNOTATED BIBLIOGRAPHY

NEUROSCIENCE AND INTUITIONISM

The neurological mechanisms that account for how one intuits the world through constructive mental exercises are an evolving framework that I only began to discover while writing the final drafts of this book—and one I hope to more fully explore in the future. The 2023 study involving infants in Barcelona is Bohus et al. (2023). The study on self-assessment of math ability based on math scores is Wan et al. (2022). The urban coyotes study was described in a public presentation by Stan Gehrt at a New Horizons in Science meeting in Columbus, Ohio, in 2014. The work of researchers at the University of Tübingen in Germany studying crows is described in Liao et al. (2022) and Kirschhock and Nieder (2023). The Okinawa Institute of Science and Technology studies on the clown fish (anemonefish *Amphiprion ocellaris*) is Hayashi et al. (2024).

SOURCES

"The Achievement of the Army Medical Department in the World War." *JAMA*, vol. 325, no. 11, March 19, 1921, p. 1115, https://doi.org/10.1001/jama.2020.17846.

Adams, John W. "The Influences Affecting Naval Shipbuilding Legislation, 1910–1916." *Naval War College Review*, vol. 22, no. 4, 1969, pp. 41–70. JSTOR, http://www.jstor.org/stable/44639502.

Adams, L. J. "Mathematical World News." *National Mathematics Magazine*, vol. 12, no. 2, 1937, pp. 90–93. JSTOR, http://www.jstor.org/stable/3028476.

Adams, R. J. Q. "Asquith's Choice: The May Coalition and the Coming of Conscription, 1915–1916." *Journal of British Studies*, vol. 25, no. 3, 1986, pp. 243–263. JSTOR, http://www.jstor.org/stable/175463.

Alexanderson, Gerald L. "About the Cover: Hilbert and the Paris ICM." *Bulletin of the American Mathematical Society*, vol. 51, no. 2, January 27, 2014, pp. 329–334.

Alexanderson, Gerald L. "An Interview with Constance Reid." *The Two-Year College Mathematics Journal*, vol. 11, no. 4, 1980, pp. 226–238. JSTOR, https://doi.org/10.2307/3027201.

Alford, Roger P. "The Nobel Effect." *Proceedings of the American Society of International Law*, vol. 103, 2009, pp. 467–468. JSTOR, http://www.jstor.org/stable/10.5305/procannmeetasil.103.1.0467.

Allen, Ann Taylor. *The Journal of Modern History*, vol. 78, no. 1, 2006, pp. 255–257. JSTOR, https://doi.org/10.1086/502761.

Allen, Frederick Lewis. *Only Yesterday: An Informal History of the 1920s*. Harper & Row, 1964.

ANNOTATED BIBLIOGRAPHY

Allen, Garland E. "Is a New Eugenics Afoot?" *Science*, vol. 294, no. 5540, 2001, pp. 59–61. JSTOR, http://www.jstor.org/stable/3084765.

Allen, Henry T. "Present Franco-German Situation." *The Annals of the American Academy of Political and Social Science*, vol. 126, 1926, pp. 15–18. JSTOR, http://www.jstor.org/stable/1015515.

Alsberg, Henry G. "War Aims." *The Antioch Review*, vol. 1, no. 1, 1941, pp. 21–34. JSTOR, https://doi.org/10.2307/4608817.

Altick, Richard D. "Victorian Biography: Eminent Victorianism: What Lytton Strachey Hath Wrought." *The American Scholar*, vol. 64, no. 1, 1995, pp. 81–89. JSTOR, http://www.jstor.org/stable/41212289.

Anderson, Gary M., and Robert D. Tollison. "Ideology, Interest Groups, and the Repeal of the Corn Laws." *Journal of Institutional and Theoretical Economics*, vol. 141, no. 2, 1985, pp. 197–212. JSTOR, http://www.jstor.org/stable/40750831.

"Armored Trains." *Scientific American (1845–1908)*, vol. 82, January 13, 1900, p. 18.

"The Armored Trains in South Africa." *Scientific American Supplement*, vol. 1255, January 20, 1900, p. 20211.

Armytage, W. H. "The 1870 Education Act." *British Journal of Educational Studies*, vol. 18, no. 2, June 1970, pp. 121–133, https://doi.org/10.2307/3120304.

"Asquith: The Master Statesman." *The North American Review*, vol. 198, no. 695, 1913, pp. 433–443. JSTOR, http://www.jstor.org/stable/25120104.

Avery, Todd. "'The Historian of the Future': Lytton Strachey and Modernist Historiography Between the Two Cultures." *ELH*, vol. 77, no. 4, 2010, pp. 841–866. JSTOR, http://www.jstor.org/stable/40963111.

Avigad, Jeremy. "Review: *Gnomes in the Fog. The Reception of Brouwer's Intuitionism in the 1920s*." *The Mathematical Intelligencer*, vol. 28, no. 4, September 2006, pp. 71–74, https://doi.org/10.1007/bf02984712.

Avigad, Jeremy, and Erich H. Reck. "'Clarifying the Nature of the Infinite': The Development of Metamathematics and Proof Theory." Carnegie Mellon University, 2001, https://philpapers.org/rec/AVICTN.

Ayer, A. J. "Bertrand Russell on Meaning and Truth." *Nature*, vol. 148, no. 3747, August 23, 1941, pp. 206–207, https://doi.org/10.1038/148206a0.

Ayres, Leonard P. *The Official Record of the United States' Part in the Great War*. Government Printing Office, 1920.

Badsey, Stephen. "The Boer War (1899–1902) and British Cavalry Doctrine: A Re-evaluation." *The Journal of Military History*, vol. 71, no. 1, 2007, pp. 75–97. JSTOR, http://www.jstor.org/stable/4138030.

Ball, Stuart R. "Asquith's Decline and the General Election of 1918." *The Scottish Historical Review*, vol. 61, no. 171, 1982, pp. 44–61. JSTOR, http://www.jstor.org/stable/25529447.

"The Barbed Wire Patent Declared Invalid." *Scientific American (1845–1908)*, vol. 58, January 14, 1888, p. 16.

ANNOTATED BIBLIOGRAPHY

Bardi, Jason Socrates. "Peace, Love, and Mathiness: Considering Bertrand Russell's Relevance." *The Humanist*, March–April 2020, published online February 25, 2020, https://thehumanist.com/magazine/march-april-2020/features/peace-love-and-mathiness/.

Barnes, Marcus Emmanuel. "John Charles Fields and the Fields Medal." *Pi Mu Epsilon Journal*, vol. 12, no. 2, 2005, pp. 65–70. JSTOR, http://www.jstor.org/stable/24340684.

Barrett, Deborah, and Charles Kurzman. "Globalizing Social Movement Theory: The Case of Eugenics." *Theory and Society*, vol. 33, no. 5, 2004, pp. 487–527. JSTOR, http://www.jstor.org/stable/4144884.

Barton, Clara. "Our Work and Observations in Cuba." *The North American Review*, vol. 166, no. 498, 1898, pp. 552–559. JSTOR, http://www.jstor.org/stable/25118997.

Bates, Gordon. "What Social Hygiene Means." *The Public Health Journal*, vol. 16, no. 8, 1925, pp. 383–386. JSTOR, http://www.jstor.org/stable/41973351.

Bauer, Andrej. "Five Stages of Accepting Constructive Mathematics." *Bulletin of the American Mathematical Society*, vol. 54, no. 3, July 2017, pp. 481–498, https://doi.org/10.1090/bull/1556.

Bayly, Martin. "Fatalism and an Absence of Public Grief: How British Society Dealt with the 1918 Flu." *British Politics and Policy at LSE*, October 28, 2020, https://blogs.lse.ac.uk/politicsandpolicy/public-memory-1918-flu/.

Beadnell, C. Marsh. "The War in South Africa." *The British Medical Journal*, vol. 1, no. 2037, 1900, pp. 99–104. JSTOR, http://www.jstor.org/stable/20263135.

Bederson, Benjamin. "Fritz Reiche and the Emergency Committee in Aid of Displaced Foreign Scholars." *Physics in Perspective*, vol. 7, no. 4, December 2005, pp. 453–472, https://doi.org/10.1007/s00016-005-0245-3.

"Before and After the British Election." *Advocate of Peace Through Justice*, vol. 84, no. 12, 1922, pp. 430–431. JSTOR, http://www.jstor.org/stable/20660160.

Bell, Robert H. "Confession and Concealment in 'The Autobiography of Bertrand Russell.'" *Biography*, vol. 8, no. 4, 1985, pp. 318–335. JSTOR, http://www.jstor.org/stable/23539390.

Bell, Vanessa. *Helen Dudley*. Oil paint on canvas, 1915. Tate Collection, https://www.tate.org.uk/art/artworks/bell-helen-dudley-t01123.

Bernand, Francis, editor. *Punch, or the London Charivari*, vol. 103. Bradbury & Evans, 1892, via Project Gutenberg, https://www.gutenberg.org/files/20759/20759-h/20759-h.htm.

Berns, Walter. "Buck v. Bell: Due Process of Law?" *The Western Political Quarterly*, vol. 6, no. 4, 1953, pp. 762–775. JSTOR, https://doi.org/10.2307/443203.

Bernstein, B. A. "Whitehead and Russell's *Principia Mathematica*." *Bulletin of the American Mathematical Society*, vol. 32, no. 6, 1926, pp. 711–713.

ANNOTATED BIBLIOGRAPHY

"Bertrand Russell Again." *The Nation*, vol. 100, no. 2597, April 8, 1915, p. 385.

"Bertrand Russell / The Reviewer Replies." *The Nation*, vol. 100, no. 2593, March 11, 1915, pp. 274–275.

"Bertrand Russell Prize of the AMS." American Mathematical Society, accessed January 29, 2025, https://www.ams.org/prizes-awards/paview.cgi?parent_id=40.

Binkley, Robert C. "Ten Years of Peace Conference History." *The Journal of Modern History*, vol. 1, no. 4, 1929, pp. 607–629. JSTOR, http://www.jstor.org/stable/1871103.

Bishop, Errett. "The Crisis in Contemporary Mathematics." *Historia Mathematica*, vol. 2, no. 4, November 1975, pp. 507–517, https://doi.org/10.1016/0315-0860(75)90113-5.

Bisschop, W. R. "The Locarno Pact, October 15–December 1, 1925." *Transactions of the Grotius Society*, vol. 11, 1925, pp. 79–115. JSTOR, http://www.jstor.org/stable/742835.

Blackburn, Sheila. "Ideology and Social Policy: The Origins of the Trade Boards Act." *The Historical Journal*, vol. 34, no. 1, 1991, pp. 43–64. JSTOR, http://www.jstor.org/stable/2639707.

Boas, Franz. "Eugenics." *The Scientific Monthly*, vol. 3, no. 5, 1916, pp. 471–478. JSTOR, http://www.jstor.org/stable/6055.

Boas, R. P. "Award for Distinguished Service to Otto Neugebauer." *The American Mathematical Monthly*, vol. 86, no. 2, 1979, pp. 77–78. JSTOR, http://www.jstor.org/stable/2321939.

"The Boer Army." *Scientific American Supplement*, no. 1275, June 9, 1900, p. 20439.

"The Boer War." *The British Medical Journal*, vol. 2, no. 2079, 1900, pp. 1322–1323. JSTOR, http://www.jstor.org/stable/20266256.

Bohus, Kinga Anna, et al. "The Scope and Role of Deduction in Infant Cognition." *Current Biology*, vol. 33, no. 18, September 25, 2023, https://doi.org/10.1016/j.cub.2023.08.028.

Bolduan, Charles. "Medical Science in the Service of War." *Scientific American*, vol. 111, no. 19, 1914, pp. 382–396. JSTOR, http://www.jstor.org/stable/26015318.

Boolos, George. "Frege's Theorem and the Peano Postulates." *Bulletin of Symbolic Logic*, vol. 1, no. 3, September 1995, pp. 317–326, https://doi.org/10.2307/421158.

Boolos, George. *Logic, Logic, and Logic*. Harvard University Press, 1998.

Bossenbroek, Martin. *The Boer War*. Translated by Yvette Rosenberg. Seven Stories Press, 2018.

Boughton, Clement R. "Jaundice & War: Viral Hepatitis and Other Causes of Jaundice in Times of War." *Health and History*, vol. 4, no. 2, 2002, pp. 41–56. JSTOR, https://doi.org/10.2307/40111437.

Boulter, Michael. "Old Habits Die Hard, 1901–14." *Bloomsbury Scientists: Science and Art in the Wake of Darwin*. UCL Press, 2017, pp. 115–134. JSTOR, https://doi.org/10.2307/j.ctt1vxm8sr.13.

ANNOTATED BIBLIOGRAPHY

Boulter, Michael. "The Rise of Eugenics, 1901–14." *Bloomsbury Scientists: Science and Art in the Wake of Darwin.* UCL Press, 2017, pp. 102–114. JSTOR, https://doi.org/10.2307/j.ctt1vxm8sr.12.

Boulter, Michael. "Time Passes, 1914–18." *Bloomsbury Scientists: Science and Art in the Wake of Darwin.* UCL Press, 2017, pp. 135–145. JSTOR, https://doi.org/10.2307/j.ctt1vxm8sr.14.

Bowlby, Anthony A., and Frederick Treves. "The War in South Africa." *The British Medical Journal*, vol. 1, no. 2061, 1900, pp. 1610–1612. JSTOR, http://www.jstor.org/stable/20265027.

Braslow, Joel T. "In the Name of Therapeutics: The Practice of Sterilization in a California State Hospital." *Journal of the History of Medicine and Allied Sciences*, vol. 51, no. 1, January 1996, pp. 29–51. JSTOR, https://www.jstor.org/stable/24624084.

Brawley, Mark R. "Tariff Reform, Taxes and Land: Trade-Based Cleavages in Pre–World War I Britain." *Review of International Political Economy*, vol. 16, no. 5, 2009, pp. 827–853. JSTOR, http://www.jstor.org/stable/27756196.

Brightman, R. "The Scientific Spirit in Education." *Nature*, vol. 130, no. 3293, December 10, 1932, pp. 863–865.

Brink, Andrew. "Lady Ottoline Morrell's Life." *Russell: The Journal of Bertrand Russell Studies*, no. 25–28, December 31, 1977, pp. 75–83, https://doi.org/10.15173/russell.v0i1.1478.

Brink, Andrew. "Russell to Lady Ottoline Morrell: The Letters of Transformation." *Russell: The Journal of Bertrand Russell Archives*, vol. 21, 1976, pp. 3–15. Project MUSE, https://muse.jhu.edu/article/882546.

Brinkle, Lydle, and Paul Griesacker. "Euler's Solution (or Discoveries) in Relation to the Problem of the Königsberg Bridges: The Origin of the Mathematics of Topology." *Journal of the Pennsylvania Academy of Science*, vol. 64, no. 2, 1990, pp. 103–107. JSTOR, http://www.jstor.org/stable/44149364.

Broad, William J. "Findings Back Einstein in a Plagiarism Dispute." *The New York Times*, November 18, 1997, https://www.nytimes.com/1997/11/18/science/findings-back-einstein-in-a-plagiarism-dispute.html.

Broadbent, T. A. A. "George Boole (1815–1864)." *The Mathematical Gazette*, vol. 48, no. 366, December 1964, pp. 373–378, https://doi.org/10.2307/3611693.

Brooks, David. "Lloyd George, for and Against." *The Historical Journal*, vol. 24, no. 1, 1981, pp. 223–230. JSTOR, http://www.jstor.org/stable/2638915.

Brouwer, L. E. J. "Intuitionism and Formalism." Translated by Arnold Dresden. *Bulletin of the American Mathematical Society*, vol. 20, no. 2, 1913, pp. 81–96, https://doi.org/10.1090/s0002-9904-1913-02440-6.

Browder, Felix. "Reflections on the Future of Mathematics." *Notices of the American Mathematical Society*, vol. 49, no. 6, 2002, pp. 658–662.

ANNOTATED BIBLIOGRAPHY

Buckland, P. J. "The Southern Irish Unionists, the Irish Question, and British Politics 1906–14." *Irish Historical Studies*, vol. 15, no. 59, 1967, pp. 228–255. JSTOR, http://www.jstor.org/stable/30004963.

Buckley, Maureen. "Irish Easter Rising of 1916." *Social Science*, vol. 31, no. 1, 1956, pp. 49–55. JSTOR, http://www.jstor.org/stable/41884424.

Burke, Chloe S., and Christopher J. Castaneda. "The Public and Private History of Eugenics: An Introduction." *The Public Historian*, vol. 29, no. 3, 2007, pp. 5–17. JSTOR, https://doi.org/10.1525/tph.2007.29.3.5.

Calder, Allan. "Constructive Mathematics." *Scientific American*, vol. 241, no. 4, October 1979, pp. 146–171.

Cantor, Georg. *Contributions to the Founding of the Theory of Transfinite Numbers*. Dover Edition, 1955.

Carden, Lieutenant Godfrey L. "British and Boer Guns—a Lesson from the South African War." *Scientific American* (1845–1908), vol. 82, January 20, 1900, p. 35.

Carlson, Elof Axel. "Commentary: R. L. Dugdale and the Jukes Family: A Historical Injustice Corrected." *BioScience*, vol. 30, no. 8, August 1980, pp. 535–539, https://doi.org/10.2307/1307974.

Carnap, Rudolf, Arend Heyting, and Johann von Neumann. "Symposium on the Foundations of Mathematics." *Philosophy of Mathematics: Selected Readings*. Ed. Paul Benacerraf and Hilary Putnam. Cambridge University Press, 1984.

Carr, Avery. "Russell Was Not Naive." AMS Blogs, June 27, 2013, https://blogs.ams.org/mathgradblog/2013/06/27/russell-naive/.

Carrasco, Isabel. "The Cuban Holocaust No One Talks About That Inspired the Nazis." *Cultura Colectiva*, November 16, 2017, https://www.culturacolectiva.com/en/history/cuban-concentration-camp-inspired-hitler/.

Cassels, J. W. S. "Review of Mathematics Without Borders: A History of the International Mathematical Union." *Notices of the American Mathematical Society*, vol. 46, no. 10, November 1999, pp. 1230–1232.

Castelvecchi, Davide. "Machine Learning Leads Mathematicians to Unsolvable Problem." *Nature*, vol. 565, no. 7739, January 2019, p. 277, https://doi.org/10.1038/d41586-019-00083-3.

Cawood, Ian, and Tom Crook, editors. *The Many Lives of Corruption: The Reform of Public Life in Modern Britain, c. 1750–1950*. Manchester University Press, 2022. JSTOR, http://www.jstor.org/stable/j.ctv2jsh3z5.

Chaitin, Gregory. "Computers, Paradoxes and the Foundations of Mathematics: Some Great Thinkers of the 20th Century Have Shown That Even in the Austere World of Mathematics, Incompleteness and Randomness Are Rife." *American Scientist*, vol. 90, no. 2, 2002, pp. 164–171, https://doi.org/10.1511/2002.2.164.

Chamberlain, J. P. "Current Legislation: Eugenics and Limitations of Marriage." *American Bar Association Journal*, vol. 9, no. 7, 1923, pp. 429–430. JSTOR, http://www.jstor.org/stable/25711334.

ANNOTATED BIBLIOGRAPHY

Chamberlain, J. P. "Eugenics and Limitations of Marriage." *Journal of Comparative Legislation and International Law*, vol. 5, no. 4, 1923, pp. 253–257. JSTOR, http://www.jstor.org/stable/752972.

Chang, Kenneth. "Vladimir Arnold Dies at 72; Pioneering Mathematician." *The New York Times*, June 11, 2010, http://www.nytimes.com/2010/06/11/science/11arnold.html.

Cheyne, W. Watson, et al. "The War in South Africa." *The British Medical Journal*, vol. 1, no. 2053, 1900, pp. 1093–1103. JSTOR, http://www.jstor.org/stable/20264453.

"Christiane Rousseau to Receive the 2018 Bertrand Russell Prize of the AMS." American Mathematical Society, November 16, 2017, https://www.ams.org/news?news_id=3821.

Church, Alonzo. "On the Law of the Excluded Middle." *Bulletin of the American Mathematical Society*, vol. 34, January 1, 1928, pp. 75–78, https://doi.org/10.1090/S0002-9904-1928-04516-0.

Clark, Ronald. *The Life of Bertrand Russell*. Bloomsbury Reader, 2012.

Clarke, Peter. "Bertrand Russell and the Dimensions of Edwardian Liberalism." *Russell: The Journal of Bertrand Russell Archives*, vol. 4, no. 1, 1984, pp. 207–221. Project MUSE, https://muse.jhu.edu/article/882251.

Clausewitz, Carl von. *Clausewitz on War*. Edited by Anatol Rapoport. Penguin Books, 1968.

Clemens, Detlev. "The 'Bavarian Mussolini' and His 'Beerhall Putsch': British Images of Adolf Hitler, 1920–24." *The English Historical Review*, vol. 114, no. 455, 1999, pp. 64–84. JSTOR, http://www.jstor.org/stable/579915.

Clement, Kevin C. "A New Century in the Life of a Paradox. Review-Essay of One Hundred Years of Russell's Paradox." *The Review of Modern Logic*, vol. 11, no. 1–2, 2007, pp. 7–30.

Cleveland, Richard. "The Axioms of Set Theory." *Mathematics Magazine*, vol. 52, no. 4, 1979, pp. 256–257. JSTOR, https://doi.org/10.2307/2689426.

"Clown Anemonefish Seem to Be Counting Bars and Laying Down the Law." Eurekalert!, February 1, 2024, https://www.eurekalert.org/news-releases/1032510.

Coetzee, Franz, and Marilyn Shevin-Coetzee. *World War One: A History and Documents*. Oxford University Press, 2002.

Cohen, Paul J., and Reuben Hersh. "Non-Cantorian Set Theory." *Scientific American*, vol. 217, no. 6, 1967, pp. 104–117. JSTOR, http://www.jstor.org/stable/24925924.

Cohrs, Patrick O. "The First 'Real' Peace Settlements After the First World War: Britain, the United States and the Accords of London and Locarno, 1923–1925." *Contemporary European History*, vol. 12, no. 1, 2003, pp. 1–31. JSTOR, http://www.jstor.org/stable/20081138.

Cole, G. D. H. "Recent Developments in the British Labor Movement." *The American Economic Review*, vol. 8, no. 3, 1918, pp. 485–504. JSTOR, http://www.jstor.org/stable/894.

ANNOTATED BIBLIOGRAPHY

Collodi, Carlo. *Pinocchio: The Adventures of a Marionette*. Translated by Walter S. Cramp. Heritage Edition, 1937.

"Conference on Birth Control." *The British Medical Journal*, vol. 2, no. 3212, 1922, pp. 132–133. JSTOR, http://www.jstor.org/stable/20420630.

Connelly, James. "Russell and Wittgenstein on Logical Form and Judgement: What Did Wittgenstein Try That Wouldn't Work?" *Theoria*, vol. 80, no. 3, August 19, 2013, pp. 232–254, https://doi.org/10.1111/theo.12029.

"Conquest of Anemia One of Medicine's Great Epics." *The Science News-Letter*, vol. 26, no. 708, 1934, pp. 275–276. JSTOR, https://doi.org/10.2307/3910077.

Copi, Irving M. "The Burali-Forti Paradox." *Philosophy of Science*, vol. 25, no. 4, 1958, pp. 281–286. JSTOR, http://www.jstor.org/stable/185640.

"Correspondence: Sighting Smokeless Flashes." *Scientific American*, February 10, 1900, p. 86.

Corry, Leo. "David Hilbert Between Mechanical and Electromagnetic Reductionism (1910–1915)." *Archive for History of Exact Sciences*, vol. 53, no. 6, 1999, pp. 489–527. JSTOR, http://www.jstor.org/stable/41134068.

Corry, Leo, et al. "Belated Decision in the Hilbert-Einstein Priority Dispute." *Science*, vol. 278, no. 5341, 1997, pp. 1270–1273. JSTOR, http://www.jstor.org/stable/2894186.

Cosgrove, Richard A. "The Boer War and the Modernization of British Martial Law." *Military Affairs*, vol. 44, no. 3, 1980, pp. 124–127. JSTOR, https://doi.org/10.2307/1987436.

"The Cost of War." *The Advocate of Peace (1894–1920)*, vol. 69, no. 3, 1907, pp. 66–68. JSTOR, http://www.jstor.org/stable/25752864.

Councell, Clara E. "War and Infectious Disease." *Public Health Reports (1896–1970)*, vol. 56, no. 12, 1941, pp. 547–573. JSTOR, https://doi.org/10.2307/4583663.

Courant, Richard. "Mathematics in the Modern World." *Scientific American*, vol. 211, no. 3, September 1964, pp. 41–49, https://doi.org/10.1038/scientificamerican0964-40.

Cregier, Don M. "The Murder of the British Liberal Party." *The History Teacher*, vol. 3, no. 4, 1970, pp. 27–36. JSTOR, https://doi.org/10.2307/3054322.

Crosby, Oscar T. "Locarno." *Advocate of Peace Through Justice*, vol. 88, no. 4, 1926, pp. 223–236. JSTOR, http://www.jstor.org/stable/20661236.

Crozier, Andrew J. "The Colonial Question in Stresemann's Locarno Policy." *The International History Review*, vol. 4, no. 1, 1982, pp. 37–54. JSTOR, http://www.jstor.org/stable/40105792.

"Cuba." *The Journal of Education*, vol. 55, no. 25 (1384), 1902, pp. 395–398. JSTOR, http://www.jstor.org/stable/44055253.

Cubitt, Toby S., et al. "The Unsolvable Problem." *Scientific American*, October 2018, https://www.scientificamerican.com/article/the-unsolvable-problem/.

Cunningham, Daniel W. *Set Theory*. Cambridge University Press, 2016.

ANNOTATED BIBLIOGRAPHY

Cunningham, Hugh. "The Language of Patriotism, 1750–1914." *History Workshop*, no. 12, 1981, pp. 8–33. JSTOR, http://www.jstor.org/stable/4288376.

Curbera, Guillermo P. "The International Congress of Mathematicians: A Human Endeavour." *Current Science*, vol. 99, no. 3, 2010, pp. 2–7. JSTOR, http://www.jstor.org/stable/24108275.

Czech, Herwig, et al. "The Lancet Commission on Medicine, Nazism, and the Holocaust: Historical Evidence, Implications for Today, Teaching for Tomorrow." *The Lancet*, vol. 402, no. 10415, November 8, 2023, pp. 1867–1940, https://doi.org/10.1016/s0140-6736(23)01845-7.

Dangerfield, George. *The Strange Death of Liberal England, 1910–1914*. Pedigree Books, 1980.

Darroch, Sandra Jobson. *The Life of Lady Ottoline Morrell*. Coward, McCann & Geoghegan, 1975.

Darwin, Leonard. "The Aims and Methods of Eugenical Societies." *Science*, vol. 54, no. 1397, 1921, pp. 313–323. JSTOR, http://www.jstor.org/stable/1646320.

Darwin, Leonard. "The Field of Eugenic Reform." *The Scientific Monthly*, vol. 13, no. 5, 1921, pp. 385–398. JSTOR, http://www.jstor.org/stable/6520.

Dauben, Joseph W. "Georg Cantor and Pope Leo XIII: Mathematics, Theology, and the Infinite." *Journal of the History of Ideas*, vol. 38, no. 1, 1977, pp. 85–108. JSTOR, https://doi.org/10.2307/2708842.

Dauben, Joseph W. "Georg Cantor and the Origins of Transfinite Set Theory." *Scientific American*, vol. 248, no. 6, 1983, pp. 122–131. JSTOR, http://www.jstor.org/stable/24968925.

Dauben, Joseph W. *Georg Cantor: His Mathematics and Philosophy of the Infinite*. Princeton University Press, 1979.

Dauben, Joseph W. "Georg Cantor: The Personal Matrix of His Mathematics." *Isis*, vol. 69, no. 4, 1978, pp. 534–550. JSTOR, http://www.jstor.org/stable/231091.

Dauben, Joseph W. "Review." *Isis*, vol. 79, no. 4, 1988, pp. 700–702. JSTOR, http://www.jstor.org/stable/234775.

Dauben, Joseph W. "The Trigonometric Background to Georg Cantor's Theory of Sets." *Archive for History of Exact Sciences*, vol. 7, no. 3, 1971, pp. 181–216. JSTOR, http://www.jstor.org/stable/41133323.

Davenport, Chas. B. "Harry Hamilton Laughlin." *Science*, vol. 97, no. 2513, 1943, pp. 194–195. JSTOR, http://www.jstor.org/stable/1669981.

Davenport, Charles B. "Research in Eugenics." *Science*, vol. 54, no. 1400, 1921, pp. 391–397. JSTOR, http://www.jstor.org/stable/1645486.

David, Edward. "The Liberal Party Divided 1916–1918." *The Historical Journal*, vol. 13, no. 3, 1970, pp. 509–532. JSTOR, http://www.jstor.org/stable/2637886.

Davies, Brian. "Whither Mathematics?" *Notices of the American Mathematical Society*, vol. 52, no. 11, December 2005, pp. 1350–1356.

ANNOTATED BIBLIOGRAPHY

Davis, Martin. "Gödel's Universe." *Nature*, vol. 435, May 2005, pp. 19–20, https://doi.org/10.1038/435019a.

Davis, Martin. "The Incompleteness Theorem." *Notices of the American Mathematical Society*, vol. 53, no. 4, April 2006, pp. 414–418.

Davis, Philip J. "Otto Neugebauer: Reminiscences and Appreciation." *The American Mathematical Monthly*, vol. 101, no. 2, 1994, pp. 129–131. JSTOR, https://doi.org/10.2307/2324359.

Davis, Philip J., and David Mumford. "Henri's Crystal Ball." *Notices of the American Mathematical Society*, vol. 55, no. 4, April 2008, pp. 458–466.

Dawson, John W. "Gödel and the Limits of Logic." *Scientific American*, vol. 280, no. 6, 1999, pp. 76–81. JSTOR, http://www.jstor.org/stable/26058291.

Dawson, John W. "The Reception of Gödel's Incompleteness Theorems." *PSA: Proceedings of the Biennial Meeting of the Philosophy of Science Association*, vol. 2, 1984, pp. 253–271. JSTOR, http://www.jstor.org/stable/192508.

Dawson, John W. "The Vienna Circle and the Epic Quest for the Foundations of Science," review of *Exact Thinking in Demented Times: The Vienna Circle and the Epic Quest for the Foundations of Science*, by Karl Sigmund. *Notices of the American Mathematical Society*, vol. 65, no. 7, September 2018, pp. 1002–1005.

Dawson, William Harbutt. "Reviewed Work: German Social Democracy by Bertrand Russell, Alys Russell." *The Economic Journal*, vol. 7, no. 26, 1897, pp. 248–250. JSTOR, https://doi.org/10.2307/2957248.

"Death of Mr. Gladstone." *Scientific American Supplement*, no. 1169, May 28, 1898, p. 18702.

Dedekind, Richard. *Essays on the Theory of Numbers*. Translated by Wooster Woodruff Beman, Open Court Publishing Company, 1901, via Project Gutenberg, https://www.gutenberg.org/files/21016/21016-pdf.pdf.

DeLong, Howard. *A Profile of Mathematical Logic*. Addison-Wesley, 1970.

de Reuck, Jenny. "Social Suffering and the Politics of Pain: Observations on the Concentration Camps in the Anglo-Boer War 1899–1902." *English in Africa*, vol. 26, no. 2, 1999, pp. 69–88. JSTOR, http://www.jstor.org/stable/40238883.

DeSua, Frank C. "Metamathematics: A Non-technical Exposition." *American Scientist*, vol. 42, no. 3, 1954, pp. 488–495. JSTOR, http://www.jstor.org/stable/27826561.

Dick, Auguste. *Emmy Noether, 1882–1935*. Translated by H. I. Blocher. Birkhäuser, 1981.

Dickson, L. E. "Book Review: Essays on the Theory of Numbers: I. Continuity and Irrational Numbers. II. The Nature and Meaning of Numbers by Richard Dedekind." *Bulletin of the American Mathematical Society*, vol. 8, no. 6, March 1, 1902, pp. 259–261, https://doi.org/10.1090/s0002-9904-1902-00891-4.

"Dietetic Treatment of Pernicious Anaemia." *The British Medical Journal*, vol. 2, no. 3431, 1926, pp. 650–651. JSTOR, http://www.jstor.org/stable/25325966.

ANNOTATED BIBLIOGRAPHY

Di Lellio, Anna. "Introduction." *The Battle of Kosovo 1389*. Translated by Robert Elsie. Bloomsbury Publishing, 2009.

"The Diseases of South Africa." *The British Medical Journal*, vol. 1, no. 2052, 1900, pp. 1044–1045. JSTOR, http://www.jstor.org/stable/20264399.

Donaldson, Peter. "Introduction" and "Conclusion." *Remembering the South African War: Britain and the Memory of the Anglo-Boer War, from 1899 to the Present*. Liverpool University Press, 2013, pp. 1–10, 170–174. JSTOR, https://doi.org/10.2307/j.ctt5vjmmh.12.

Doxiadis, Apostolos, and Christos Papadimitriou. *LOGICOMIX*. Bloomsbury USA, 2009.

Doyle, Peter. *World War I in 100 Objects*. Penguin Books, 2014.

Dresden, Arnold. "Brouwer's Contributions to the Foundations of Mathematics." *Bulletin of the American Mathematical Society*, vol. 30, no. 1, 1924, pp. 31–40, https://doi.org/10.1090/s0002-9904-1924-03844-0.

Dresden, Arnold. "Some Philosophical Aspects of Mathematics." *Bulletin of the American Mathematical Society*, vol. 34, no. 4, 1928, pp. 438–452, https://doi.org/10.1090/s0002-9904-1928-04560-3.

Dresden, Arnold. "The Migration of Mathematicians." *The American Mathematical Monthly*, vol. 49, no. 7, 1942, pp. 415–429. JSTOR, https://doi.org/10.2307/2303266.

Duggan, Stephen P. "The New Conception of War." *The Annals of the American Academy of Political and Social Science*, vol. 144, 1929, pp. 29–31. JSTOR, http://www.jstor.org/stable/1017327.

Dunbabin, J. P. D. "Parliamentary Elections in Great Britain, 1868–1900: A Psephological Note." *The English Historical Review*, vol. 81, no. 318, 1966, pp. 82–99. JSTOR, http://www.jstor.org/stable/559902.

Duncan, R. G., et al. "The Sociopolitical in Human Genetics Education." *Science*, vol. 383, no. 6685, February 23, 2024, pp. 826–828, https://doi.org/10.1126/science.adi8227.

Dunnington, G. Waldo. "Emile Picard." *National Mathematics Magazine*, vol. 16, no. 4, 1942, pp. 186–187. JSTOR, https://doi.org/10.2307/3028268.

du Sautoy, Marcus. "'The Music of the Primes.'" *The New York Times*, July 6, 2003, https://www.nytimes.com/2003/07/06/books/chapters/the-music-of-the-primes.html.

Džamonja, Mirna. "A New Foundational Crisis in Mathematics: Is It Really Happening?" *Synthese Library*, vol. 407, 2019, pp. 255–269, https://doi.org/10.1007/978-3-030-15655-8_11.

Edel, Leon. "The Group and the Salon." *The American Scholar*, vol. 46, no. 1, 1977, pp. 116–124. JSTOR, http://www.jstor.org/stable/41207458.

Edwards, David A. "Letters to the Editor: Response to Quinn." *Notices of the American Mathematical Society*, vol. 59, no. 3, 2011, p. 366.

ANNOTATED BIBLIOGRAPHY

Einstein, Albert. "Fundamental Ideas and Problems of the Theory of Relativity." Nordic Assembly of Naturalists at Gothenburg, July 11, 1923, Gothenburg, Sweden.

"Elementary Education Act 1870." 33 & 34 Vict., c. 75. Education UK, UK Government, 1870, https://education-uk.org/documents/acts/1870-elementary-education-act.html.

Elton, Lewis. "Einstein, General Relativity, and the German Press, 1919–1920." *Isis*, vol. 77, no. 1, 1986, pp. 95–103. JSTOR, http://www.jstor.org/stable/232505.

"The Emergency Committee in Aid of Displaced German Scholars." *Science*, vol. 78, no. 2012, 1933, pp. 52–53. JSTOR, http://www.jstor.org/stable/1659382.

"The Emergency Committee in Aid of Displaced German Scholars." *Science*, vol. 79, no. 2042, 1934, pp. 153–154. JSTOR, http://www.jstor.org/stable/1659536.

"Epimenides." *Encyclopedia Britannica*, February 29, 2024, https://www.britannica.com/biography/Epimenides.

"Eugenics." *The British Medical Journal*, vol. 2, no. 2747, 1913, pp. 508–509. JSTOR, http://www.jstor.org/stable/25302615.

Ewald, William. "Review." *Bulletin of the American Mathematical Society*, vol. 40, no. 1, 2003, pp. 125–129, https://www.ams.org/journals/bull/2003-40-01/S0273-0979-02-00959-X/S0273-0979-02-00959-X.pdf.

F., A. R. "Karl Weierstrass." *Nature*, vol. 55, no. 1428, March 1897, p. 443, https://doi.org/10.1038/055443a0.

Fabricant, Mona, et al. "Why Women Succeed in Mathematics." *The Mathematics Teacher*, vol. 83, no. 2, 1990, pp. 150–154. JSTOR, http://www.jstor.org/stable/27966564.

Fairchild, Henry Pratt. "The Immigration Law of 1924." *The Quarterly Journal of Economics*, vol. 38, no. 4, 1924, pp. 653–665. JSTOR, https://doi.org/10.2307/1884595.

Farwell, Byron. *Eminent Victorian Soldiers: Seekers of Glory*. W. W. Norton & Company, 1985.

Farwell, Byron. *The Great Boer War*. Pen & Sword, 1976.

Feferman, Solomon. "Lieber Herr Bernays! Lieber Herr Gödel! Gödel on Finitism, Constructivity and Hilbert's Program." *Dialectica*, vol. 62, no. 2, 2008, pp. 179–203. JSTOR, http://www.jstor.org/stable/42971217.

Ferguson, Niall. *Empire: The Rise and Demise of the British World Order and the Lessons for Global Power*. Basic Books, 2002.

Ferreirós, José. "The Crisis in the Foundations of Mathematics." *The Princeton Companion to Mathematics*. Edited by Timothy Gowers, June Barrow-Green, and Imre Leader. Princeton University Press, 2009b, pp. 142–156, https://doi.org/10.1515/9781400830398.142.

Ferreirós, José. "Hilbert, Logicism, and Mathematical Existence." *Synthese*, vol. 170, no. 1, 2009a, pp. 33–70. JSTOR, http://www.jstor.org/stable/40271343.

ANNOTATED BIBLIOGRAPHY

Ferreirós, José. "Traditional Logic and the Early History of Sets, 1854–1908." *Archive for History of Exact Sciences*, vol. 50, no. 1, 1996, pp. 5–71. JSTOR, http://www.jstor.org/stable/41134015.

Field, James A. "The Progress of Eugenics." *The Quarterly Journal of Economics*, vol. 26, no. 1, 1911, pp. 1–67. JSTOR, https://doi.org/10.2307/1884524.

Fierlinger, Zdenek. "Central Europe and the Balkans." *Proceedings of the Academy of Political Science in the City of New York*, vol. 12, no. 1, 1926, pp. 276–281. JSTOR, https://doi.org/10.2307/1180391.

"Fifth International Congress of Philosophy, London, 1915." *The Monist*, vol. 24, no. 4, 1914, pp. 636–638. JSTOR, http://www.jstor.org/stable/27900513.

Fink, Carole. "Stresemann's Minority Policies, 1924–29." *Journal of Contemporary History*, vol. 14, no. 3, 1979, pp. 403–422. JSTOR, http://www.jstor.org/stable/260014.

Fischer, Conan. "Scoundrels Without a Fatherland? Heavy Industry and Transnationalism in Post–First World War Germany." *Contemporary European History*, vol. 14, no. 4, 2005, pp. 441–464. JSTOR, http://www.jstor.org/stable/20081279.

"The Folly of Human Sterilization." *Scientific American*, vol. 151, no. 4, 1934, pp. 188–190. JSTOR, http://www.jstor.org/stable/24968632.

Foy, Michael. "Ulster Unionist Propaganda Against Home Rule 1912–14." *History Ireland*, vol. 4, no. 1, 1996, pp. 49–53. JSTOR, http://www.jstor.org/stable/27724315.

"France Ratifies the Locarno Treaties." *Advocate of Peace Through Justice*, vol. 88, no. 4, 1926, pp. 211–213. JSTOR, http://www.jstor.org/stable/20661227.

"Francis Galton." *Journal of the Royal Statistical Society*, vol. 85, no. 2, 1922, pp. 293–298. JSTOR, https://doi.org/10.2307/2341167.

Frankel, Richard. "From the Beer Halls to the Halls of Power: The Cult of Bismarck and the Legitimization of a New German Right, 1898–1945." *German Studies Review*, vol. 26, no. 3, 2003, pp. 543–560. JSTOR, https://doi.org/10.2307/1432746.

Frege, Gottlob. "The Fundamental Laws of Arithmetic. [Introductory Note]." *The Monist*, vol. 25, no. 4, 1915, pp. 481–494. JSTOR, http://www.jstor.org/stable/27900555.

French, David. "Spy Fever in Britain, 1900–1915." *The Historical Journal*, vol. 21, no. 2, 1978, pp. 355–370. JSTOR, http://www.jstor.org/stable/2638264.

Fromkin, David. *Europe's Last Summer: Who Started the Great War in 1914?* Random House, 2004.

Fry, Michael. "Political Change in Britain, August 1914 to December 1916: Lloyd George Replaces Asquith: The Issues Underlying the Drama." *The Historical Journal*, vol. 31, no. 3, 1988, pp. 609–627. JSTOR, http://www.jstor.org/stable/2639759.

Frye, Bruce B. "The German Democratic Party 1918–1930." *The Western Political Quarterly*, vol. 16, no. 1, 1963, pp. 167–179. JSTOR, https://doi.org/10.2307/445966.

ANNOTATED BIBLIOGRAPHY

"The Funeral of M. Poincaré." *Science*, vol. 36, no. 919, 1912, pp. 167–168. JSTOR, http://www.jstor.org/stable/1637838.

"The Future of International Science." *The Geographical Journal*, vol. 51, no. 1, 1918, pp. 33–35. JSTOR, https://doi.org/10.2307/1779518.

Fyfe, Herbert C. "The Automobile in Warfare: Experiences in the South African Campaign." *Scientific American* (1845–1908), vol. 88, April 11, 1903, p. 268.

Gaab, Jeffrey. "Hitler's Beer Hall Politics: A Reassessment Based on New Historical Scholarship." *International Journal of Humanities and Social Science*, vol. 1, no. 20, December 2011, pp. 35–41.

Galbraith, John Kenneth. *The Great Crash, 1929*. Houghton Mifflin, 1961.

Galison, Peter Louis. "Minkowski's Space-Time: From Visual Thinking to the Absolute World." *Historical Studies in the Physical Sciences*, vol. 10, 1979, pp. 85–121. JSTOR, https://www.jstor.org/stable/27757388.

Galton, Francis. "Eugenics: Its Definition, Scope, and Aims." *American Journal of Sociology*, vol. 10, no. 1, 1904, pp. 1–25. JSTOR, http://www.jstor.org/stable/2762125.

Galton, Francis. "Studies in Eugenics." *American Journal of Sociology*, vol. 11, no. 1, 1905, pp. 11–25. JSTOR, http://www.jstor.org/stable/2762356.

García, Guadalupe. "Urban Guajiros: Colonial Reconcentración, Rural Displacement and Criminalisation in Western Cuba, 1895–1902." *Journal of Latin American Studies*, vol. 43, no. 2, 2011, pp. 209–235. JSTOR, http://www.jstor.org/stable/23030619.

Gårding, Lars. "Review." *American Scientist*, vol. 75, no. 1, 1987, pp. 88–89. JSTOR, http://www.jstor.org/stable/27854500.

Garnett, David, editor. *Carrington Letters and Extracts from Her Diaries*. Jonathan Cape, 1975.

Gathorne-Hardy, Robert, editor. *Memoirs of Lady Ottoline Morrell: A Study in Friendship, 1873–1915*. Alfred A. Knopf, 1964.

Gathorne-Hardy, Robert, editor. *Ottoline at Garsington: Memoirs of Lady Ottoline Morrell, 1915–1918*. Alfred A. Knopf, 1975.

Gay, Frederick P. "The Contribution of Medical Science to Medical Art as Shown in the Study of Typhoid Fever." *Science*, vol. 44, no. 1126, 1916, pp. 109–124. JSTOR, http://www.jstor.org/stable/1644145.

Gehrt, Stan. "Coyotes in the Loop: A Close-up View of Survival in the Urban Core." Oral presentation at New Horizons in Science meeting, ScienceWriters2014, Columbus, Ohio, October 20, 2014, https://casw.org/wp-content/uploads/CASWNewHorizons2014Program-forweb-1.pdf.

"General Pact for the Renunciation of War." *The American Journal of International Law*, vol. 22, no. 4, 1928, pp. 171–176. JSTOR, https://doi.org/10.2307/2213112.

"General Valeriano Weyler 1838–1930." *Library of Congress Research Guides*, accessed August 30, 2019, https://guides.loc.gov/world-of-1898.

ANNOTATED BIBLIOGRAPHY

A German Naval Expert. "The German Navy and England." *The North American Review*, vol. 190, no. 645, 1909, pp. 250–257. JSTOR, http://www.jstor.org/stable/25106433.

Gibson, Robin. "Simon Bussy's 'Portrait of Lytton Strachey', 1904 (National Portrait Gallery)." *The Burlington Magazine*, vol. 111, no. 799, 1969, pp. 617–619. JSTOR, http://www.jstor.org/stable/876089.

Gill, Rebecca. "'The Rational Administration of Compassion': The Origins of British Relief in War." *Le Mouvement Social*, no. 227, 2009, pp. 9–26. JSTOR, http://www.jstor.org/stable/40538264.

Gill, Richard. "Invitation to Garsington." *The Virginia Quarterly Review*, vol. 50, no. 2, 1974, pp. 198–214. JSTOR, http://www.jstor.org/stable/26435479.

Godby, Michael. "Confronting Horror: Emily Hobhouse and the Concentration Camp Photographs of the South African War." *Kronos*, no. 32, 2006, pp. 34–48. JSTOR, http://www.jstor.org/stable/41056558.

Gödel, Kurt. "What Is Cantor's Continuum Problem?" *The American Mathematical Monthly*, vol. 54, no. 9, 1947, pp. 515–525. JSTOR, https://doi.org/10.2307/2304666.

"The Gold Fields of the Transvaal." *Scientific American Supplement*, no. 1051, February 22, 1896.

Goldstein, Rebecca. *Incompleteness*. W. W. Norton & Company, 2005.

Gölz, Sabine I., et al. "Hypnotism and Medicine in 1888 Paris: Contemporary Observations by Sofia Kovalevskaya." *SubStance*, vol. 25, no. 1, 1996, pp. 3–23. JSTOR, https://www.jstor.org/stable/3685226.

Gomme, Alice B. "Boer Folk-Medicine and Some Parallels." *Folklore*, vol. 13, no. 1, 1902, pp. 69–75. JSTOR, http://www.jstor.org/stable/1254379.

Gomme, Alice B., and Edward Peacock. "Boer Folk-Medicine and Some Parallels. II (Continued)." *Folklore*, vol. 13, no. 2, 1902, pp. 181–183. JSTOR, http://www.jstor.org/stable/1254665.

Goodwin, Craufurd D. "'Roger Fry: Art and Commerce': Introduction." *Journal of Cultural Economics*, vol. 22, no. 1, 1998, pp. 43–47. JSTOR, http://www.jstor.org/stable/41810651.

Gordon, Donald C. "The Admiralty and Dominion Navies, 1902–1914." *The Journal of Modern History*, vol. 33, no. 4, 1961, pp. 407–422. JSTOR, http://www.jstor.org/stable/1877217.

Gouvêa, Fernando Q. "Was Cantor Surprised?" *The American Mathematical Monthly*, vol. 118, no. 3, March 2011, pp. 198–209, https://doi.org/10.4169/amer.math.monthly.118.03.198.

Graham, Colin C. "Review." *Philosophy of Science*, vol. 47, no. 1, 1980, pp. 159–160. JSTOR, http://www.jstor.org/stable/187157.

ANNOTATED BIBLIOGRAPHY

Graham, Loren, and Jean-Michel Kantor. "A Comparison of Two Cultural Approaches to Mathematics: France and Russia, 1890–1930." *Isis*, vol. 97, no. 1, 2006, pp. 56–74. JSTOR, https://doi.org/10.1086/501100.

Grainger, John D. *The British Navy in the Baltic*. Boydell & Brewer, 2014, ch. 12. JSTOR, http://www.jstor.org/stable/10.7722/j.ctt6wpbsq.

Grant, Madison, and Henry Fairfield Osborn. *The Passing of the Great Race or the Racial Basis of European History*. Charles Scribner's Sons, 1923. https://www.gutenberg.org/files/68185/68185-h/68185-h.htm.

Grathwol, Robert. "Gustav Stresemann: Reflections on His Foreign Policy." *The Journal of Modern History*, vol. 45, no. 1, 1973, pp. 52–70. JSTOR, http://www.jstor.org/stable/1877593.

Grattan-Guinness, Ivor. "Bertrand Russell (1872–1970), Man of Dissent." *Notes and Records of the Royal Society of London*, vol. 63, no. 4, May 20, 2009, pp. 365–379, https://doi.org/10.1098/rsnr.2009.0020.

Grattan-Guinness, Ivor. "Bertrand Russell and Lady Ottoline Morrell." *Russell: The Journal of Bertrand Russell Studies*, vol. 21, no. 1, 2001, pp. 87–94. Project MUSE, https://doi.org/10.1353/rss.2001.0017.

Grattan-Guinness, Ivor. "Book Review: *Mystic, Geometer, and Intuitionist. The Life of L. E. J. Brouwer*, Volume 1: *The Dawning Revolution*." *Bulletin of the American Mathematical Society*, vol. 36, no. 4, July 19, 1999, pp. 529–533, https://doi.org/10.1090/s0273-0979-99-00794-6.

Grattan-Guinness, Ivor. "Missing Materials Concerning the Life and Work of Georg Cantor." *Isis*, vol. 62, no. 4, 1971, pp. 516–517. JSTOR, http://www.jstor.org/stable/229823.

Graves, Robert. *Goodbye to All That*. Random House, 1929.

Gray, Jeremy. "Mathematicians as Philosophers of Mathematics: Part 1." *For the Learning of Mathematics*, vol. 18, no. 3, 1998, pp. 20–24. JSTOR, http://www.jstor.org/stable/40248274.

Gray, Jeremy. *Plato's Ghost: The Modernist Transformation of Mathematics*. Princeton University Press, 2008.

"Great Britain and Locarno." *Advocate of Peace Through Justice*, vol. 88, no. 2, 1926, pp. 112–118. JSTOR, http://www.jstor.org/stable/20661170.

Greco, Pietro. "Mileva Marić." *Lettera Matematica International*, vol. 5, no. 1, April 2017, pp. 43–48, https://doi.org/10.1007/s40329-017-0158-4.

Green, E. H. H. "Radical Conservatism: The Electoral Genesis of Tariff Reform." *The Historical Journal*, vol. 28, no. 3, 1985, pp. 667–692. JSTOR, http://www.jstor.org/stable/2639144.

Greer, Jeff. "Famous Rhodes Scholars." *U.S. News and World Report*, November 19, 2009, https://www.usnews.com/education/slideshows/famous-rhodes-scholars.

ANNOTATED BIBLIOGRAPHY

Gregory, Adrian. "Peculiarities of the English? War, Violence and Politics: 1900–1939." *Journal of Modern European History*, vol. 1, no. 1, 2003, pp. 44–59. JSTOR, https://www.jstor.org/stable/26265778.

Gregory, William King. "Henry Fairfield Osborn." *Proceedings of the American Philosophical Society*, vol. 76, no. 3, 1936, pp. 395–408. JSTOR, http://www.jstor.org/stable/984553.

Griffin, Nicholas. "That Obscure Object of Desire." *Russell: The Journal of Bertrand Russell Archives*, vol. 13, no. 2, 1993, pp. 209–224. Project MUSE, https://muse.jhu.edu/article/881128.

Griffith, Benjamin. "An Edwardian Progress." *The Sewanee Review*, vol. 105, no. 2, 1997, pp. 271–273. JSTOR, http://www.jstor.org/stable/27548350.

Griffiths, Andrew. "Winston Churchill, the 'Morning Post', and the End of the Imperial Romance." *Victorian Periodicals Review*, vol. 46, no. 2, 2013, pp. 163–183. JSTOR, http://www.jstor.org/stable/43663687.

Grun, George A. "Locarno: Idea and Reality." *International Affairs* (Royal Institute of International Affairs 1944–), vol. 31, no. 4, 1955, pp. 477–485. JSTOR, https://doi.org/10.2307/2604823.

"Guns of Position and Siege Guns for War." *Scientific American Supplement*, no. 1263, March 17, 1900, pp. 20250–20251.

H., S. A. "Note on the Franco-Soviet Pact and the Locarno Treaty." *Bulletin of International News*, vol. 12, no. 18, 1936, pp. 8–13. JSTOR, http://www.jstor.org/stable/25639512.

Hadamard, J. "Emile Picard. 1856–1941." *Obituary Notices of Fellows of the Royal Society*, vol. 4, no. 11, 1942, pp. 129–150. JSTOR, http://www.jstor.org/stable/769154.

Hall, G. Stanley. "Can the Masses Rule the World?" *The Scientific Monthly*, vol. 18, no. 5, 1924, pp. 456–466. JSTOR, http://www.jstor.org/stable/7382.

Hallstein, Christian W. "'Ohm Krüger': The Genesis of a Nazi Propaganda Film." *Literature/Film Quarterly*, vol. 30, no. 2, 2002, pp. 133–139. JSTOR, http://www.jstor.org/stable/43797083.

Halmos, Paul R. "Has Progress in Mathematics Slowed Down? *The American Mathematical Monthly*, vol. 97, no. 7, August 1990, pp. 561–588, https://www.jstor.org/stable/2324635.

Halmos, Paul R. *Naïve Set Theory*. Martino Publishing, 2011.

Hampton, Mark. "The Press, Patriotism, and Public Discussion: C. P. Scott, the 'Manchester Guardian', and the Boer War, 1899–1902." *The Historical Journal*, vol. 44, no. 1, 2001, pp. 177–197. JSTOR, http://www.jstor.org/stable/3133666.

Hankins, Frank H. "Individual Differences: The Galton-Pearson Approach." *Social Forces*, vol. 4, no. 2, 1925, pp. 272–281. JSTOR, https://doi.org/10.2307/3004575.

Hanks, Peter W. "How Wittgenstein Defeated Russell's Multiple Relation Theory of Judgment." *Synthese*, vol. 154, no. 1, 2007, pp. 121–146. JSTOR, http://www.jstor.org/stable/27653445.

350

ANNOTATED BIBLIOGRAPHY

Hardie, J. Keir. "Federated Labor as a New Factor in British Politics." *The North American Review*, vol. 177, no. 561, 1903, pp. 233–241. JSTOR, http://www.jstor.org/stable/25119435.

Hare, William. "Ideas for Teachers: Russell's Legacy." *Oxford Review of Education*, vol. 28, no. 4, December 2002, pp. 491–507, https://doi.org/10.1080/0305498022000013634.

Harper, Tyler Austin. "The 100-Year Extinction Panic Is Back, Right on Schedule." *The New York Times*, January 26, 2024, https://www.nytimes.com/2024/01/26/opinion/polycrisis-doom-extinction-humanity.html.

Hart, Bradley W. "Watching the 'Eugenic Experiment' Unfold: The Mixed Views of British Eugenicists Toward Nazi Germany in the Early 1930s." *Journal of the History of Biology*, vol. 45, no. 1, 2012, pp. 33–63. JSTOR, http://www.jstor.org/stable/41488441.

Hastings, Charles J. "Democracy and Public Health Administration." *The Public Health Journal*, vol. 10, no. 3, 1919, pp. 97–112. JSTOR, http://www.jstor.org/stable/41975703.

Hayashi, Kina, et al. "Counting Nemo: Anemonefish *Amphiprion Ocellaris* Identify Species by Number of White Bars." *Journal of Experimental Biology*, vol. 227, no. 2, January 15, 2024, https://doi.org/10.1242/jeb.246357.

Hazlehurst, Cameron. "Asquith as Prime Minister, 1908–1916." *The English Historical Review*, vol. 85, no. 336, 1970, pp. 502–531. JSTOR, http://www.jstor.org/stable/563193.

Heck, Richard Kimberly. "Frege's Theorem: An Introduction." *Manuscripto*, vol. 26, no. 2, 2003, pp. 471–503.

Heilbron, Johan, et al. "Toward a Transnational History of the Social Sciences." *Journal of the History of the Behavioral Sciences*, vol. 44, no. 2, March 2008, pp. 146–160, https://doi.org/10.1002/jhbs.20302.

Hentschel, Klaus. "Erwin Finlay Freundlich and Testing Einstein's Theory of Relativity." *Archive for History of Exact Sciences*, vol. 47, no. 2, 1994, pp. 143–201. JSTOR, http://www.jstor.org/stable/41133977.

Hepburn, James. "Ottoline the Terrible." *The Sewanee Review*, vol. 84, no. 3, 1976, pp. 517–522. JSTOR, http://www.jstor.org/stable/27543144.

Hertzman, Lewis. "Gustav Stresemann: The Problem of Political Leadership in the Weimar Republic." *International Review of Social History*, vol. 5, no. 3, 1960, pp. 361–377. JSTOR, http://www.jstor.org/stable/44583585.

Hesseling, Dennis E. *Gnomes in the Fog: The Reception of Brouwer's Intuitionism in the 1920s*. Birkhäuser Verlag, 2003.

Heung, Marina. "'Breaker Morant' and the Melodramatic Treatment of History." *Film Criticism*, vol. 8, no. 2, 1984, pp. 3–13. JSTOR, http://www.jstor.org/stable/44018752.

ANNOTATED BIBLIOGRAPHY

Heyningen, Elizabeth van. "Costly Mythologies: The Concentration Camps of the South African War in Afrikaner Historiography." *Journal of Southern African Studies*, vol. 34, no. 3, 2008, pp. 495–513. JSTOR, http://www.jstor.org/stable/40283165.

Hilbert, David. *The Foundations of Geometry*. Translated by E. J. Townsend. Altus Classics, 2021.

Hilbert, D., and W. Ackerman. *Principles of Mathematical Logic*. Chelsea Publishing Co., 1950.

Hille, Einar. "Mathematics and Mathematicians from Abel to Zermelo." *Mathematics Magazine*, vol. 26, no. 3, 1953, pp. 127–146. JSTOR, https://doi.org/10.2307/3029614.

Hillier, Alfred. "The Geography and Climate of South Africa." *The British Medical Journal*, vol. 2, no. 2031, 1899, pp. 1537–1542. JSTOR, http://www.jstor.org/stable/20262667.

Himmelfarb, Gertrude. "The Politics of Democracy: The English Reform Act of 1867." *Journal of British Studies*, vol. 6, no. 1, 1966, pp. 97–138. JSTOR, http://www.jstor.org/stable/175195.

Hirsch, Felix E. "Stresemann in Historical Perspective." *The Review of Politics*, vol. 15, no. 3, 1953, pp. 360–377. JSTOR, http://www.jstor.org/stable/1405174.

Hirshfield, Claire. "Liberal Women's Organizations and the War Against the Boers, 1899–1902." *Albion: A Quarterly Journal Concerned with British Studies*, vol. 14, no. 1, 1982, pp. 27–49. JSTOR, https://doi.org/10.2307/4048484.

Hirshfield, Claire. "Working Class Peace Activities in Victorian England." *Peace Research*, vol. 18, no. 3, 1986, pp. 17–80. JSTOR, http://www.jstor.org/stable/23609860.

Hobhouse, Emily. "Appeal of Miss Hobhouse to Mr. Broderick." *South Africa Conciliation Committee*, no. 86, October 15, 1901.

Hobhouse, Emily. "Dust-Women." *The Economic Journal*, vol. 10, no. 39, 1900, pp. 411–420. JSTOR, http://www.jstor.org/stable/2957231.

Hobhouse, Emily. *Report of a Visit to the Camps of Women and Children in the Cape and Orange River Colonies: Addressed to the Committee of the Distress Fund for South African Women and Children*. Friars Printing Association, 1909.

Hochberg, Herbert. "Peano, Russell, and Logicism." *Analysis*, vol. 16, no. 5, April 1956, pp. 118–120. JSTOR, http://www.jstor.org/stable/3327052.

Hodges, Andrew. "In Retrospect: Gödel's Proof." *Nature*, vol. 454, August 2008, pp. 829, https://doi.org/10.1038/454829a.

Hodges, Wilfrid, and Wilfried Sieg. "A Symposium on Hilbert's Program." *The Journal of Symbolic Logic*, vol. 53, no. 2, 1988, p. 337. JSTOR, http://www.jstor.org/stable/2274506.

Holton, Gerald. "Of Love, Physics and Other Passions: The Letters of Albert and Mileva." *Physics Today*, vol. 47, no. 8, August 1, 1994, pp. 23–29, https://doi.org/10.1063/1.881398.

ANNOTATED BIBLIOGRAPHY

Hopkins, Albert A. "Our Latest Science—Eugenics." *Scientific American*, vol. 125, no. 16, 1921, pp. 273-279. JSTOR, http://www.jstor.org/stable/24980680.

Horowitz, Irving Louis. "Bertrand Russell on War and Peace." *Science & Society*, vol. 21, no. 1, 1957, pp. 30-51. JSTOR, http://www.jstor.org/stable/40400481.

Hutch, Richard A. "Strategic Irony and Lytton Strachey's Contribution to Biography." *Biography*, vol. 11, no. 1, 1988, pp. 1-15. JSTOR, http://www.jstor.org/stable/23539315.

Huxley, Aldous. *Crome Yellow*. 1921.

"Immigration Act of 1924." U.S. Statutes at Large, vol. 43, 1924, p. 153.

"The Influenza: Dr. Niven's Warning." *The Guardian*, November 2, 1918, www.theguardian.com/society/1918/nov/02/health.lifeandhealth.

"The International Commission of Eugenics." *Science*, vol. 56, no. 1457, 1922, pp. 626-627. JSTOR, http://www.jstor.org/stable/1648566.

"The International Congress of Mathematicians." *Nature*, vol. 62, no. 1609, August 30, 1900, pp. 418-420, https://doi.org/10.1038/062418a0.

"International Scientific Organization." *Science*, vol. 48, no. 1247, 1918, pp. 509-510. JSTOR, http://www.jstor.org/stable/1642414.

Irvine, Andrew D., and Peter Simons. "FORMALISM in Philosophy of Mathematics (Handbook of the Philosophy of Science)." *Philosophia Mathematica*, vol. 18, no. 1, November 5, 2009, pp. 291-310, https://doi.org/10.1093/philmat/nkp018.

Irwin, Douglas A. "The Political Economy of Free Trade: Voting in the British General Election of 1906." *The Journal of Law & Economics*, vol. 37, no. 1, 1994, pp. 75-108. JSTOR, http://www.jstor.org/stable/725605.

Jackman, Steven D. "Shoulder to Shoulder: Close Control and 'Old Prussian Drill' in German Offensive Infantry Tactics, 1871-1914." *The Journal of Military History*, vol. 68, no. 1, 2004, pp. 73-104. JSTOR, http://www.jstor.org/stable/3397249.

Jackson, Alvin. "The Larne Gun Running of 1914." *History Ireland*, vol. 1, no. 1, 1993, pp. 35-38. JSTOR, http://www.jstor.org/stable/27724046.

Jalland, Patricia. "United Kingdom Devolution 1910-14: Political Panacea or Tactical Diversion?" *The English Historical Review*, vol. 94, no. 373, 1979, pp. 757-785. JSTOR, http://www.jstor.org/stable/565552.

Jalland, Patricia, and John Stubbs. "The Irish Question After the Outbreak of War in 1914: Some Unfinished Party Business." *The English Historical Review*, vol. 96, no. 381, 1981, pp. 778-807. JSTOR, http://www.jstor.org/stable/569840.

"James Niven, M.A., LL.D., M.B., B.Ch." *British Medical Journal*, vol. 2, no. 3380, October 10, 1925, pp. 673-674, https://doi.org/10.1136/bmj.2.3380.673.

Johnson, Dale M. "Review." *The British Journal for the History of Science*, vol. 14, no. 1, 1981, pp. 101-103. JSTOR, http://www.jstor.org/stable/4026086.

Johnson, Dale M. "Review of L. E. J. Brouwer—Topologist, Intuitionist, Philosopher: How Mathematics Is Rooted in Life." *Notices of the American Mathematical Society*, vol. 61, June 1, 2014, pp. 607-610.

ANNOTATED BIBLIOGRAPHY

Johnson, Dale M. "The Problem of the Invariance of Dimension in the Growth of Modern Topology, Part I." *Archive for History of Exact Sciences*, vol. 20, no. 2, 1979, pp. 97–188. JSTOR, http://www.jstor.org/stable/41133541.

Johnson, Donald. "The Political Career of A. Mitchell Palmer." *Pennsylvania History: A Journal of Mid-Atlantic Studies*, vol. 25, no. 4, 1958, pp. 345–370. JSTOR, http://www.jstor.org/stable/27769836.

Johnson, G. E. W. "Something New in Peace Machinery." *The North American Review*, vol. 238, no. 4, 1934, pp. 312–322. JSTOR, http://www.jstor.org/stable/25114515.

Johnson, Phillip E. "The Early Beginnings of Set Theory." *The Mathematics Teacher*, vol. 63, no. 8, 1970, pp. 690–692. JSTOR, http://www.jstor.org/stable/27958491.

Johnson, Roswell H. "Eugenics and So-Called Eugenics." *American Journal of Sociology*, vol. 20, no. 1, 1914, pp. 98–103. JSTOR, http://www.jstor.org/stable/2762976.

Jones, David S., et al. "Explaining Health Inequities—the Enduring Legacy of Historical Biases." *New England Journal of Medicine*, vol. 390, no. 5, February 1, 2024, pp. 389–395, https://doi.org/10.1056/nejmp2307312.

Jones, Greta. "Eugenics in Ireland: The Belfast Eugenics Society, 1911–15." *Irish Historical Studies*, vol. 28, no. 109, 1992, pp. 81–95. JSTOR, http://www.jstor.org/stable/30008006.

Jones, R. V. "Winston Leonard Spencer Churchill, 1874–1965." *Biographical Memoirs of Fellows of the Royal Society*, vol. 12, 1966, pp. 35–105. JSTOR, http://www.jstor.org/stable/769525.

Jourdain, Philip E. B. "Introduction to Mathematical Philosophy." *The Mathematical Gazette*, vol. 10, no. 145, March 1920, p. 46, https://doi.org/10.2307/3603237.

Jourdain, Philip E. B. "Mr. Bertrand Russell's First Work on the Principles of Mathematics." *The Monist*, vol. 22, no. 1, 1912, pp. 149–158. JSTOR, http://www.jstor.org/stable/27900364.

Jourdain, Philip E. B. "The Philosophy of Mr. B*rtr*nd R*ss*ll." *The Monist*, vol. 26, no. 1, 1916b, pp. 24–62. JSTOR, http://www.jstor.org/stable/27900570.

Jourdain, Philip E. B. "Review: Our Knowledge of the External World as a Field for Scientific Method in Philosophy by Bertrand Russell." *The Mathematical Gazette*, vol. 7, no. 113, October 1914, pp. 404–406, https://doi.org/10.2307/3604842.

Jourdain, Philip E. B. "Richard Dedekind (1833–1916)." *The Monist*, vol. 26, no. 3, 1916a, pp. 415–427. JSTOR, http://www.jstor.org/stable/27900599.

Jourdain, Philip E. B. "Some Modern Advances in Logic." *The Monist*, vol. 21, no. 4, 1911, pp. 564–566. JSTOR, http://www.jstor.org/stable/27900347.

Jourdain, Philip E. B. "The Study of Mathematics." *The Mathematical Gazette*, vol. 4, no. 73, 1908, pp. 306–307. JSTOR, https://doi.org/10.2307/3604856.

Jourdain, Philip E. B. "Transfinite Numbers and the Principles of Mathematics: Part I." *The Monist*, vol. 20, no. 1, 1910, pp. 93–118. JSTOR, http://www.jstor.org/stable/27900235.

ANNOTATED BIBLIOGRAPHY

K. "The Paris Academy of Sciences." *Science*, vol. 49, no. 1269, 1919, p. 404. JSTOR, http://www.jstor.org/stable/1642889.

Kalai, Gil. "Gödel, Hilbert and Brouwer." *Combinatorics and More* (blog), December 2, 2008, https://gilkalai.wordpress.com/2008/12/02/godel-hilbert-and-brouwer/.

Kamareddine, Fairouz, et al. "Types in Logic and Mathematics Before 1940." *The Bulletin of Symbolic Logic*, vol. 8, no. 2, June 2002, pp. 185–245, https://doi.org/10.2178/bsl/1182353871.

Kanamori, Akihiro. "The Mathematical Development of Set Theory from Cantor to Cohen." *The Bulletin of Symbolic Logic*, vol. 2, no. 1, 1996, pp. 1–71. JSTOR, https://doi.org/10.2307/421046.

Kanamori, Akihiro. "Zermelo and Set Theory." *The Bulletin of Symbolic Logic*, vol. 10, no. 4, 2004, pp. 487–553. JSTOR, http://www.jstor.org/stable/3216738.

Kant, Immanuel. *The Critique of Pure Reason*. Translated by J. M. D. Meiklejohn, via Project Gutenberg, https://www.gutenberg.org/files/4280/4280-h/4280-h.htm.

Kelley, Loretta. "Why Were So Few Mathematicians Female?" *The Mathematics Teacher*, vol. 89, no. 7, 1996, pp. 592–596. JSTOR, http://www.jstor.org/stable/27969922.

Kennedy, Hubert C. "An Appreciation of Giuseppe Peano." *Pi Mu Epsilon Journal*, vol. 3, no. 3, 1960, pp. 107–113. JSTOR, http://www.jstor.org/stable/24338285.

Kennedy, Paul M. "The Tradition of Appeasement in British Foreign Policy 1865–1939." *British Journal of International Studies*, vol. 2, no. 3, 1976, pp. 195–215. JSTOR, http://www.jstor.org/stable/20096775.

Kennedy, Steve. "Emmy Noether." *Math Horizons*, vol. 4, no. 2, 1996, p. 17. JSTOR, http://www.jstor.org/stable/25678088.

Kennedy, Thomas C. "Nourishing Life: Russell and the 20th-Century British Peace Movement, 1900–18." *Russell: The Journal of Bertrand Russell Studies*, vol. 4, no. 1, June 30, 1984, https://doi.org/10.15173/russell.v4i1.1613.

Kerr, Fergus. "Russell vs Lawrence and/or Wittgenstein." *New Blackfriars*, vol. 63, no. 748, 1982, pp. 430–440. JSTOR, http://www.jstor.org/stable/43248720.

Kershaw, J. J. "Will Bacteria Be Used in War?" *Scientific American*, vol. 165, no. 2, 1941, pp. 56–58. JSTOR, http://www.jstor.org/stable/24966974.

Kevles, Daniel J. *In the Name of Eugenics*. Harvard University Press, 1995.

Keynes, John Maynard. *The Economic Consequences of the Peace*. East India Publishing Co., 2023.

Kiester, Edwin. *An Incomplete History of World War I*. Pier 9, Murdoch Books, 2007.

Kimberling, Clark. "Emmy Noether, Greatest Woman Mathematician." *The Mathematics Teacher*, vol. 75, no. 3, 1982, pp. 246–249. JSTOR, http://www.jstor.org/stable/27962871.

Kimberling, Clark. "Emmy Noether." *The American Mathematical Monthly*, vol. 79, no. 2, 1972, pp. 136–149. JSTOR, https://doi.org/10.2307/2316534.

"Kimberly Cave Dwellings." *Scientific American Supplement*, no. 1301, December 8, 1900, p. 20854.

ANNOTATED BIBLIOGRAPHY

Kirschhock, Maximilian E., and Andreas Nieder. "Association Neurons in the Crow Telencephalon Link Visual Signs to Numerical Values." *Proceedings of the National Academy of Sciences*, vol. 120, no. 45, October 30, 2023, https://doi.org/10.1073/pnas.2313923120.

Klement, Kevin C. "A New Century in the Life of a Paradox: Review of One Hundred Years of Russell's Paradox, edited by G. Link." *Review of Modern Logic*, vol. 11, no. 1–2, 2007, pp. 7–30.

Klement, Kevin C. "Putting Form Before Function: Logical Grammar in Frege, Russell, and Wittgenstein." *Philosopher's Imprint*, vol. 4, no. 2, August 2004, pp. 1–47, http://hdl.handle.net/2027/spo.3521354.0004.002.

Klug, Adam. "Why Chamberlain Failed and Bismarck Succeeded: The Political Economy of Tariffs in British and German Elections." *European Review of Economic History*, vol. 5, no. 2, 2001, pp. 219–250. JSTOR, http://www.jstor.org/stable/41377896.

Knowles, Owen. "Conrad and Bertrand Russell: New Light on Their Relationship." *The Conradian*, vol. 13, no. 2, 1988, pp. 192–202. JSTOR, http://www.jstor.org/stable/20873936.

Koblitz, Ann Hibner. "Science, Women, and the Russian Intelligentsia: The Generation of the 1860s." *Isis*, vol. 79, no. 2, 1988, pp. 208–226. JSTOR, http://www.jstor.org/stable/233605.

Koss, Stephen E. "The Destruction of Britain's Last Liberal Government." *The Journal of Modern History*, vol. 40, no. 2, 1968, pp. 257–277. JSTOR, http://www.jstor.org/stable/1876732.

Kovalevskaya, Sofya. *A Russian Childhood*. Translated by Beatrice Stillman. Springer-Verlag, 1978.

Krebs, Paula M. "'The Last of the Gentlemen's Wars': Women in the Boer War Concentration Camp Controversy." *History Workshop*, no. 33, 1992, pp. 38–56. JSTOR, http://www.jstor.org/stable/4289138.

Kreisel, G. "Bertrand Arthur William Russell, Earl Russell, 1872–1970." *Biographical Memoirs of Fellows of the Royal Society*, vol. 19, 1973, pp. 583–620. JSTOR, http://www.jstor.org/stable/769574.

Kreisel, G. "Review of Wittgenstein's Lectures on the Foundations of Mathematics, Cambridge 1939." *Bulletin of the American Mathematical Society*, vol. 84, no. 1, January 1978, pp. 79–90.

Krieg, Joann P. "'Don't Let Us Talk of That Anymore': Whitman's Estrangement from the Costelloe-Smith Family." *Walt Whitman Quarterly Review*, vol. 17, no. 3, January 1, 2000, pp. 91–120, https://doi.org/10.13008/2153-3695.1579.

Krivine, Jean-Louis. *Introduction to Axiomatic Set Theory*. D. Reidel Publishing Co., 1971.

Krugman, Paul. "Will Putin Kill the Global Economy?" *The New York Times*, March 31, 2022, www.nytimes.com/2022/03/31/opinion/putin-global-economy.html.

ANNOTATED BIBLIOGRAPHY

Kuitenbrouwer, Vincent. "'All Will Be Well!': Pro-Boer Propaganda, June 1900–June 1902." *War of Words: Dutch Pro-Boer Propaganda and the South African War (1899–1902)*. Amsterdam University Press, 2012a, pp. 215–252. JSTOR, http://www.jstor.org/stable/j.ctt46mxtq.9.

Kuitenbrouwer, Vincent. "'Dum-Dums of Public Opinion': Pro-Boer Propaganda, October 1899–June 1900." *War of Words: Dutch Pro-Boer Propaganda and the South African War (1899–1902)*. Amsterdam University Press, 2012b, pp. 179–214. JSTOR, http://www.jstor.org/stable/j.ctt46mxtq.8.

Kunen, Kenneth. *The Foundations of Mathematics*. College Publications, 2009.

"Labor Government in England." *Advocate of Peace Through Justice*, vol. 91, no. 5, 1929, pp. 271–273. JSTOR, http://www.jstor.org/stable/20681334.

Lally, Erica. "Race and Racism: British Responses to Civilian Prison Camps in the Boer War and the Kenya Emergency." *UCLA Historical Journal*, vol. 26, 2015, https://escholarship.org/uc/item/0h5760fh.

Lamberti, Marjorie. "The Reception of Refugee Scholars from Nazi Germany in America: Philanthropy and Social Change in Higher Education." *Jewish Social Studies*, vol. 12, no. 3, 2006, pp. 157–192. JSTOR, http://www.jstor.org/stable/4467750.

"The Land of the Boers." *Scientific American Supplement*, vol. 48, no. 1247, November 25, 1899.

Langer, Susanne K. "Review." *The Journal of Symbolic Logic*, vol. 3, no. 4, 1938, pp. 156–157. JSTOR, https://doi.org/10.2307/2267779.

Lasher, Pamela. "Review." *The Mathematics Teacher*, vol. 93, no. 8, 2000, p. 726. JSTOR, http://www.jstor.org/stable/27971560.

Lasley, J. W. "The Revolt Against Aristotle." *American Scientist*, vol. 30, no. 4, 1942, pp. 275–287. JSTOR, http://www.jstor.org/stable/27825957.

Lawrence, D. H. *Women in Love*. Bygone Media Publishing, 2022.

Lee, John Thomas. "The Philosophy of Bertrand Russell." *The Nation*, vol. 98, no. 2538, February 19, 1914, pp. 180–183.

"The Legality of Sterilisation." *Nature*, vol. 130, no. 3293, December 10, 1932, p. 863, https://doi.org/10.1038/130863a0.

Le Neve Foster, C. "The Progress in the Art of Mining." *Scientific American Supplement*, no. 805, June 6, 1891, p. 12860.

Leon, Sharon M. "'Hopelessly Entangled in Nordic Pre-Suppositions': Catholic Participation in the American Eugenics Society in the 1920s." *Journal of the History of Medicine and Allied Sciences*, vol. 59, no. 1, 2004, pp. 3–49. JSTOR, http://www.jstor.org/stable/24623991.

"The Lesson of Geneva." *Advocate of Peace Through Justice*, vol. 88, no. 4, 1926, pp. 197–198. JSTOR, http://www.jstor.org/stable/20661217.

Liao, Diana A., et al. "Recursive Sequence Generation in Crows." *Science Advances*, vol. 8, no. 44, November 4, 2022, https://doi.org/10.1126/sciadv.abq3356.

ANNOTATED BIBLIOGRAPHY

Lifton, Robert Jay. *The Nazi Doctors: Medical Killing and the Psychology of Genocide.* Basic Books, 1986.

Lindley, D. V. "A Brief History of Statistics in the Last 100 Years." *The Mathematical Gazette*, vol. 80, no. 487, March 1996, pp. 92–100, https://doi.org/10.2307/3620336.

Linnebo, Øystein. *Philosophy of Mathematics.* Princeton University Press, 2017.

Lisle, John. "Einstein up in Smoke." *Physics in Perspective*, vol. 17, no. 4, January 2016, pp. 354–360, https://doi.org/10.1007/s00016-015-0171-y.

Lobell, Steven E. "Britain's Paradox: Cooperation or Punishment Prior to World War I." *Review of International Studies*, vol. 27, no. 2, 2001, pp. 169–186. JSTOR, http://www.jstor.org/stable/20097726.

"Locarno." *Advocate of Peace Through Justice*, vol. 87, no. 10/11, 1925, pp. 578–580. JSTOR, http://www.jstor.org/stable/20661058.

"Locarno." *Bulletin of International News*, vol. 1, no. 23, 1925, pp. 2–10. JSTOR, http://www.jstor.org/stable/25637938.

Lombardo, Paul A. "Facing Carrie Buck." *The Hastings Center Report*, vol. 33, no. 2, 2003, pp. 14–17. JSTOR, https://doi.org/10.2307/3528148.

Lombardo, Paul A. "'Ridding the Race of His Defective Blood'—Eugenics in the Journal, 1906–1948." *New England Journal of Medicine*, vol. 390, no. 10, March 2, 2024, pp. 869–873, https://doi.org/10.1056/nejmp2307346.

Lord Tennyson, Alfred. "The Charge of the Light Brigade." *Poetry Foundation*, November 1, 2017, www.poetryfoundation.org/poems/45319/the-charge-of-the-light-brigade.

Louçã, Francisco. "Emancipation Through Interaction—How Eugenics and Statistics Converged and Diverged." *Journal of the History of Biology*, vol. 42, no. 4, 2009, pp. 649–684. JSTOR, http://www.jstor.org/stable/25650625.

Lowe, Victor. "A. N. Whitehead on His Mathematical Goals: A Letter of 1912." *Annals of Science*, vol. 32, no. 2, March 1975, pp. 85–101, https://doi.org/10.1080/00033797500200161.

Lucas, J. R. "Minds, Machines and Gödel." *Philosophy*, vol. 36, no. 137, 1961, pp. 112–127. JSTOR, http://www.jstor.org/stable/3749270.

Ludmerer, Kenneth M. "American Geneticists and the Eugenics Movement: 1905–1935." *Journal of the History of Biology*, vol. 2, no. 2, 1969, pp. 337–362, https://doi.org/10.1007/bf00125023.

Ludmerer, Kenneth M. "Genetics, Eugenics, and the Immigration Restriction Act of 1924." *Bulletin of the History of Medicine*, vol. 46, no. 1, 1972, pp. 59–81. JSTOR, http://www.jstor.org/stable/44447480.

Luft, Eric V. D. "The Foundations of Mathematics: Hilbert's Formalism vs. Brouwer's Intuitionism." Encyclopedia.com, www.encyclopedia.com/science/encyclopedias-almanacs-transcripts-and-maps/foundations-mathematics-hilberts-formalism-vs-brouwers-intuitionism.

ANNOTATED BIBLIOGRAPHY

Lutskanov, Rosen. "Whitehead's Early Philosophy of Mathematics and the Development of Formalism." *Logique et Analyse*, vol. 54, no. 214, 2011, pp. 161–172. JSTOR, http://www.jstor.org/stable/44085002.

Lutz, H. L. "Inter-Allied Debts, Reparations, and National Policy." *Journal of Political Economy*, vol. 38, no. 1, 1930, pp. 29–61. JSTOR, http://www.jstor.org/stable/1823216.

Lyon, John M. "'Poor Ottoline!'" *The Cambridge Quarterly*, vol. 23, no. 1, 1994, pp. 92–97. JSTOR, http://www.jstor.org/stable/42967311.

MacColl, Hugh. "Reviewed Work: *A Treatise on Universal Algebra with Applications*, Alfred North Whitehead." *Mind*, vol. 8, no. 29, 1899, pp. 108–113. JSTOR, http://www.jstor.org/stable/2247747.

Macdonald, Arthur. "Education and Eugenics." *The Journal of Education*, vol. 102, no. 17 (2553), 1925, pp. 451–454. JSTOR, http://www.jstor.org/stable/42832005.

Macfarlane, Alexander. *Science*, vol. 9, no. 218, 1899, pp. 324–328. JSTOR, http://www.jstor.org/stable/1626993.

MacFarlane, John. "Frege, Kant, and the Logic in Logicism." *The Philosophical Review*, vol. 111, no. 1, 2002, pp. 25–65. JSTOR, https://doi.org/10.2307/3182569.

Makins, G. H., and a South African Campaigner. "The War in South Africa." *The British Medical Journal*, vol. 2, no. 2035, 1899, pp. 1809–1812. JSTOR, http://www.jstor.org/stable/20262982.

Makins, G. H., and George Ashton. "The War in South Africa." *The British Medical Journal*, vol. 1, no. 2041, 1900, pp. 343–348. JSTOR, http://www.jstor.org/stable/20263421.

Mallard, Timothy S., and Nathan H. White. *A Persistent Fire: The Strategic Ethical Impact of World War I on the Global Profession of Arms*. National Defense University Press, 2019.

Mallion, Roger. "A Contemporary Eulerian Walk over the Bridges of Kaliningrad." *BSHM Bulletin: Journal of the British Society for the History of Mathematics*, vol. 23, no. 1, January 2008, pp. 24–36, https://doi.org/10.1080/17498430701799183.

Mancosu, Paolo. "Between Russell and Hilbert: Behmann on the Foundations of Mathematics." *The Bulletin of Symbolic Logic*, vol. 5, no. 3, 1999a, pp. 303–330. JSTOR, https://doi.org/10.2307/421183.

Mancosu, Paolo. "Between Vienna and Berlin: The Immediate Reception of Gödel's Incompleteness Theorems." *History and Philosophy of Logic*, vol. 20, no. 1, January 1999b, pp. 33–45, https://doi.org/10.1080/014453499298174.

Mancosu, Paolo. "Book Review: Matthias Baaz, Christos H. Papadimitriou, Hilary W. Putnam, Dana S. Scott, Charles L. Harper Jr. (Editors). *Kurt Gödel and the Foundations of Mathematics: Horizons of Truth*." *Isis*, vol. 103, no. 2, 2012, pp. 383–384. JSTOR, https://doi.org/10.1086/667472.

Mancosu, Paolo. "The Russellian Influence on Hilbert and His School." *Synthese*, vol. 137, no. 1/2, 2003, pp. 59–101. JSTOR, http://www.jstor.org/stable/20118352.

ANNOTATED BIBLIOGRAPHY

Mancosu, Paolo. *From Brouwer to Hilbert: The Debate on the Foundations of Mathematics in the 1920s.* Oxford University Press, 1998.

Mancosu, Paolo. *The Adventure of Reason: Interplay Between Philosophy of Mathematics and Mathematical Logic, 1900–1940.* Oxford University Press, 2010a.

Mancosu, Paolo. "Reviewed Work: *LOGICOMIX* by Apostolos Doxiadis, Christos H. Papadimitriou, Alecos Papadatos, Annie di Donna." *The Bulletin of Symbolic Logic*, vol. 16, no. 3, 2010b, pp. 419–420. JSTOR, http://www.jstor.org/stable/20749629.

Mancosu, Paolo. "Reviewed Work: *Wittgenstein, Finitism, and the Foundations of Mathematics* by Mathieu Marion." *The Philosophical Review*, vol. 110, no. 2, 2001, pp. 286–289. JSTOR, https://doi.org/10.2307/2693685.

Mancosu, Paolo, et al. *An Introduction to Proof Theory.* Oxford University Press, 2021.

Manning, Frederic. *Her Privates We.* G. P. Putnam's Sons, 1930.

March, Lucien. "The Consequences of War and the Birth Rate in France." *The Scientific Monthly*, vol. 13, no. 5, 1921, pp. 399–419. JSTOR, http://www.jstor.org/stable/6521.

Marion, Mathieu. "Wittgenstein and Brouwer." *Synthese*, vol. 137, no. 1/2, 2003, pp. 103–127. JSTOR, http://www.jstor.org/stable/20118353.

Marks, Shula, and Stanley Trapido. "Lord Milner and the South African State." *History Workshop*, no. 8, 1979, pp. 50–80. JSTOR, http://www.jstor.org/stable/4288258.

Martel, Gordon, et al. *The Origins of the First World War*, 3rd ed. Routledge, 2013.

Martin, Donald A. "Gödel's Conceptual Realism." *Bulletin of Symbolic Logic*, vol. 11, no. 2, June 2005, pp. 207–224, https://doi.org/10.2178/bsl/1120231631.

Martínez, Alberto. "Getting to Know Mileva Marić." *Physics Today*, vol. 72, no. 7, July 1, 2019, p. 53, https://doi.org/10.1063/pt.3.4251.

Marx, Fritz Morstein. "Germany's New Civil Service Act." *American Political Science Review*, vol. 31, no. 5, October 1937, pp. 878–883, https://doi.org/10.2307/1947915.

Mashaal, Maurice. *Bourbaki: A Secret Society of Mathematicians.* American Mathematical Society, 2006.

Massie, Robert K. *Dreadnought: Britain, Germany, and the Coming of the Great War.* Ballantine Books, 1991.

McCready, H. W. "Home Rule and the Liberal Party, 1899–1906." *Irish Historical Studies*, vol. 13, no. 52, 1963, pp. 316–348. JSTOR, http://www.jstor.org/stable/30005014.

McEwen, John M. "The Liberal Party and the Irish Question During the First World War." *Journal of British Studies*, vol. 12, no. 1, 1972, pp. 109–131. JSTOR, http://www.jstor.org/stable/175330.

McEwen, John M. "The Press and the Fall of Asquith." *The Historical Journal*, vol. 21, no. 4, 1978a, pp. 863–883. JSTOR, http://www.jstor.org/stable/2638972.

ANNOTATED BIBLIOGRAPHY

McEwen, John M. "The Struggle for Mastery in Britain: Lloyd George Versus Asquith, December 1916." *Journal of British Studies*, vol. 18, no. 1, 1978b, pp. 131–156. JSTOR, http://www.jstor.org/stable/175459.

McGuinness, Brian. *Wittgenstein: A Life: Young Ludwig, 1889–1921.* University of California Press, 1988.

"The Medical Aspects of the War: XIV." *The British Medical Journal*, vol. 1, no. 2043, 1900, pp. 476–478. JSTOR, http://www.jstor.org/stable/20263566.

"The Medical Aspects of the War: XXIII." *The British Medical Journal*, vol. 1, no. 2059, 1900, p. 1496. JSTOR, http://www.jstor.org/stable/20264883.

Mehlberg, Henryk. "The Present Situation in the Philosophy of Mathematics." *Synthese*, vol. 12, no. 4, 1960, pp. 380–414. JSTOR, http://www.jstor.org/stable/20114358.

Mels, Edgar. "The Future of South Africa—I." *Scientific American* (1845–1908), vol. 81, November 25, 1899, p. 342.

Menzel, Christopher. "Cantor and the Burali-Forti Paradox." *The Monist*, vol. 67, no. 1, 1984, pp. 92–107. JSTOR, http://www.jstor.org/stable/27902845.

"Metallic Fences." *Scientific American* (1845–1908), vol. 41, October 25, 1879, p. 257.

Meyer, Karl E. "An Edwardian Warning: The Unraveling of a Colossus." *World Policy Journal*, vol. 17, no. 4, 2000, pp. 47–57. JSTOR, http://www.jstor.org/stable/40209718.

"Military Service Act 1916." 6 & 7 Geo. 5, c. 104. UK Government, 1916, https://www.legislation.gov.uk/ukpga/1916/104/contents/enacted.

"The Military Service Act." *The British Medical Journal*, vol. 1, no. 2875, 1916, p. 211. JSTOR, http://www.jstor.org/stable/25316023.

Miller, David. "Gödel: Incomplete Success." *Nature*, vol. 323, October 30, 1986, pp. 766–767, https://doi.org/10.1038/323766b0.

Miller, G. A. "A Popular Account of Some New Fields of Thought in Mathematics." *The American Mathematical Monthly*, vol. 7, no. 4, 1900, pp. 91–99. JSTOR, https://doi.org/10.2307/2969415.

Miller, G. A. "Professor Charles Emile Picard." *The Scientific Monthly*, vol. 22, no. 5, 1926, pp. 464–467. JSTOR, http://www.jstor.org/stable/7661.

Miller, Stephen M. "In Support of the 'Imperial Mission'? Volunteering for the South African War, 1899–1902." *The Journal of Military History*, vol. 69, no. 3, 2005, pp. 691–711. JSTOR, http://www.jstor.org/stable/3397115.

"A Mobile Hospital in Natal." *The British Medical Journal*, vol. 1, no. 2050, 1900, pp. 914–915. JSTOR, http://www.jstor.org/stable/20264251.

Monk, Ray. *Bertrand Russell: The Ghost of Madness, 1921–1970.* Free Press Edition, 2001.

Monk, Ray. *Bertrand Russell: The Spirit of Solitude.* Random House, 1996.

Monk, Ray C. *Ludwig Wittgenstein: The Duty of Genius.* The Free Press, 1990.

ANNOTATED BIBLIOGRAPHY

Moody, T. W. "Michael Davitt and the British Labour Movement 1882–1906." *Transactions of the Royal Historical Society*, vol. 3, 1953, pp. 53–76. JSTOR, https://doi.org/10.2307/3678709.

Moore, A. W. "A Brief History of Infinity." *Scientific American*, April 1, 1995, https://www.scientificamerican.com/article/a-brief-history-of-infinity/.

Moore, C. L. "The Fourth International Congress of Mathematicians." *Bulletin of the American Mathematical Society*, vol. 14, no. 10, July 1, 1908, pp. 481–498, https://doi.org/10.1090/s0002-9904-1908-01656-2.

Moore, Gregory H. "Early History of the Generalized Continuum Hypothesis: 1878–1938." *The Bulletin of Symbolic Logic*, vol. 17, no. 4, 2011, pp. 489–532. JSTOR, http://www.jstor.org/stable/41302100.

Moore, Gregory H. "The Origins of Zermelo's Axiomatization of Set Theory." *Journal of Philosophical Logic*, vol. 7, no. 1, 1978, pp. 307–329. JSTOR, http://www.jstor.org/stable/30226178.

Moore, Gregory H. "Review." *The Review of Modern Logic*, vol. 9, no. 1 & 2, 2001, pp. 215–220.

Moore, Paul. "'And What Concentration Camps Those Were!': Foreign Concentration Camps in Nazi Propaganda, 1933–9." *Journal of Contemporary History*, vol. 45, no. 3, 2010, pp. 649–674. JSTOR, http://www.jstor.org/stable/20753619.

Moorhouse, H. F. "The Political Incorporation of the British Working Class: An Interpretation." *Sociology*, vol. 7, no. 3, 1973, pp. 341–359. JSTOR, http://www.jstor.org/stable/42853020.

Moran, Margaret. "Bertrand Russell Meets His Muse: The Impact of Lady Ottoline Morrell (1911–12)." *Russell: The Journal of Bertrand Russell Archives*, vol. 11, no. 2, 1991, pp. 180–192. Project MUSE, https://muse.jhu.edu/article/881539.

Mordell, L. J. *The Mathematical Gazette*, vol. 16, no. 220, 1932, pp. 279–280. JSTOR, https://doi.org/10.2307/3605935.

Morgenstern, Oskar. "Thirteen Critical Points in Contemporary Economic Theory: An Interpretation." *Journal of Economic Literature*, vol. 10, no. 4, 1972, pp. 1163–1189. JSTOR, http://www.jstor.org/stable/2721542.

Moriconi, Enrico. "On the Meaning of Hilbert's Consistency Problem (Paris, 1900)." *Synthese*, vol. 137, no. 1/2, 2003, pp. 129–139. JSTOR, http://www.jstor.org/stable/20118354.

Mowry, George E., editor. *The Twenties: Fords, Flappers & Fanatics*. Prentice-Hall, 1965.

"Mr. Russell's Confession of Faith," *The Nation*, vol. 104, no. 2700, March 29, 1917, pp. 367–368.

Muhadri, Bedri. "The Invasion of Kosovo from the Ottomans in the XIV Century." *European Journal of Social Sciences Studies*, vol. 2, no. 6, August 2017, https://dx.doi.org/10.5281/zenodo.841841.

ANNOTATED BIBLIOGRAPHY

Mulligan, William. "From Case to Narrative: The Marquess of Lansdowne, Sir Edward Grey, and the Threat from Germany, 1900–1906." *The International History Review*, vol. 30, no. 2, 2008, pp. 273–302. JSTOR, http://www.jstor.org/stable/41220101.

Mullin, Emily. "How Tuberculosis Shaped Victorian Fashion." *Smithsonian Magazine*, May 10, 2016, https://www.smithsonianmag.com/science-nature/how-tuberculosis-shaped-victorian-fashion-180959029/.

Murray, Gilbert, et al. "Address on the Locarno Pact." *Transactions of the Grotius Society*, vol. 12, 1926, pp. xxii–xliv. JSTOR, http://www.jstor.org/stable/742671.

Musgrave, Alan. "Logicism Revisited." *The British Journal for the Philosophy of Science*, vol. 28, no. 2, 1977, pp. 99–127. JSTOR, http://www.jstor.org/stable/686628.

Mycielski, Jan. "The Meaning of Pure Mathematics." *Journal of Philosophical Logic*, vol. 18, no. 3, 1989, pp. 315–320. JSTOR, http://www.jstor.org/stable/30227216.

Nagel, Ernest, and James R. Newman. *Gödel's Proof*. Routledge & Kegan Paul, 1958.

Nagel, Ernest, and James R. Newman. "Gödel's Proof." *Scientific American*, vol. 194, no. 6, 1956, pp. 71–90. JSTOR, http://www.jstor.org/stable/24943884.

"Nature and Treatment of Pernicious Anaemia." *The British Medical Journal*, vol. 2, no. 3850, 1934, pp. 726–727. JSTOR, http://www.jstor.org/stable/25342063.

Nelson, Edward. "Review of *Gnomes in the Fog: The Reception of Brouwer's Intuitionism in the 1920s*, by Dennis E. Hesseling." *Bulletin of the American Mathematical Society*, vol. 41, no. 4, June 17, 2004, pp. 545–549.

"Nemo Can Count!" Eurekalert!, February 1, 2024, https://www.eurekalert.org/news-releases/1032398.

Neugebauer, Otto. "Biographical Sketch." *National Mathematics Magazine*, vol. 11, no. 1, 1936, pp. 14–16. JSTOR, https://doi.org/10.2307/3028163.

"New Cabinet in Germany." *Advocate of Peace Through Justice*, vol. 87, no. 2, 1925, pp. 79–81. JSTOR, http://www.jstor.org/stable/20660823.

"The New Rifle for the British Army." *Scientific American Supplement*, vol. 55, no. 1421, March 28, 1903, p. 22767, https://doi.org/10.1038/scientificamerican03281903-22767asupp.

Newman, M. H. "Hermann Weyl." *Journal of the London Mathematical Society*, vol. 1–33, no. 4, October 1958, pp. 500–511, https://doi.org/10.1112/jlms/s1-33.4.500.

"The Nobel Prize in Physiology and Medicine." *The Scientific Monthly*, vol. 39, no. 6, 1934, pp. 565–567. JSTOR, http://www.jstor.org/stable/15843.

Notestein, Wallace. "The Career of Mr. Asquith." *Political Science Quarterly*, vol. 31, no. 3, 1916, pp. 361–379. JSTOR, https://doi.org/10.2307/2141650.

Novaes, Catarina Dutilh. "Review." *Mind*, vol. 122, no. 486, 2013, pp. 571–575. JSTOR, http://www.jstor.org/stable/24489573.

Nunemacher, Jeffrey L. "Review." *The American Mathematical Monthly*, vol. 110, no. 6, 2003, pp. 554–557. JSTOR, https://doi.org/10.2307/3647929.

ANNOTATED BIBLIOGRAPHY

Nye, Mary Jo. "The Scientific Periphery in France: The Faculty of Sciences at Toulouse (1880–1930)." *Minerva*, vol. 13, no. 3, 1975, pp. 374–403. JSTOR, http://www.jstor.org/stable/41820241.

Oberschelp, Arnold. *The Journal of Symbolic Logic*, vol. 47, no. 2, 1982, pp. 456–457. JSTOR, https://doi.org/10.2307/2273171.

Offner, John L. "McKinley and the Spanish-American War." *Presidential Studies Quarterly*, vol. 34, no. 1, 2004, pp. 50–61. JSTOR, http://www.jstor.org/stable/27552563.

Offner, John. "Why Did the United States Fight Spain in 1898?" *OAH Magazine of History*, vol. 12, no. 3, 1998, pp. 19–23. JSTOR, http://www.jstor.org/stable/25163215.

Ogg, F. A. "The British Representation of the People Act." *American Political Science Review*, vol. 12, no. 3, August 1918, pp. 498–503, https://doi.org/10.2307/1946102.

"On the Theory of the Transfinite: Correspondence of Georg Cantor and J.B. Cardinal Franzelin (1885–1886)." *Fidelio*, vol. 3, no. 3, 1994.

"Organum Novissimum: Review of *Our Knowledge of the External World as a Field for Scientific Method in Philosophy*, by Bertrand Russell." *The Nation*, vol. 100, no. 2586, January 21, 1915, pp. 83–84.

Osborn, Henry Fairfield. "Eugenics—the American and Norwegian Programs." *Science*, vol. 54, no. 1403, 1921, pp. 482–484. JSTOR, http://www.jstor.org/stable/1645650.

Overbye, Dennis. "A Century Ago, Einstein's Theory of Relativity Changed Everything." *The New York Times*, November 24, 2015, https://www.nytimes.com/2015/11/24/science/a-century-ago-einsteins-theory-of-relativity-changed-everything.html.

"Overthrow of the Barb Fence Patents." *Scientific American* (1845–1908), vol. 48., June 16, 1883, p. 368.

Pakenham, Thomas. *The Scramble for Africa*. Harper Perennial, 2003.

Parker, R. A. C. "The First Capitulation: France and the Rhineland Crisis of 1936." *World Politics*, vol. 8, no. 3, 1956, pp. 355–373. JSTOR, https://doi.org/10.2307/2008855.

"Patent Perfidy." *Scientific American* (1845–1908), vol. 44, May 7, 1881, p. 288.

"Patents in Congress." *Scientific American* (1845–1908), vol. 50, April 12, 1884, p. 224.

Paton, Stewart. "Democracy's Opportunity." *The Scientific Monthly*, vol. 11, no. 3, 1920, pp. 254–262. JSTOR, http://www.jstor.org/stable/6597.

Paul, Harry W. "The Issue of Decline in Nineteenth-Century French Science." *French Historical Studies*, vol. 7, no. 3, 1972, pp. 416–450. JSTOR, https://doi.org/10.2307/286222.

Peckhaus, Volker. "The Pragmatism of Hilbert's Programme." *Synthese*, vol. 137, November 2003, pp. 141–156, https://doi.org/10.1023/a:1026235118657.

Peckhaus, Volker. "The Way of Logic into Mathematics." *Theoria: An International Journal for Theory, History and Foundations of Science*, vol. 12, no. 1 (28), 1997, pp. 39–64. JSTOR, http://www.jstor.org/stable/23917977.

ANNOTATED BIBLIOGRAPHY

"Pernicious Anaemia." *Canadian Public Health Journal*, vol. 27, no. 9, 1936, p. 459. JSTOR, http://www.jstor.org/stable/41977488.

Pfeiffer, G. A. "Book Review: Introduction to Mathematical Philosophy." *Bulletin of the American Mathematical Society*, vol. 27, no. 2, November 1, 1920, pp. 81–90, https://doi.org/10.1090/s0002-9904-1920-03365-3.

"[Photograph]: Emile Picard." *American Journal of Mathematics*, vol. 17, no. 1, 1895. JSTOR, http://www.jstor.org/stable/2369704.

Pierpont, James. "The History of Mathematics in the Nineteenth Century." *Bulletin of the American Mathematical Society*, vol. 37, no. 1, December 21, 1999, pp. 9–24.

Pingree, David. "Eloge: Otto Neugebauer, 26 May 1899–19 February 1990." *Isis*, vol. 82, no. 1, 1991, pp. 87–88. JSTOR, http://www.jstor.org/stable/233516.

Pitzer, Andrea. "Concentration Camps Existed Long Before Auschwitz." *Smithsonian Magazine*, November 2, 2017, https://www.smithsonianmag.com/history/concentration-camps-existed-long-before-Auschwitz-180967049/.

Poirier, Sylvain. "Set Theory and Foundations of Mathematics." Settheory.net, accessed October 22, 2024.

Pólya, G. "A Story with a Moral." *The Mathematical Gazette*, vol. 57, no. 400, 1973, pp. 86–87. JSTOR, https://doi.org/10.2307/3615343.settheory.net/foundations/1.

Pólya, G. "As Their Students See Them." *The Two-Year College Mathematics Journal*, vol. 7, no. 2, 1976, p. 54. JSTOR, http://www.jstor.org/stable/3027007.

Pope, Arthur Upham. "Roger Fry." *Bulletin of the American Institute for Persian Art and Archaeology*, vol. 3, no. 7, 1934, pp. 53–54. JSTOR, http://www.jstor.org/stable/44235954.

"Portable Shields for Infantry." *Scientific American (1845–1908)*, vol. 82, June 2, 1900, p. 340.

Potts, C. S. "World Chaos Once More." *The Southwestern Social Science Quarterly*, vol. 16, no. 2, 1935, pp. 1–10. JSTOR, http://www.jstor.org/stable/42879635.

Powell, Julie M. "Making 'The Case Against the "Reds"': Racializing Communism, 1919–1920." In *Historicizing Fear: Ignorance, Vilification, and Othering*, edited by Travis D. Boyce and Winsome M. Chunnu. University Press of Colorado, 2019, pp. 102–121. JSTOR, http://www.jstor.org/stable/j.ctvwh8d12.9.

Power, Paul F. "Gandhi in South Africa." *The Journal of Modern African Studies*, vol. 7, no. 3, October 1969, pp. 441–455, https://doi.org/10.1017/s0022278x00018590.

"The Prevention of Waterborne Typhoid in Armies in the Field." *The British Medical Journal*, vol. 1, no. 2091, 1901, pp. 242–243. JSTOR, http://www.jstor.org/stable/20267189.

Puleston, W. D. "Asquith and Kitchener." *Scientific American*, vol. 145, no. 2, August 1931, pp. 116–117.

Puleston, W. D. "Viscount Grey and Lord Haldane." *Scientific American*, vol. 146, no. 5, 1932, pp. 276–315. JSTOR, http://www.jstor.org/stable/24965921.

ANNOTATED BIBLIOGRAPHY

Putnam, Hilary. "Mathematics Without Foundations." *The Journal of Philosophy*, vol. 64, no. 1, 1967, pp. 5–22. JSTOR, https://doi.org/10.2307/2024603.

Pyenson, Lewis. "Einstein's Natural Daughter." *History of Science*, vol. 28, no. 4, December 1990, pp. 365–379, https://doi.org/10.1177/007327539002800402.

Pyenson, Lewis. "Hermann Minkowski and Einstein's Special Theory of Relativity." *Archive for History of Exact Sciences*, vol. 17, no. 1, 1977, pp. 71–95. JSTOR, http://www.jstor.org/stable/41133480.

Quick, Jonathan R. "Virginia Woolf, Roger Fry and Post-Impressionism." *The Massachusetts Review*, vol. 26, no. 4, 1985, pp. 547–570. JSTOR, http://www.jstor.org/stable/25089694.

Quill, R. H. "Airborne Typhoid." *The British Medical Journal*, vol. 1, no. 2146, 1902, pp. 383–384. JSTOR, http://www.jstor.org/stable/20271241.

Quinn, Frank. "'A Revolution in Mathematics? What Really Happened a Century Ago and Why It Matters Today.'" *Notices of the American Mathematical Society*, vol. 59, no. 1, January 1, 2012, pp. 31–37.

R., L. G. "Louis Courturat (1868–1914)." *The Monist*, vol. 25, no. 3, July 1915, pp. 476–477. JSTOR, http://www.jstor.org/stable/27900551.

Raatikainen, Panu. "Hilbert's Program Revisited." *Synthese*, vol. 137, no. 1/2, 2003, pp. 157–177. JSTOR, http://www.jstor.org/stable/20118356.

Rahn, Werner. "German Navies from 1848 to 2016: Their Development and Courses from Confrontation to Cooperation." *Naval War College Review*, vol. 70, no. 4, 2017, pp. 12–47. JSTOR, http://www.jstor.org/stable/26398064.

Rantavaara, Irma. "On Lytton Strachey's Personality and Style." *Neuphilologische Mitteilungen*, vol. 73, no. 1/3, 1972, pp. 326–339. JSTOR, http://www.jstor.org/stable/43345363.

Rappaport, Karen D. "S. Kovalevsky: A Mathematical Lesson." *The American Mathematical Monthly*, vol. 88, no. 8, 1981, pp. 564–574. JSTOR, https://doi.org/10.2307/2320506.

Rappaport, Karen D. "Women Mathematicians: A Bibliography." *Women's Studies Newsletter*, vol. 6, no. 4, 1978, pp. 15–17. JSTOR, http://www.jstor.org/stable/40042445.

Rawlins, F. I. G. "Gödel's Theorems in English." *Nature*, vol. 201, January 11, 1964, p. 117, https://www.nature.com/articles/201117a0.

Readman, Paul. "The Conservative Party, Patriotism, and British Politics: The Case of the General Election of 1900." *Journal of British Studies*, vol. 40, no. 1, 2001, pp. 107–145. JSTOR, http://www.jstor.org/stable/3070771.

Reamer, Kevin. "The Importance of Intelligence in Combating a Modern Insurgency." *Journal of Strategic Security*, vol. 2, no. 2, 2009, pp. 73–90. JSTOR, http://www.jstor.org/stable/26462960.

Redfield, Caspar L. "Eugenics." *Canadian Journal of Public Health*, vol. 7, no. 3, March 1916, pp. 149–151. JSTOR, http://www.jstor.org/stable/41996966.

ANNOTATED BIBLIOGRAPHY

"Regulating Eugenics." *Harvard Law Review*, vol. 121, no. 6, 2008, pp. 1578–1599. JSTOR, http://www.jstor.org/stable/40042704.

Reid, Constance. "Book Review: *The Emergence of the American Mathematical Research Community, 1876–1900* by J. J. Sylvester, Felix Klein, and E. H. Moore, by Karen Hunger Parshall and David E. Rowe." *Bulletin of the American Mathematical Society*, vol. 32, no. 3, July 1, 1995, pp. 349–354, https://doi.org/10.1090/s0273-0979-1995-00595-1.

Reid, Constance. *Courant*. Copernicus, 1976.

Reid, Constance. *Hilbert*. Springer-Verlag, 1970.

Reilly, Philip R. "Involuntary Sterilization in the United States: A Surgical Solution." *The Quarterly Review of Biology*, vol. 62, no. 2, June 1987, pp. 153–170, https://doi.org/10.1086/415404.

Reingold, Nathan. "Refugee Mathematicians in the United States of America, 1933–1941: Reception and Reaction." *Annals of Science*, vol. 38, no. 3, May 1981, pp. 313–338, https://doi.org/10.1080/00033798100200251.

"'Remember the Maine!'" U.S. History, 2008, https://www.ushistory.org/us/44c.asp.

Rempel, Richard. "British Quakers and the South African War." *Quaker History*, vol. 64, no. 2, 1975, pp. 75–95. JSTOR, http://www.jstor.org/stable/41947626.

Rempel, Richard. "From Imperialism to Free Trade: Couturat, Halévy and Russell's First Crusade." *Journal of the History of Ideas*, vol. 40, no. 3, 1979, pp. 423–443. JSTOR, https://doi.org/10.2307/2709246.

Renna, Thomas. "Peace in the Humanitarian Tradition." *Peace Research*, vol. 10, no. 4, 1978, pp. 155–158. JSTOR, http://www.jstor.org/stable/23609476.

"Review: *Justice in War Time* by Bertrand Russell." *The Advocate of Peace* (1894–1920), vol. 78, no. 8, 1916, p. 250. JSTOR, http://www.jstor.org/stable/20667590.

Richards, Noel J. "Political Nonconformity at the Turn of the Twentieth Century." *Journal of Church and State*, vol. 17, no. 2, 1975, pp. 239–258. JSTOR, http://www.jstor.org/stable/23914736.

Richardson, Robert P., and Edward H. Landis. "Numbers, Variables and Mr. Russell's Philosophy." *The Monist*, vol. 25, no. 3, 1915, pp. 321–364. JSTOR, http://www.jstor.org/stable/27900544.

Robič, Borut. "Chapter 2: The Foundational Crisis in Mathematics." *The Foundations of Computability Theory*. Springer-Verlag, 2015, pp. 9–30.

Robinson, Abraham. "From a Formalist's Point of View." *Dialectica*, vol. 23, no. 1, 1969, pp. 45–49. JSTOR, http://www.jstor.org/stable/42968450.

Ross, Robert S. "Nationalism, Geopolitics, and Naval Expansionism: From the Nineteenth Century to the Rise of China." *Naval War College Review*, vol. 71, no. 4, 2018, pp. 10–44. JSTOR, https://www.jstor.org/stable/26607088.

Rosser, Barkley. "The Burali-Forti Paradox." *The Journal of Symbolic Logic*, vol. 7, no. 1, 1942, pp. 1–17. JSTOR, https://doi.org/10.2307/2267550.

ANNOTATED BIBLIOGRAPHY

Rostker, Bernard. "The World War." *Providing for the Casualties of War: The American Experience Through World War II*. RAND Corporation, 2013, pp. 123–174. JSTOR, http://www.jstor.org/stable/10.7249/j.ctt2tt90p.16.

Roth, Joseph. *What I Saw: Reports from Berlin 1920–1933*. Translated by Michael Hofmann. W. W. Norton & Company, 2004.

Rowe, David E. "Book Review: Plato's Ghost: The Modernist Transformation of Mathematics." *Bulletin of the American Mathematical Society*, vol. 50, no. 3, July 2013, pp. 513–521, https://doi.org/10.1090/s0273-0979-2012-01403-9.

Rowe, David E. "'Jewish Mathematics' at Göttingen in the Era of Felix Klein." *Isis*, vol. 77, no. 3, 1986, pp. 422–449. JSTOR, http://www.jstor.org/stable/231607.

Russell, Bertrand. *The Autobiography of Bertrand Russell*, vol. 1, *1872–1914*. Little, Brown and Company, 1967a.

Russell, Bertrand. *The Autobiography of Bertrand Russell*, vol. 2, *1914–1944*. Little, Brown and Company, 1968.

Russell, Bertrand. *The Autobiography of Bertrand Russell*, vol. 3, *1944–1969*. Little, Brown and Company, 1969.

Russell, Bertrand. *The Basic Writings of Bertrand Russell*. Clarion Books, 1967b.

Russell, Bertrand. "Definitions and Methodological Principles in Theory of Knowledge." *The Monist*, vol. 24, no. 4, 1914, pp. 582–593. JSTOR, http://www.jstor.org/stable/27900507.

Russell, Bertrand. "The Ethics of War." *International Journal of Ethics*, vol. 25, no. 2, 1915, pp. 127–142. JSTOR, http://www.jstor.org/stable/2376578.

Russell, Bertrand. "Free Speech in Childhood." *The Nation*, vol. 133, no. 3443, July 1, 1931, pp. 12–13.

Russell, Bertrand. *Icarus, or the Future of Science*. Wilder Publications, 2021.

Russell, Bertrand. *Introduction to Mathematical Philosophy*. Martino Fine Books, 2017.

Russell, Bertrand. "Mathematical Logic as Based on the Theory of Types." *American Journal of Mathematics*, vol. 30, no. 3, 1908, pp. 222–262. JSTOR, https://doi.org/10.2307/2369948.

Russell, Bertrand. "New Morals for Old Styles in Ethics." *The Nation*, vol. 118, no. 3069, April 30, 1924, pp. 497–498.

Russell, Bertrand. "On the Experience of Time." *The Monist*, vol. 25, no. 2, 1915, pp. 212–233. JSTOR, http://www.jstor.org/stable/27900529.

Russell, Bertrand. *Our Knowledge of the External World as a Field for Scientific Method in Philosophy*. Allen & Unwin, 1926.

Russell, Bertrand. *The Philosophy of Leibniz*. Routledge, 1992a.

Russell, Bertrand. "The Philosophy of Logical Atomism." *The Monist*, vol. 29, no. 1, 1919a, pp. 32–63. JSTOR, http://www.jstor.org/stable/27900724.

Russell, Bertrand. "The Philosophy of Logical Atomism." *The Monist*, vol. 29, no. 2, 1919b, pp. 190–222. JSTOR, http://www.jstor.org/stable/27900737.

ANNOTATED BIBLIOGRAPHY

Russell, Bertrand. "The Philosophy of Logical Atomism." *The Monist*, vol. 29, no. 3, 1919c, pp. 345–380. JSTOR, http://www.jstor.org/stable/27900748.

Russell, Bertrand. *The Selected Letters of Bertrand Russell*, vol. 1, *The Private Years (1884–1914)*. Houghton Mifflin Company, 1992b.

Russell, Bertrand. *The Selected Letters of Bertrand Russell*, vol. 2: *The Public Years (1914–1970)*. Routledge, 2001.

Russell, Bertrand. "Soviet Russia—1920." *The Nation*, vol. 111, no. 2874, July 31, 1920, pp. 121–122.

Russell, Bertrand. "The Superior Virtue of the Oppressed." *The Nation*, vol. 144, 1937.

Russell, Bertrand. "The Teaching of Euclid." *The Mathematical Gazette*, vol. 2, no. 33, 1902, pp. 165–167. JSTOR, https://doi.org/10.2307/3604768.

Russell, Bertrand. "What I Believe." *The Nation*, vol. 132, no. 3434, April 29, 1931, pp. 469–470.

Russell, Bertrand. "A World I'd Like / An Unprophetic Vision." *The Nation*, vol. 132, no. 3434, November 7, 1953.

Rygiel, Mary Ann. "Sofya Kovalevskaya's 'A Russian Childhood' as Poetic Autobiography." *Biography*, vol. 10, no. 3, 1987, pp. 208–224. JSTOR, http://www.jstor.org/stable/23539381.

Sambourne, Edward Linley. "The Rhodes Colossus." Cornell Digital Library, April 14, 2017, https://digital.library.cornell.edu/catalog/ss:19343183.

Sanders, Charles Richard. "Lytton Strachey's Conception of Biography." *PMLA*, vol. 66, no. 4, 1951, pp. 295–315. JSTOR, https://doi.org/10.2307/459477.

Schaffer, Gavin. "'Like a Baby with a Box of Matches': British Scientists and the Concept of 'Race' in the Inter-war Period." *The British Journal for the History of Science*, vol. 38, no. 3, 2005, pp. 307–324. JSTOR, http://www.jstor.org/stable/4028672.

Schamel, Ray, et al. "Glidden's Patent Application for Barbed Wire." *Social Education*, vol. 61, no. 1, January 1997, pp. 52–55.

Schweitzer, Arthur. *Journal of Political Economy*, vol. 54, no. 1, 1946, pp. 84–86. JSTOR, http://www.jstor.org/stable/1824940.

Schwerin, Alan. "A Lady, Her Philosopher and a Contradiction." *Russell: The Journal of Bertrand Russell Archives*, vol. 19, no. 1, 1999, pp. 5–28. Project MUSE, https://muse.jhu.edu/article/880959.

Scott, Charlotte Angas. "The International Congress of Mathematicians in Paris." *Bulletin of the American Mathematical Society*, vol. 7, no. 2, November 1, 1900, pp. 57–80, https://doi.org/10.1090/s0002-9904-1900-00768-3.

Seidel, Brina, and Laurence Chandy. "Is Globalization's Second Wave About to Break?" Brookings, October 4, 2016, https://www.brookings.edu/articles/is-globalizations-second-wave-about-to-break/.

Segal, S. L. "Helmut Hasse in 1934." *Historia Mathematica*, vol. 7, no. 1, February 1980, pp. 46–56, https://doi.org/10.1016/0315-0860(80)90063-4.

ANNOTATED BIBLIOGRAPHY

Seymour, Miranda. *Ottoline Morrell: Life on the Grand Scale*. Farrar, Straus & Giroux, 1992.

Sheffield, Gary, editor. *War on the Western Front*. Hachette, 2007.

Sieg, Wilfred. "Hilbert's Programs: 1917–1922." *The Bulletin of Symbolic Logic*, vol. 5, no. 1, March 1999, pp. 1–44.

Shepherdson, J. C. "Mathematical Logic." *Nature*, vol. 220, no. 5164, October 1968, p. 310, https://doi.org/10.1038/220310b0.

Shields, Brit. "Mathematics, Peace, and the Cold War: Scientific Diplomacy and Richard Courant's Scientific Identity." *Historical Studies in the Natural Sciences*, vol. 46, no. 5, 2016, pp. 556–591. JSTOR, https://www.jstor.org/stable/26413628.

Sieg, Wilfried. "Hilbert's Program Sixty Years Later." *The Journal of Symbolic Logic*, vol. 53, no. 2, 1988, pp. 338–348. JSTOR, https://doi.org/10.2307/2274507.

Sieg, Wilfried. *Hilbert's Programs and Beyond*. Oxford University Press, 2019.

Sieg, Wilfried. "On Tait on Kant and Finitism." *The Journal of Philosophy*, vol. 113, no. 5/6, 2016, pp. 274–285. JSTOR, https://www.jstor.org/stable/48568255.

Siegmund-Schultze, Reinhard. "The Institute Henri Poincaré and Mathematics in France Between the Wars." *Revue d'Histoire des Sciences*, vol. 62, no. 1, 2009, pp. 247–283. JSTOR, http://www.jstor.org/stable/23634493.

Siegmund-Schultze, Reinhard. "'Mathematics Knows No Races': A Political Speech That David Hilbert Planned for the ICM in Bologna in 1928." *The Mathematical Intelligencer*, vol. 38, no. 1, October 28, 2016, pp. 56–66, https://doi.org/10.1007/s00283-015-9559-4.

Sigmund, Karl. "Deciphering an Enigma." *Nature*, vol. 387, no. 6631, May 1997, pp. 362–363, https://doi.org/10.1038/387362a0.

"The Significance of Locarno." *Bulletin of International News*, vol. 1, no. 24, 1925, pp. 5–6. JSTOR, http://www.jstor.org/stable/25637948.

Simpson, Stephen G. "Partial Realizations of Hilbert's Program." *The Journal of Symbolic Logic*, vol. 53, no. 2, 1988, pp. 349–363. JSTOR, https://doi.org/10.2307/2274508.

Singh, Simon. "Mathematics 'Proves' What the Grocer Always Knew." *The New York Times*, August 25, 1998.

Smale, Steve. "Mathematical Problems for the Next Century." *The Mathematical Intelligencer*, vol. 20, no. 2, March 1998, pp. 7–15, https://doi.org/10.1007/bf03025291.

Smith, Jeremy. "Federalism, Devolution and Partition: Sir Edward Carson and the Search for a Compromise on the Third Home Rule Bill, 1913–14." *Irish Historical Studies*, vol. 35, no. 140, 2007, pp. 496–518. JSTOR, http://www.jstor.org/stable/20547491.

Smoryński, C. "Review." *The Journal of Symbolic Logic*, vol. 44, no. 1, 1979, pp. 116–119. JSTOR, https://doi.org/10.2307/2273711.

Snapper, Ernst. "The Three Crises in Mathematics: Logicism, Intuitionism and Formalism." *Mathematics Magazine*, vol. 52, no. 4, 1979a, pp. 207–216. JSTOR, https://doi.org/10.2307/2689412.

ANNOTATED BIBLIOGRAPHY

Snapper, Ernst. "What Is Mathematics?" *The American Mathematical Monthly*, vol. 86, no. 7, 1979b, pp. 551–557. JSTOR, https://doi.org/10.2307/2320582.

Snyder, Virgil. "The Fifth International Congress of Mathematicians, Cambridge, 1912." *Bulletin of the American Mathematical Society*, vol. 19, no. 3, December 1, 1912, pp. 107–130, https://doi.org/10.1090/s0002-9904-1912-02309-1.

South African Campaigner. "The Medical Aspects of the Boer War. III." *The British Medical Journal*, vol. 2, no. 2031, 1899, pp. 1556–1557. JSTOR, http://www.jstor.org/stable/20262681.

Sparrow, Robert. "A Not-So-New Eugenics: Harris and Savulescu on Human Enhancement." *The Hastings Center Report*, vol. 41, no. 1, 2011, pp. 32–42. JSTOR, http://www.jstor.org/stable/41058988.

Spiers, Edward M. "Re-engaging the Boers." *The Victorian Soldier in Africa*. Manchester University Press, 2004, pp. 159–179. JSTOR, http://www.jstor.org/stable/j.ctt155jj67.17.

Spies, S. B. *Methods of Barbarism*. Jonathan Ball Publishers, 2001.

Stahl, Stanley H. "Book Review: *Georg Cantor, His Mathematics and Philosophy of the Infinite*." *Bulletin of the American Mathematical Society*, vol. 2, no. 1, January 1, 1980, pp. 214–216, https://doi.org/10.1090/s0273-0979-1980-14725-4.

Stambrook, F. G. "'Das Kind': Lord D'Abernon and the Origins of the Locarno Pact." *Central European History*, vol. 1, no. 3, 1968, pp. 233–263. JSTOR, http://www.jstor.org/stable/4545496.

"Star of South Africa." *Encyclopedia Britannica*, July 20, 1998, https://www.britannica.com/topic/Star-of-South-Africa.

Stephens, Hugh W. "Party Realignment in Britain, 1900–1925: A Preliminary Analysis." *Social Science History*, vol. 6, no. 1, 1982, pp. 35–66. JSTOR, https://doi.org/10.2307/1170846.

Stewart, Herbert L. "Freedom of Speech in War Time." *The Nation*, vol. 105, no. 2722, August 30, 1917, pp. 219–220.

Stillwell, Devon. "Eugenics Visualized: The Exhibit of the Third International Congress of Eugenics, 1932." *Bulletin of the History of Medicine*, vol. 86, no. 2, 2012, pp. 206–236. JSTOR, https://www.jstor.org/stable/26305847.

Stillwell, John. "The Continuum Problem." *The American Mathematical Monthly*, vol. 109, no. 3, 2002, pp. 286–297. JSTOR, https://doi.org/10.2307/2695360.

St. John, Ronald B. "European Naval Expansion and Mahan, 1889–1906." *Naval War College Review*, vol. 23, no. 7, 1971, pp. 74–83. JSTOR, http://www.jstor.org/stable/44641219.

Stokesbury, James. *A Short History of World War One*. William Morrow Paperbacks, 1981.

Strachan, Hew. *The Oxford Illustrated History of the First World War*. Oxford University Press, 2014.

Stratford, Jenny. "Eminent Victorians." *The British Museum Quarterly*, vol. 32, no. 3/4, 1968, pp. 93–96. JSTOR, https://doi.org/10.2307/4422997.

ANNOTATED BIBLIOGRAPHY

Stresemann, Gustav. "The Economic Restoration of the World." *Foreign Affairs*, vol. 2, no. 4, 1924, pp. 552–557. JSTOR, https://doi.org/10.2307/20028328.

Strogatz, Steven. "The Hilbert Hotel." *The New York Times*, May 9, 2010, https://archive.nytimes.com/opinionator.blogs.nytimes.com/2010/05/09/the-hilbert-hotel/.

Studenski, Paul. "Armament Expenditures in Principal Countries." *The Annals of the American Academy of Political and Social Science*, vol. 214, 1941, pp. 29–37. JSTOR, http://www.jstor.org/stable/1024149.

"Study Shows That Students Compare Their Math Performance with Their Own Reading Performance to Determine Whether They Are a 'Math Person' or 'Reading Person.'" Eurekalert!, October 12, 2022, https://www.eurekalert.org/news-releases/967098.

"The Supremacy of the Modern Magazine Rifle." *Scientific American (1845–1908)*, vol. 82, March 24, 1900, p. 178.

Suter, Sonia M. "A Brave New World of Designer Babies?" *Berkeley Technology Law Journal*, vol. 22, no. 2, 2007, pp. 897–969. JSTOR, http://www.jstor.org/stable/24117430.

T., B. F. "Cuba Since the War." *The Advocate of Peace (1894–1920)*, vol. 62, no. 1, 1900, pp. 12–15. JSTOR, http://www.jstor.org/stable/25751493.

Tahta, Dick. "Recounting Cantor." *For the Learning of Mathematics*, vol. 27, no. 3, 2007, pp. 8–11. JSTOR, http://www.jstor.org/stable/40248577.

Tait, W. W. "Finitism." *The Journal of Philosophy*, vol. 78, no. 9, 1981, pp. 524–546. JSTOR, https://doi.org/10.2307/2026089.

Tames, Richard. *Bloomsbury Past*. Historical Publications, 1993.

Tanner, Jakob. "Eugenics Before 1945." *Journal of Modern European History*, vol. 10, no. 4, 2012, pp. 458–479. JSTOR, https://www.jstor.org/stable/26266044.

Taussky, Olga. "Prof. David Hilbert, For.Mem.R.S." *Nature*, vol. 152, no. 3850, August 1943, pp. 182–183, https://doi.org/10.1038/152182a0.

Taylor, David G. "The Aesthetic Theories of Roger Fry Reconsidered." *The Journal of Aesthetics and Art Criticism*, vol. 36, no. 1, 1977, pp. 63–72. JSTOR, https://doi.org/10.2307/430750.

Taylor, R. Gregory. "Reviewed Works: *Zermelo: Definiteness and the Universe of Definable Sets*; Heinz-Dieter Ebbinghaus, *Zermelo in the Mirror of the Baer Correspondence, 1930–1931*," *The Bulletin of Symbolic Logic*, vol. 10, no. 4, 2004, pp. 590–592. JSTOR, http://www.jstor.org/stable/3216749.

Taylor, William L. "The Debate over Changing Cavalry Tactics and Weapons, 1900–1914." *Military Affairs*, vol. 28, no. 4, 1964, pp. 173–183. JSTOR, https://doi.org/10.2307/1984387.

"Text-Book on Gunnery: Review of Mackinlay, Major G." *Nature*, vol. 37, no. 946, December 15, 1887, pp. 148–149, https://doi.org/10.1038/037148a0.

ANNOTATED BIBLIOGRAPHY

Thiele, Rüdger. "Hilbert's Twenty-Fourth Problem." *The American Mathematical Monthly*, vol. 110, no. 1, 2003, pp. 1–24. JSTOR, https://doi.org/10.2307/3072340.

Thompson, David Croal, editor. "The Paris Exhibition, 1900." *Art Journal*, 1901.

Thompson, Warren S. "Eugenics and the Social Good." *The Journal of Social Forces*, vol. 3, no. 3, 1925, pp. 414–419. JSTOR, https://doi.org/10.2307/3004978.

Thompson, Warren S. "Eugenics as Viewed by a Sociologist." *Monthly Labor Review*, vol. 18, no. 2, 1924, pp. 11–23. JSTOR, http://www.jstor.org/stable/41828847.

Thomson, William, and Surgeon-Lieutenant-Colonel Hartley. "The War in South Africa." *The British Medical Journal*, vol. 2, no. 2064, 1900, pp. 181–185. JSTOR, http://www.jstor.org/stable/20265230.

Thorning, Joseph F. "Franco-Germanic Relations." *Social Science*, vol. 9, no. 3, 1934, pp. 315–318. JSTOR, http://www.jstor.org/stable/41885566.

"Titus 1:12, New International Version Bible." Bible Gateway, Biblica, 2011, https://www.biblegateway.com/passage/?search=Titus%201%3A12&version=NIV.

Tonelli, L. "Report on the 1928 International Congress of Mathematicians." *Bulletin of the American Mathematical Society*, vol. 35, no. 2, 1929, pp. 201–205, https://doi.org/10.1090/s0002-9904-1929-04700-1.

"Training Field Workers in Eugenics." *Scientific American*, vol. 120, no. 15, 1919, p. 373. JSTOR, http://www.jstor.org/stable/26039323.

"The Transvaal." *Scientific American Supplement*, vol. 51, no. 1049, February 8, 1896, pp. 16759–16761.

Travers, T. H. E. "Technology, Tactics, and Morale: Jean de Bloch, the Boer War, and British Military Theory, 1900–1914." *The Journal of Modern History*, vol. 51, no. 2, 1979, pp. 264–286. JSTOR, http://www.jstor.org/stable/1879217.

Trent, Sydney. "World's Largest Body of Human Geneticists Apologizes for Eugenics Role." *Washington Post*, January 24, 2023, https://www.washingtonpost.com/dc-md-va/2023/01/24/geneticists-eugenics-apology/.

Trimble, Donald E. "The Geologic Story of the Great Plains." *Geological Survey Bulletin*, no. 1493, 2006, https://www.nps.gov/parkhistory/online_books/geology/publications/bul/1493/intro.htm.

Tuchman, Barbara W. *The Guns of August*. Random House, 1962.

Tuchman, Barbara W. *The Zimmermann Telegram*. Macmillan, 1966.

Tur, J. Soliveres, and J. Climent Vidal. "The Modernity of Dedekind's Anticipations Contained in *What Are Numbers and What Are They Good For?*" *Archive for History of Exact Sciences*, vol. 72, no. 2, January 17, 2018, pp. 99–141, https://doi.org/10.1007/s00407-018-0202-6.

Turing, A. M. "Computability and λ-Definability." *The Journal of Symbolic Logic*, vol. 2, no. 4, 1937, pp. 153–163. JSTOR, https://doi.org/10.2307/2268280.

ANNOTATED BIBLIOGRAPHY

Turing, A. M. "Computing Machinery and Intelligence." *Mind*, vol. 59, October 1, 1950, pp. 433–460, https://doi.org/10.1093/mind/lix.236.433.

Turner, Raymond. "Locarno." *The Virginia Quarterly Review*, vol. 2, no. 4, 1926, pp. 481–500. JSTOR, http://www.jstor.org/stable/26441398.

"Two Famous Boer Guns." *Scientific American Supplement*, no. 1258, February 10, 1900.

Tyler, H. W. "The International Congress of Mathematicians at Heidelberg." *Bulletin of the American Mathematical Society*, vol. 11, no. 4, January 1, 1905, pp. 191–205, https://doi.org/10.1090/s0002-9904-1905-01199-x.

Urquhart, Alasdair. "Russell and Gödel." *The Bulletin of Symbolic Logic*, vol. 22, no. 4, December 2016, pp. 504–520, https://doi.org/10.1017/bsl.2016.35.

van Atten, Mark. *The Bulletin of Symbolic Logic*, vol. 10, no. 3, 2004, pp. 423–427. JSTOR, http://www.jstor.org/stable/3185194.

van Dalen, Dirk. "Brouwer and Fraenkel on Intuitionism." *The Bulletin of Symbolic Logic*, vol. 6, no. 3, 2000, pp. 284–310. JSTOR, https://doi.org/10.2307/421057.

van Dalen, Dirk. "Hermann Weyl's Intuitionistic Mathematics." *The Bulletin of Symbolic Logic*, vol. 1, no. 2, 1995, pp. 145–169. JSTOR, https://doi.org/10.2307/421038.

van Dalen, Dirk. *L. E. J. Brouwer—Topologist, Intuitionist, Philosopher: How Mathematics Is Rooted in Life*. Springer London, 2013.

van Dalen, Dirk. *Mystic, Geometer, and Intuitionist: The Life of L. E. J. Brouwer*, vol. 1, *The Dawning Revolution*. Oxford University Press, 1999.

van Dalen, Dirk. *The Selected Correspondence of L. E. J. Brouwer*. Springer London, 2011.

van Dalen, Dirk. "The War of the Frogs and the Mice, or the Crisis of the Mathematische Annalen." *Mathematical Conversations*, 2001, pp. 445–465, https://doi.org/10.1007/978-1-4613-0195-0_40.

van Heijenoort, Jean, editor. *From Frege to Gödel: A Source Book in Mathematical Logic*. Harvard University Press, 1967.

von Plato, Jan. "Kurt Gödel's First Steps in Logic: Formal Proofs in Arithmetic and Set Theory Through a System of Natural Deduction." *The Bulletin of Symbolic Logic*, vol. 24, no. 3, September 2018, pp. 319–335, https://doi.org/10.1017/bsl.2017.42.

von Plato, Jan. "Mystic, Geometer, and Intuitionist: The Life of L. E. J. Brouwer. Volume 1. The Dawning Revolution." *The Bulletin of Symbolic Logic*, vol. 7, no. 1, March 2001, pp. 62–65, https://doi.org/10.2307/2687824.

Vytniorgu, Richard. "Ottoline Morrell: Personalist Thinker." *The Modern Language Review*, vol. 113, no. 1, 2018, pp. 57–79. JSTOR, https://doi.org/10.5699/modelangrevi.113.1.0057.

W., A. S. "Prof. Henry Fairfield Osborn, For. Mem. R.S." *Nature*, vol. 136, November 16, 1935, pp. 784–785, https://doi.org/10.1038/136784a0.

Wan, Sirui, et al. "Developmental Changes in Students' Use of Dimensional Comparisons to Form Ability Self-Concepts in Math and Verbal Domains." *Child*

Development, vol. 94, no. 1, October 12, 2022, pp. 272–287, https://doi.org/10.1111/cdev.13856.

"The War in South Africa: British Prisoners at Pretoria." *The British Medical Journal*, vol. 2, no. 2072, 1900, pp. 775–776. JSTOR, http://www.jstor.org/stable/20265729.

"The War in South Africa: The Battle of Colenso." *The British Medical Journal*, vol. 1, no. 2038, 1900, pp. 161–164. JSTOR, http://www.jstor.org/stable/20263208.

"The War in South Africa: The Medical Aspects of the War." *The British Medical Journal*, vol. 1, no. 2040, 1900, pp. 280–283. JSTOR, http://www.jstor.org/stable/20263349.

"The War in the Transvaal." *Scientific American Supplement*, no. 1243, October 28, 1899, p. 19920.

"The Warfare of the Future." *Scientific American Supplement*, no. 1266, April 7, 1900, p. 20297.

Warnock, G. J. *English Philosophy Since 1900*. Oxford University Press, 1966.

Wavre, Rolin. "Is There a Crisis in Mathematics?" *The American Mathematical Monthly*, vol. 41, no. 8, 1934, pp. 488–499. JSTOR, https://doi.org/10.2307/2300414.

Webb, Judson C. "Hilbert's Formalism and Arithmetization of Mathematics." *Synthese*, vol. 110, no. 1, 1997, pp. 1–14. JSTOR, http://www.jstor.org/stable/20117583.

Weber, Bruce. "Hilary Putnam, Giant of Modern Philosophy, Dies at 89." *The New York Times*, March 17, 2016.

Weikart, Richard. "Progress Through Racial Extermination: Social Darwinism, Eugenics, and Pacifism in Germany, 1860–1918." *German Studies Review*, vol. 26, no. 2, 2003, pp. 273–294. JSTOR, https://doi.org/10.2307/1433326.

Welch, Philip, and Leon Horsten. "Reflecting on Absolute Infinity." *The Journal of Philosophy*, 2016, pp. 89–111.

West, Rebecca. *1900*. Viking Press, 1982.

Weston, Corinne Comstock. "The Liberal Leadership and the Lords' Veto, 1907–1910." *The Historical Journal*, vol. 11, no. 3, 1968, pp. 508–537. JSTOR, http://www.jstor.org/stable/2638166.

Weyl, Hermann. "David Hilbert and His Mathematical Work." *Bulletin of the American Mathematical Society*, vol. 50, no. 9, September 1944, pp. 612–654, https://projecteuclid.org/journals/bulletin-of-the-american-mathematical-society/volume-50/issue-9/David-Hilbert-and-his-mathematical-work/bams/1183506085.full.

Weyl, Hermann. "Emmy Noether." *Emmy Noether 1882–1935*, 1981, pp. 112–152, https://doi.org/10.1007/978-1-4684-0535-4_6.

Wheeler, John Archibald. "Hermann Weyl and the Unity of Knowledge: In the Linkage of Four Mysteries—the 'How Come' of Existence, Time, the Mathematical Continuum, and the Discontinuous Yes-or-No of Quantum Physics—May Lie the Key to Deep New Insight." *American Scientist*, vol. 74, no. 4, 1986, pp. 366–375. JSTOR, http://www.jstor.org/stable/27854250.

ANNOTATED BIBLIOGRAPHY

White, Trumball. *Our War with Spain for Cuba's Freedom*. Monarch Book Company, 1898, via Project Gutenberg, https://www.gutenberg.org/cache/epub/4210/pg4210-images.html.

Whitehead, Alfred North, and Bertrand Russell. *Principia Mathematica*, 3 vols. Rough Draft Printing, 2011.

Whitman, Alden. "Bertrand Russell Is Dead; British Philosopher, 97." *The New York Times*, February 3, 1970.

Whittaker, Edmund T. "Alfred North Whitehead, 1861–1947." *Obituary Notices of Fellows of the Royal Society*, vol. 6, no. 17, 1948, pp. 281–296. JSTOR, http://www.jstor.org/stable/768923.

Whittaker, Edmund T. "The Philosophy of Bertrand Russell." *Nature*, vol. 155, no. 3927, February 3, 1945, pp. 128–131, https://doi.org/10.1038/155128a0.

Wiener, Norbert. "The Relation of Space and Geometry to Experience." *The Monist*, vol. 32, no. 1, 1922, pp. 12–60. JSTOR, http://www.jstor.org/stable/27900892.

Wiest, Lynda R. "Female Mathematicians as Role Models for All Students." *Feminist Teacher*, vol. 19, no. 2, 2009, pp. 162–167. JSTOR, http://www.jstor.org/stable/40546095.

Wikler, Daniel. "Can We Learn from Eugenics?" *Journal of Medical Ethics*, vol. 25, no. 2, 1999, pp. 183–194. JSTOR, http://www.jstor.org/stable/27718281.

Wilder, Raymond Louis. *Introduction to the Foundations of Mathematics*. Dover Publications, 2019.

Willis, Kirk. "'This Place Is Hell': Bertrand Russell at Harvard, 1914." *The New England Quarterly*, vol. 62, no. 1, 1989, pp. 3–26. JSTOR, https://doi.org/10.2307/366207.

Wilkinson, S. "'Eugenics Talk' and the Language of Bioethics." *Journal of Medical Ethics*, vol. 34, no. 6, 2008, pp. 467–471. JSTOR, http://www.jstor.org/stable/27720112.

Wilson, Janelle L., and Carmen M. Latterell. "Nerds? Or Nuts? Pop Culture Portrayals of Mathematicians." *ETC: A Review of General Semantics*, vol. 58, no. 2, 2001, pp. 172–178. JSTOR, http://www.jstor.org/stable/42578095.

Wilson, Trevor. *The Downfall of the Liberal Party, 1914–1935*. Faber & Faber, 2011.

"Wireless Telegraphy." *Nature*, vol. 61, no. 1581, February 15, 1900, pp. 377–380, https://doi.org/10.1038/061377a0.

"Wisconsin Eugenics Laws." *Journal of the American Institute of Criminal Law and Criminology*, vol. 9, no. 4, 1919, pp. 597–598. JSTOR, http://www.jstor.org/stable/1134135.

Wittgenstein, Ludwig. *Notebooks 1914–1916*, 2nd ed. Edited by G. H. Wright and G. E. M. Anscombe. University of Chicago Press, 1984.

Wittgenstein, Ludwig, and B. F. McGuinness. *Tractatus Logico-Philosophicus*. Translated by D. F. Pears. Routledge & Kegan Paul, 1969.

ANNOTATED BIBLIOGRAPHY

"Work of the Corps of Engineers of the Army." *The Scientific Monthly*, vol. 5, no. 6, 1917, pp. 568–570. JSTOR, http://www.jstor.org/stable/22503.

"A World's Record in Bridge Building." *Scientific American* (1845–1908), vol. 82, February 24, 1900, p. 116.

Wright, Jonathan. "Locarno: A Democratic Peace?" *Review of International Studies*, vol. 36, no. 2, 2010, pp. 391–411. JSTOR, http://www.jstor.org/stable/40783204.

Wright, Jonathan. "Stresemann and Locarno." *Contemporary European History*, vol. 4, no. 2, 1995, pp. 109–131. JSTOR, http://www.jstor.org/stable/20068657.

Wrinch, D. M. "Review." *The Mathematical Gazette*, vol. 17, no. 226, 1933, pp. 332–333. JSTOR, https://doi.org/10.2307/3606521.

Young, J. W. A. "The Fifth International Congress of Mathematicians." *The American Mathematical Monthly*, vol. 19, no. 10/11, 1912, pp. 161–166. JSTOR, https://doi.org/10.2307/2971877.

Zach, Richard. "The Practice of Finitism: Epsilon Calculus and Consistency Proofs in Hilbert's Program." *Synthese*, vol. 137, no. 1/2, 2003, pp. 211–259. JSTOR, http://www.jstor.org/stable/20118359.

INDEX

"absolute" proofs, 209
absolute space and time, 212
Achilles, 37
Acta Mathematica, 50
Aesthetes, The (Turner), 192
airplanes, 134, 257
Aiyangar, Srinivasa Ramanujan, 135
Albert, Prince, 189
Albert Einstein Prize, 290
aleph-naught, 47–48, 50, 89
aleph-one, 48, 50, 89, 101–102, 219
Alexandrina Victoria. *See* Victoria, Queen
algebra, fundamental theorem of, 243
Aliens Restriction Act (Great Britain), 188
Allen, Frederick Lewis, 258
American Association for the Advancement of Sciences, 257
American Mathematical Society, 6, 10–11
anemonefish, 317
Annals of Mathematics, 12
antinomies, 94
anti-Semitism, 294–299
antiwar activism, 76–77, 84, 109, 164, 167, 182–185, 188, 197
Aquinas, Thomas, 38
Archimedes, 175
Aristotelian Society, 136
Aristotle, 35–36, 38, 40, 98, 233

Arnold, Vladimir Igorevich, 11
Asquith, Herbert Henry, 120–121, 155–156, 171, 182, 184, 189, 190, 195
assassinations, political, 160, 162–164
Austrian First Army, 170
Austro-Hungarian Empire, 160
aviation, 133–134, 257
Axiom of Choice, 102–104
axiomatic method, 175, 209–210, 232, 287

Bacon, Francis, 59, 62
Balfour, Arthur James, 199
barbed wire, 85
Barrymore, Ethel, 3
Basic Laws of Arithmetic (*Grundgesetze der Arithmetik*; Frege), 88–89
Batrachomyomachia, 263*fig*
Begriffsschrift (Concept notation; Frege), 65–67, 210
Behmann, Heinrich, 198
beliefs, Russell on, 97
Bell, Clive, 116
Bell, Vanessa, 116, 125
Berlin Academy, 181
Berners-Lee, Tim, 314
Bernhardt, Sarah, 3
Bethe, Hans, 301
Bieberbach, Ludwig Georg Elias Moses, 260, 273–274, 276

big bang, concept of, 213, 240
Black Tuesday, 282
Black Week, 75
Blaricum, 140
Bloch, Felix, 301
blockhouses, 85
Bloemfontein, 76
Bloomsbury Group, 116, 121–122, 192
Blumenthal, Ludwig Otto, 264, 265, 268–269, 271, 272–273, 274–276, 277–278, 279
Boer War, 63*fig*, 69–70, 72–78, 80–81, 82–86
Bol, Manute, 105
Bolshevism, 246
Boltzmann, Ludwig Eduard, 174
Boole, George, 65, 67, 98, 179
Boolean logic, 67
Boolos, George, 96
Boon (Wells), 192
Borel, Félix Édouard Justin Émile, 103
Born, Max, 261, 301
Botha, Louis, 75
Brahms, Johannes, 131
Broadbent, T. A. A., 28
Brontë, Charlotte, 23
Brouwer, Lize, 141–142, 145, 146, 259, 309
Brouwer, Luitzen Egbertus Jan "L. E. J."
  approach taken by, 211–215, 216–217, 234–235
  attempt at reconciliation from, 258–259
  background of, 138–140
  Compositio Mathematica and, 309
  death of Scheltema and, 248
  Descartes and, 212
  ego of, 267–268
  German scientists' boycott and, 260–261
  Göttingen University and, 302
  Hilbert and, 142–143, 144–145, 176, 206–208, 209, 210–211, 215, 236, 239, 240–241
  Hilbert's dismissal of, 264–267, 270–279
  impact of, 317–318
  Kant and, 212–213, 214
  later life of, 309–310
  Life, Art, and Mysticism and, 142–143
  military language and, 257
  Noether and, 256
  publications of, 242–243
  quotation from, 263
  rejection of intuitionism and, 246–247, 248–250, 255
  reputation of, 137–138
  squabbles involving, 269–270
  stepdaughter of, 145–146
  topology and, 144
  Weyl and, 218–219, 221–224, 228, 258, 308
  wife of, 141–142
Buck, Carrie, 303
Buck v. Bell, 303
Bulletin of the American Mathematical Society, 5
Bunsen, Robert, 25
Burali-Forti, Cesare, 61
Burali-Forti Paradox, 61, 89
burrowing duke, 112–113

Calculus Wars, 139
Calder, Allan, 89, 242
Campbell-Bannerman, Henry, 120, 121
Cannan, Gilbert, 193
Cantor, Constantin, 61
Cantor, Georg Sr. (father), 33–34
Cantor, Georg Ferdinand Ludwig Philipp

INDEX

background of, 34–35
belief in divine intellect and, 55–56
Brouwer and, 143
Catholic church and, 57–59
continuum hypothesis and, 49–51, 53–54
criticism of, 46
Hilbert and, 270
infinity and, 36–37, 39–44, 219–220
International Congress of Mathematicians and, 100–101
König and, 101–102
Kronecker and, 46–47, 48–49, 50–51, 52–53, 223
Mathematiker Vereinigung and, 60–61
mental health of, 51–52, 62
paradoxes and, 89
Principia Mathematica and, 118
quotation from, 293
set theory and, 32–34, 60–61
Shakespeare and, 59, 62
transfinite numbers and, 47–48
Zermelo and, 102–103
See also set theory
Cantor, Rudolph (son), 62
Cantor's Paradox, 61, 90
Cape Colony, 70
Capone, Al, 281*fig*
Carathéodory, Constantin, 264, 265–267, 271–272, 273, 274–275, 278
cardinality, 89–90, 100
Carnap, Rudolf, 288, 289
Cartesian coordinates, 42
Catholic church, 57–59
Cavendish-Bentinck, Ottoline Violet Anne. *See* Morrell, Ottoline
Cézanne, Paul, 4
Chaitin, Gregory, 232
Chekhov, Anton, 3

chess, 229–230, 233
choice functions, 102
choice sequences, 235
Chrysippus of Soli, 316
Churchill, Winston, 75, 125, 195
Cicero, Marcus Tullius, 87
Clark, Ronald, 22, 93
Clay Mathematics Institute, 12
Clement, Kevin C., 96
clown fish, 317
Cogito, ergo sum, 212
Colenso, Battle of, 74–75
Collodi, Carlo, 20
Compositio Mathematica, 309
Compton, Karl, 250
concentration camps, 83–85
Conrad, Joseph, 3
conscientious objectors, 184, 189–190
consistency proof, 240
Constantinople, capture of, 161
constructive set theory, 235
constructivism, 216–217, 223, 234, 242–243, 267, 315, 318
Contemporary Art Society (London), 121
continuum hypothesis, 48, 49–51, 53–54, 100, 101–102
Coolidge, Calvin, 281, 282
Count Ferdinand, 133
Courant, Richard, 137, 250, 251–253, 272, 273, 295, 296–297, 298, 301
Cousin, Jehan, 111*fig*
Crelle, August Leopold, 48
Crelle's Journal, 48–49
Crome Yellow (Huxley), 193
Cronjé, Pieter Arnoldus, 76
Curbera, Guillermo, 238
Cuthbert Learmont (Revermort), 192

Dada movement, 218
Daedalus, 87*fig*

INDEX

Daily Mail (London), 75
data, fallacy of rich, 204
Dauben, Joseph Warren, 32, 51, 55, 101, 118
de Fermet, Pierre, 42
de Holl, Anna Louise Elisabeth "Louise," 146, 310
de Holl, Reinharda Bernadina Frederica Elizabeth "Lize." *See* Brouwer, Lize
de Jode, P., 111*fig*
de Vries, Hugo Marie, 136–137
Debye, Peter, 174
Dedekind, Julius Wilhelm Richard, 45, 89
DeLong, Howard, 95
Descartes, René, 39, 42, 212
DeSua, Frank, 67
Deutsche Mathematiker Vereinigung (German Mathematicians Association), 60
diagonalization, 60
diamonds, 70
Dichotomy Paradox, 37–38
Dickens, Charles, 21–22
dimensional invariance of real numbers, 49
Diophantus of Alexandria, 11
diphtheria, 23
Dirac, Paul, 250
Disney, Walt, 20
Doyle, Arthur Conan, 75
Dresden, Arnold, 255
Du Bois, W. E. B., 3
du Bois-Reymond, Emil, 13–14, 241
Dudley, E. Clark, 157
Dudley, Helen, 157–158, 164–167, 194
Durkheim, Émile, 206
Džamonja, Mirna, 313

Eastman Kodak, 3
Eddington, Arthur Stanley, 205
Edwards, David A., 312
Einstein, Albert
  Brouwer and, 261, 264, 265–266, 267, 271, 272
  general relativity and, 180, 181, 205
  Great Migration and, 294, 301–302
  Hilbert and, 176–177, 181, 242
  impact of work of, 212
  Kant and, 213
  Mathematische Annalen and, 206, 270, 274–275, 276–277, 278
  mind of, 205–206
  Minkowski and, 9
  Nauheim meeting and, 236
  Noether and, 256
  special relativity and, 137
  Weyl and, 218
  See also relativity, theory of
Elements (Euclid), 7, 20–21, 175
Eliot, Thomas Stearns "T. S.," 147, 155, 191
Elizabeth II (Queen), 15
Emergency Committee in Aid of Displaced German Scholars, 299, 300–301
Eminent Victorians (Strachey), 190
Engel, Friedrich, 270
Epimenides the Cretan, 105–106, 107
Erdős, Paul, 301
Escoffier, Auguste, 2
Euclid, 7–8, 20–21, 175
eugenics, 303–304
Euler, Leonhard, 284–285
euthanasia, 304
excluded middle, principle of, 233–235, 249, 255, 259
exuberant solutionism, 12–13, 217, 228, 285, 287

Facius, G. S. & E. G., 87*fig*
fallacies

INDEX

of the illusion of desperate
    deliberateness, 130
of pure process data, 81–82, 97
of rich data, 204
of seeming, 311, 312
of unimportant results, 68
fathers and father figures, 33–35,
    46–47, 56, 62, 112–114, 130–132,
    178–179, 226–227
Ferreirós, José, 313, 315
Fields, John Charles, 247
Fields Medal, 247
Fifth International Congress of
    Mathematicians, 135–136
finitists, 239–240
First International Congress of
    Mathematicians, 60
first-order logic, 287
fixed-point theorem, 217
fluid dynamics, 134
force majeure losses, 56
formal axiomatization, 209–210
formalism, 211, 215, 228, 235–236, 239,
    241–242, 248–249, 258, 270–271,
    286–289, 313–315
Formulario Mathematico (Peano), 79
Forster, Edward Morgan "E. M.," 116, 191
Foundations of a General Theory of Sets
    (*Grundlagen einer allgemeinen*
    Mannigfaltigkeitslehre; Cantor), 32,
    40–42, 54–55, 58
Foundations of Arithmetic (*Grundlagen*
    der Arithmetik; Frege), 67
Foundations of Geometry, The
    (*Grundlagen der Geometrie*;
    Hilbert), 7, 8, 20, 142, 145, 175,
    209, 229
Franck, James, 295–296, 301
Franz Ferdinand, Archduke, 159*fig*,
    160–164

Franzelin, Johannes, 58
Frege, Friedrich Ludwig Gottlob, 65–67,
    68, 88–89, 91–96, 98, 118, 210,
    216
French, David, 189
Friends of Foreigners, 189
frogs and mice, battle of, 263*fig*, 276
Frost, Robert, 3, 96, 316
Fry, Roger, 116, 125–126, 192
futurism, 4–6, 86, 312

Galactus, 226–227
Galbraith, John Kenneth, 282
Galileo Galilei, 38–39, 40–41
Gallipoli campaign, 195
Garsington, 190–191, 192, 199
Gaudí, Antoni, 2
gauge theory, 310
Gauss, Carl Friedrich, 53–54, 179,
    243
general relativity, 176–177, 180, 181,
    205, 256
geometry
    developments in, 42
    Euclid and, 7
    Frege and, 95
    Hilbert and, 7–8, 175
George III, 15
George IV, 15
George Louis of Hanover, 189
George V, 189
German scientists, boycotts involving,
    237, 238, 247, 259–261, 264–265
German Student Association, 295
Germanophobia, 188–189
Gibbs, Josiah Willard, 174
Gladstone, William Ewart, 115, 121,
    155
Gödel, Kurt, 248, 286–291, 294,
    312, 315

383

INDEX

Gödel's Proof (Nagel and Newman), 290
gold, 70–73
Goodhart's Law, 81
Gorban, Anna, 173*fig*
Gordon, Paul, 59, 179–180
Gordon's theorem, 59
Göttingen Mathematical Society, 248
Göttingen Tageblatt, 296
Göttingen University, 208–209, 250–251, 254, 294–298, 302
Graham, Loren, 43
Grant, Duncan, 116
graph theory, 284
Graves, Robert, 190–191
gravity, general relativity and, 176–177
Great Britain
    Boer War and, 69–70, 72–78, 80–81, 82–86
    diamonds and, 70
    draft in, 184–185
    expansion of empire of, 15
    influence of, 15–16
    Irish home rule and, 155–157
    spy fever in, 188
    war and, 16–17
    World War I and, 167–168, 181–182, 188–189
Great Depression, 281*fig*, 282–283, 300
Great Migration, 298–302
greatness
    assessing, 204
    children and, 131–132
    as out of reach, 206
Greiling Paradox, 89
Griffin, Nicholas, 121
Grundgesetze der Arithmetik (*Basic Laws of Arithmetic*; Frege), 88–89
Grundlagen der Arithmetik (*The Foundations of Arithmetic*; Frege), 67
Grundlagen der Geometrie (*The Foundations of Geometry*; Hilbert), 7, 8, 20, 142, 145, 175, 209, 229
Grundlagen einer allgemeinen Mannigfaltigkeitslehre (Foundations of a general theory of sets; Cantor), 32, 40–42, 54–55, 58
Grundlagenkrise (the foundational crisis), 222
Grundzüge der theoretischen Logik (*Principles of Mathematical Logic*; Hilbert), 286–287
Guardian (newspaper), 77

Hales, Thomas Callister, 11–12, 68
Harvard University, 154–155, 157
Heisenberg, Werner, 250, 255
Henry VIII, 21
Hercules, 203*fig*
Hermite, Charles, 46, 49, 237
Hertzman, Lewis, 208
Hesseling, Dennis Edwin, 258, 260
Heyting, Arend, 213
Hilbert, David
    applied mathematics and, 175
    Brouwer and, 142, 144–145, 207, 209, 215, 216–217, 236, 239, 241, 247, 248–249, 255, 258–259, 261, 264–267, 268–269, 270–279
    Cantor and, 59, 100
    continuum hypothesis and, 48
    Courant and, 252
    death of, 308
    draw of, 250–251
    du Bois-Reymond and, 13–14, 241
    Einstein and, 264
    Euclid and, 20
    excluded middle and, 234
    as father, 227–228
    formalism and, 228–229

INDEX

foundations of math and, 209–211
general relativity and, 176–177, 180, 181
German scientists' boycott and, 260
Gödel and, 286–287, 289, 290–291
Göttingen and, 208
Great Migration and, 302
health of, 242, 253–255, 259, 261
infinity and, 239–240
journal led by, 137
kinetic theory and, 174, 175
Klein and, 250, 253
König and, 102
on math as fundamental, 316
Mathematische Annalen and, 206, 243
military language and, 257
in Nauheim, 237
Noether and, 180–181, 256
portrait of, 173*fig*
Principia Mathematica and, 176
problems presented by, 10–13
proof theory and, 230–233, 315
quotation from, 187
retirement of, 283, 285–286
reworking of geometry by, 7–8
Russell and, 79, 198–199
Russell's Paradox and, 94
at Second International Congress of Mathematicians, 6–7, 8, 9–11, 14
solutions to problems presented by, 11–12
Weyl and, 223–224, 228, 258, 308
Hilbert, Franz, 227–228, 252
Hilbert, Käthe, 227, 272
Hilbert's program, 239
Hille, Einar, 50, 65
Hitler, Adolf, 294
Hobbes, Thomas, 151
Hodges, Andrew, 290
Holmes, Oliver Wendell, 303

"home rule" bill, 155–157
homeomorphisms, 144
Hoover, Herbert, 282
Hopkins, Frederick Gowland, 134
Huxley, Aldous, 190, 193, 194
Huxley, Juliette, 192
Huygens, Christiaan, 139
Hydra, 203*fig*

Ibsen, Henrik, 3
Icarus, 62, 87*fig*, 90, 124, 277, 279, 284
Idea of Riemann Surfaces, The (Weyl), 218
ideograms, 79
immigration laws, American, 299, 300
incompleteness theorems, 288–289, 290, 294, 312, 315
inferred beliefs, 97
infinity
  actual versus potential, 35–36, 37–38
  Cantor's work on, 39–42, 219–220
  Catholic church and, 58
  concept of, 32–33, 35–37
  continuum hypothesis and, 50, 100, 101–102
  diagonalization and, 60
  Galileo on, 38–39
  Hilbert on, 239–240
  Kronecker and, 47
  large versus small, 220–221
  set theory and, 32, 35, 40–41, 42–43, 89–90, 248
  transfinite numbers and, 47–48
  Zeno of Elea and, 37–38
inflation, after World War I, 244
Institute of Technology (Zürich), 218
International Congress of Mathematicians, 4–7, 8, 9–11, 14, 60, 100–101, 135–136, 237, 238, 247, 259–261, 273

INDEX

Introduction to Mathematical Philosophy (Russell), 199
intuitionism
    Brouwer and, 211–214, 215, 216, 217, 219, 234–236, 243, 248
    criticism/rejection of, 228, 246–247, 255, 315
    Hilbert and, 223–224, 236, 258–259, 270
    Kant and, 213, 214
    Königsberg conference and, 286, 288
    as optional approach, 313–314
    potential of, 317–318
    Weyl and, 249–250, 257, 258, 308
invariants, theory of, 180
Irish home rule, 155–157, 195
Irish Volunteer Army, 156
irrational numbers, Kronecker and, 47
Isaac Newton complex, 57

Jahnke, Paul Rudolf Eugen, 269–270
James, Henry, 114, 116, 125
Jeiler, Ignatius, 58
Jogejan, Correi "Cor," 146, 310
Johannesburg, 73, 76
Jorand, Jean-Baptiste-Joseph, 307*fig*
Jourdain, Philip, 61
Journal für die reine und angewandte Mathematik (*Journal for Pure and Applied Mathematics*), 48–49
Joyce, James, 3
Judgment of Paris, 1*fig*

Kamerlingh Onnes, Heike, 137
Kanamori, Akihiro, 103, 104, 107
Kant, Immanuel, 212–213, 214, 250
Kantor, Jean-Michel, 43
Kepler, Johannes, 270
Kepler's conjecture, 11–12, 67–68
Keynes, John Maynard, 116, 134

Kiester, Edwin, 162
Kimberly (town), 70
kinetic theory of gasses, 174, 175
Kipling, Rudyard, 3
Kitchener, Horatio Herbert "H. H.," 80–81, 82–83, 84–85, 182, 187*fig*, 195
Kleene Paradox, 89
Klein, Felix, 250, 251, 253, 265, 270, 274
Klement, Kevin, 150
Klimt, Gustav, 131
König Paradox, 89
König, Gyula "Julius," 101–102
Königsberg, 283, 284–285
Kontinuum, Das (Weyl), 219
Korteweg, Diederik Johannes, 139, 142, 143
Korvin-Krukovsky, Vasily Vasilyevich, 26–27
Kosovo, First Battle of, 161
Kovalevskaya, Sofya Vasilyevna, 19, 19*fig*, 25–31, 50, 55, 177, 225
Kreisel, Georg, 118
Kronecker, Leopold, 46–47, 48–49, 50–51, 52–53, 60, 214, 223
Krüger, Stephanus Johannes Paulus "Paul," 73, 75

Ladysmith, 74, 76
land clearance, 81, 82–83
Landau, Edmund Georg Hermann, 298
language
    artificial, 211
    Brouwer and, 211–212
    constraints of, 66
    Hilbert on, 211
    inadequacy of, 210–211
    military, 257
    symbolic, 230–231

INDEX

Lasley, J. W., Jr., 36
Last Judgment: The Graves Open and the Dead Emerge, The (de Jode), 111*fig*
Law of Noncontradiction, 93
Lawrence, David Herbert "D. H.," 116, 191–192, 193–194
Lawrence, Frieda, 191, 192, 194
Le Brun, Charles, 87*fig*
Le Pautre, Jean, 293*fig*
Leaves of a Tulip Tree (Huxley), 192
Lebesgue, Henri, 268–269, 273
Leibniz, Gottfried Wilhelm, 55, 139, 284
Lenin, Vladimir, 246
Leo XIII, Pope, 57
Lewy, Hans, 301
Liar's Paradox, 105–106
Life, Art, and Mysticism (Brouwer), 142, 143, 213
Lifton, Robert Jay, 303
Lindbergh, Charles, 257
lines, as "privileged" objects, 231
"liver cure," 254
Lloyd George, David, 195
Locke, John, 39
logic
    Frege and, 66–67, 68
    predicate/first-order, 287
    Russell and, 80, 97–98
logical atomism, 79
logical inconsistency, 93, 95
logicism, 98–99, 198, 210, 216, 219, 286, 288, 289, 313–315
London, Jack, 3
London Agreement (1921), 244
Lorentz, Hendrik Antoon, 137, 174
loss, the three types of, 56–57
"Love Song of J. Alfred Prufrock" (Eliot), 191
Lowell, A. Lawrence, 154–155

Lucas, J. R., 289
Lusitania, 188

Mafeking, 76
Magersfontein, Battle of, 74–75
Mahler, Gustav, 131
Malleson, Constance (Colette O'Niel), 194
Mancosu, Paolo, 107, 118, 315
Manet, Édouard, 4
Manhattan Project, 301
Mann, Thomas, 163–164
Manning, Frederic, 245
Mannoury, Gerrit, 139
mappings, 144
Marinetti, Filippo Tommaso, 5
Massie, Robert K., 15
Mathematical Analysis of Logic, The (Boole), 67
Mathematical Association of America, 257
mathematical water nymph/mermaid problem, 31
Mathematische Annalen, 102, 137, 206, 243, 261, 264–267, 268, 270–279
Matisse, Henri, 4
Matiyasevich, Yuri, 11
McCarthy, Joseph, 299
McGuinness, Brian, 130–131
McKinley, William, 160
medical profession under Nazis, 303–305
Mehmet the Conqueror, 161
Menzel, Christopher, 90
Mercier, Désiré-Joseph, 238
metamathematics, 229, 239
Methods of Mathematical Physics (Courant), 296
Military Service Act (Great Britain), 184
Millennium Prize Problem, 12

INDEX

Minkowski, Hermann, 8–9, 144
Mittag-Leffler, Magnus Gösta, 29–30, 33, 50–51, 53–54, 55
Moder River, Battle of the, 74
Monet, Claude, 4
Monk, Ray, 21, 170
Moran, Margaret, 127
Morrell, Julian, 115
Morrell, Ottoline
  Asquith and, 121
  background of, 112–114
  children of, 115
  death of, 309
  gossip about, 192
  husband's affairs and, 196
  Lawrence and, 191, 192, 193–194
  marriage of, 114
  parodies of, 192–193
  quotation from, 159
  Russell and, 116–117, 119–120, 122–127, 128, 135, 136, 149, 152–153, 155, 157–158, 160, 164–167, 169, 194, 199, 309
  World War I and, 167, 171, 189–190
  See also Cavendish-Bentinck, Ottoline Violet Anne
Morrell, Philip Edward
  absence of, 122
  affairs of, 196
  draft and, 184
  Lawrence and, 194
  Ottoline and, 114
  in parliament, 115–116
  Russell and, 119–120, 124, 125, 153, 169
  World War I and, 167, 189–190, 194–196
Moses draws water from the rock (Le Pautre), 293*fig*
motion, Dichotomy Paradox and, 37

Munch, Edvard, 4
Murrow, Edward R., 299
Mussolini, Benito, 259

Nagel, Ernest, 118, 232, 290
Nation, The, 154, 196
Nature (journal), 9
Nauheim, 1920 conference in, 236–237
Nazis, 294–298, 302–304
Nernst, Walther, 174
Nernst equation, 174
Neugebauer, Otto, 297
"neutrality committee," 188
"New Century in the Life of a Paradox, A" (Clement), 96
new realism movement, 154
New York Times, 267
Newman, James R., 118, 232, 290
Newton, Isaac, 57, 139, 179, 270, 284
Nijinsky, Vaslav, 116
Nixon, Richard, 252
Noether, Amalie "Emmy," 177, 178–181, 255–256, 258, 297–298
Noether, Max, 179
Noether's theorem, 256
non-Euclidean geometry, 53–54
normal sets, 91–92
North American Review, 121
Northern Ireland, 156
Notes on Logic (Wittgenstein), 152
number line, infinity and, 41–42
number theory, 42
numbers, as "privileged" objects, 231
Nys, Maria, 190

Omega Workshops, 192
"On the Mapping of Manifolds" (Brouwer), 144
O'Niel, Colette (Constance Malleson), 194

INDEX

Oppenheimer, Robert, 250
Orange Free State, 69, 85
ordinary sets, 91–92
O'Reilly, John, 70
Our Knowledge of the External World (Russell), 150
Owen, Wilfred, 173

Pakenham, Thomas, 76
paradoxes
    Burali-Forti Paradox, 61, 89
    Cantor and, 89
    Cantor's Paradox, 61, 90
    Dichotomy Paradox, 37–38
    Liar's Paradox, 105–106
    Russell and, 216
    Russell's Paradox, 89, 91–96, 106, 257
    set theory and, 61, 216
    Weyl and, 222
Paris Expo, 3–5, 18
Parmenides, 93
particle physics, 256
Paul, Apostle, 105
Pauli, Wolfgang, 250
Pauling, Linus, 250
Peano, Giuseppe, 79–80, 98, 99, 103, 136
Peano postulates, 79
Pearsall Smith, Alyssa "Alys" Whitall, 64, 78, 80, 88, 96, 108–109, 119–120, 124–125
Pearsall Smith, Logan, 114, 116, 119–120, 125
Peckhaus, Volker, 67
pernicious anemia, 253–254, 259, 261
pessimism, triumphant, 13
pi, 59, 284
Picard, Charles Émile, 237–238, 247, 259
Picasso, Pablo, 2
Pied Piper, 225*fig*

Pinocchio, 20, 44
Pinsent, David, 170
Planck, Max, 174
Plato, 93
Pliny the Elder, 63
Poincaré, Jules Henri, 6, 46, 54, 103–104, 106, 143, 214, 219
political assassinations, 160, 162–164
Pólya, George, 228, 308, 315
predicate logic, 287
Pregel River bridges, 283–285
prejudice, 97
Princeton Institute for Advanced Study, 310
Princip, Gavrilo, 162, 163
Principia Mathematica (Russell and Whitehead)
    Gödel and, 286, 288
    Hilbert and, 176, 198
    impact of, 215–216
    new edition of, 308
    new realism movement and, 154
    Peano and, 136
    publication of, 117–118
    reviews of, 118–119
    success of, 127
    theory of types and, 106–108
    Weyl and, 218–219
    Wittgenstein and, 135
    writing of, 98–99, 104–105
Principles of Mathematical Logic (*Grundzüge der theoretischen Logik*; Hilbert), 286–287
Principles of Mathematics (Russell), 80, 88, 97–98
"privileged" objects, 231
Prix Bordin, 31
Prometheus, 44, 45*fig*
proof theory, 229, 231–233, 239, 240, 313–314, 315

INDEX

Prussian Academy of Sciences, 294
Prussian Officer and Other Stories, The (D. H. Lawrence), 191
Pugs and Peacocks (Cannan), 193
pure existence proofs, 59, 217, 223, 233
pure process data, fallacy of, 81–82, 97
Putnam, Hilary Whitehall, 11
pyrite, 71
Pythagorean school of thought, 37–38

quadratic polynomials, 8
quantum entanglement, 255
quantum mechanics, 255
quantum theory, 240
Quincy Market (Boston), 147*fig*

Raatikainen, Panu, 291
Raimondi, M. A., 1*fig*
Rainbow, The (D. H. Lawrence), 193–194
Raphael, 1*fig*
real numbers, 218–220
reasoning abilities of animals, 316–317
reducibility, axiom of, 198
refugee crisis, land clearance and the Boer War, 83
Reid, Constance, 8, 218, 251–252
Reingold, Nathan, 300
relativity, theory of
  general, 176–177, 180, 181, 205, 256
  impact of, 212, 240
  Nauheim meeting and, 236
  special, 9, 137
reparations, 243–244
Restoration of the Professional Civil Service Act (Germany), 294
Revermort, J. A., 192
revolutionary politics, 246–247
rich data, fallacy of, 204
Richard Paradox, 89

Riemann hypothesis, 12
Rivista di Matematica and Formulaire (Journal of Mathematics and Form), 79
Rockefeller Foundation, 251, 296, 298–299
Rodin, Auguste, 4
Rosser Paradox, 89
rotating rigid body problems, 31
Roth, Joseph, 208
rotten-apple principle, 95
Royal Society, 101, 118, 127, 237
Royal Society of Science (Göttingen), 181
Russell, Bertrand
  antiwar activism and, 84, 109, 182–185, 188, 197
  background and family of, 21–22
  Boer War and, 74, 76–77
  Brouwer and, 143, 210
  childhood of, 22, 23–25
  Euclid and, 20–21, 24–25
  expanded autobiography of, 309
  Frege and, 65–66, 68, 88–89, 91–96
  at Garsington, 190
  German mathematicians and, 64–65
  Hilbert and, 210
  inadequacy of language and, 210
  International Congress of Mathematicians and, 136
  Lawrence, D. H., and, 191–192
  logicism and, 98–99
  love triangle and, 164–167
  marriage of, 96, 108–109
  Ottoline and, 116–117, 119–120, 122–127, 135, 152–153, 157–158, 160, 164–167, 169, 194, 309
  paradoxes and, 216
  peace conversion of, 77–78, 80
  Peano and, 79–80

INDEX

photograph of, 129*fig*
Principia Mathematica and, 98–100, 104–105, 106, 108, 117–119, 127, 135, 136, 154, 176, 198, 215–216, 218–219, 286, 288, 308
Principles of Mathematics and, 80, 88, 97–98
quotation from, 129
revolution and, 246
Spalding and, 22–23
theory of types and, 106–108
trial of, 197–198
in United States, 154–155, 157
Weyl and, 218–219
Whitehead and, 78–79
wife of, 64
Wittgenstein and, 127–128, 134, 135, 148–152, 171
World War I and, 164, 167, 168–169
Zermelo and, 104
Russell, Countess, 22, 24, 64
Russell, Frank, 20–21, 23, 24
Russell, John (Lord Amberley), 21, 22, 23–24
Russell, Katharine (Lady Amberley), 21, 22, 23–24
Russell, Rachel, 23–24
Russell's Paradox, 89, 91–96, 106, 257
Russia, revolutionary fervor in, 28–29
Russian Childhood, A (Kovalevskaya), 29
Russian Imperial Academy of Sciences, 25–26
Russian Revolution, 246

Sacré Coeur, 2
Sands, Ethel, 114–115
Sanger, Margaret, 3
Scheltema, Carel Steven Adama von, 248, 267–268
Schoenflies, Arthur Moritz, 269

Scholz, Heinrich, 289
Schrödinger, Erwin, 255
Scientific American, 70, 89, 242
scientific instruments, 307*fig*
scientific revolution, 42
Scott, Charles Prestwich "C. P.," 77
Scott, Charlotte Angas, 6
Second Conference on Epistemology of the Exact Sciences, 287–288
Second International Congress of Mathematicians, 4–7, 8, 9–11, 14, 100
seeming, fallacy of, 311, 312
set theory
  axiomatic system for, 102
  Cantor's work on, 32–33, 35, 36, 39–43, 55–57
  Catholic church and, 58
  constructive, 235
  criticism of, 46, 47, 49
  infinity/infinite sets and, 32, 35, 38–39, 40–41, 42–43, 89–90, 248
  intuitionism and, 214, 235
  paradoxes and, 61, 216
  praise for, 60–61
  Principia Mathematica and, 118
  Weyl and, 219
seven bridges problem, 283–285
sexism, 177–178, 180–181. *See also* Kovalevskaya, Sofya Vasilyevna; Noether, Amalie "Emmy"
Seymour, Miranda, 126, 135
Shakespeare, William, 59, 62, 111
shapes, as "privileged" objects, 231
Sheffield, Gary, 200
Shelley, Percy Bysshe, 1
Sieg, Wilfried, 95
Siegmund-Schultze, Reinhard, 254
Simon, Max, 54
Simons, Peter, 8

INDEX

Simpson, Steve, 240
Sinclair, Upton, 3
Singleton, Henry, 245*fig*
Sitwell, Osbert, 193
Sixth International Congress of Mathematicians, 237, 238
Skolem Paradox, 89
slavery, British abolishing of, 69
Smith, Henry John Stephen, 8
Snapper, Ernest, 216, 289, 315
Society of German Scientists and Physicians, 60
Somme, Battle of the, 182, 195
Sophie, Duchess von Hohenberg, 159*fig*, 160–164
South Africa, Boer War with Great Britain in, 16–17
Spalding, Douglas, 22–23, 24
special relativity theory, 9, 137
Spinoza, Baruch, 39
Spion Kop, Battle of, 75
Springer, Ferdinand, 273–275, 277, 309
Springer-Verlag, 251
steamroller formation, 85
Stephen, Virginia. *See* Woolf, Virginia
sterilization, forced, 303, 304–305
Stone, Edward, 314
Stormberg, Battle of, 74–75
Storming of the Bastille on July 14, 1789, The, 245*fig*
Strachey, Lytton, 116, 184, 190
supersets, 90
superset/subset relational operators, 79
Sylvester Medal, 101
symbolism, 99
symmetry in nature, 256
Szilard, Leo, 301

Tale of Two Cities, A (Dickens), 22
Tarbell, Ida, 3

Teller, Edward, 301
Tertium non datur (There is no third possibility), 233–235, 249
theory of types, 106–108, 198
Third International Congress of Mathematicians, 100
Thomson, Joseph John "J. J.," 134
Thoreau, Henry David, 140
"Those Barren Leaves" (Huxley), 193
Tijdschrift voor Wijsbegeerte (Journal of philosophy), 268
Time magazine, 290
Times (London), 163
Tolstoy, Leo, 3
topology, 137, 144, 217, 268, 284, 310
Toulouse-Lautrec, Henri de, 2
Tractatus Logico-Philosophicus (Wittgenstein), 170, 258
transfinite numbers, 47–48, 58, 89
Transvaal, 69, 70, 71, 72–73, 85
Triple Fugue (Sitwell), 193
triumphant pessimism, 13
Trotsky, Leon, 246
Trowbridge, Augustus, 254
tuberculosis, 23
Turner, Walter, 192
Twain, Mark, 3

Über die neue Grundlagenkrise der Mathematik (paper) ("The new crisis in the foundations of mathematics"; Weyl), 222
Ulsters, 156–157
uncertainty principle, 255
unimportant results, fallacy of, 68
universal conservation laws, 256
unknowable unknowns, concept of, 13
US stock market crash, 282–283
US Supreme Court, 303
utilitarianism, 84, 304–305

INDEX

van Dalen, Dirk, 103, 139, 222, 243, 258, 267, 271, 278
van der Waals, Johannes Diderik, 137
van Heijenoort, Jean, 67
van't Hoff, Jacobus Henricus, 136–137
Vatican Council, 58
Verdun, Battle of, 182
Vereeniging, peace treaty signed at, 85
Versailles, Treaty of, 243
Vicious Circle Principle, 106–107
Victoria, Queen, 14–15, 16–18, 69, 74, 76, 189
von Bismarck, Otto, 205
von Laue, Max, 10
von Lindemann, Ferdinand, 59
von Plato, Jan, 286
von Schlieffen, Alfred, 239
von Treitschke, Heinrich, 177
voting rights, 21
Voyage Out, The (Woolf), 192

Waugh, Frederick Judd, 63*fig*
Wavre, Rolin, 93
Webb, Beatrice, 96–97
Weber, Max, 204–205
Weber, Werner, 295
Weierstrass, Karl Theodor Wilhelm, 28, 29
Weimar Republic, 208, 243–244, 256
well-ordering, 101–102
Wells, H. G., 3, 168, 192
Weyl, Hermann
  Brouwer and, 218–219, 221–224, 228, 258, 308
  death of, 310–311
  Einstein and, 218
  Göttingen and, 251
  Hilbert and, 223–224, 228, 236, 241, 258, 308
  intuitionism and, 249–250, 257, 258, 308
  later life of, 310
  in Nauheim, 237
  Nazis and, 297, 298
  paradoxes and, 222
  Pólya and, 308
  rejection of intuitionism and, 246–247
  Russell and, 218–219
  set theory and, 219
  on uncertainty, 314
  Whitehead and, 218–219
Weyl Paradox, 89
Wheeler, John Archibald, 310
Whitehead, Alfred North
  language and, 210
  Principia Mathematica and, 98–99, 104, 108, 117–119, 136, 143, 176, 198, 215–216, 286, 288
  Russell and, 78–79
  son of, 148
  World War I and, 169
Whitehead, Evelyn, 77–78, 169
Whittaker, Edmond T., 79, 108
Whittier, John Greenleaf, 225*fig*
William IV, 15
William of Ockham, 307
Williams, Gardner Frederick, 71–72
Wittgenstein, Gretl, 131
Wittgenstein, Hans, 132
Wittgenstein, Karl Otto Clemens, 130–131, 132
Wittgenstein, Kurt, 132
Wittgenstein, Ludwig Josef Johann, 127–128, 130–131, 133, 134–135, 148–152, 169–171, 203, 258
Wittgenstein, Paul, 132–133
Wittgenstein, Rudi, 132
Witwatersrand (White water ridge), 71–72

INDEX

Wohlordnungssatz (well-ordering theorem), 102–104
Women in Love (D. H. Lawrence), 193, 194
Woolf, Leonard, 116
Woolf, Virginia, 116, 122, 125, 191, 192
World War I
    aftereffects of, 237–238, 243–244
    casualties during, 182–183, 201
    costs of, 200–201
    Courant and, 252–253
    draft and, 184–185, 189–190
    end of, 200
    Great Britain and, 188–189
    Morrells during, 194–196
    recruitment efforts during, 181–182
    Russell and, 164, 167, 168–169
    start of, 167–168
    United States and, 197–198, 200–201
    Wittgenstein and, 169–171
    See also antiwar activism
wormholes, 310
Wright, Frank Lloyd, 2–3
Wright brothers, 134

xenophobia, 300

Yeats, William Butler, 116, 190–191

Zeeman, Pieter, 137
Zeeman effect, 137, 174
Zeno of Elea, 37–38, 93
zeppelins, 133
Zermelo, Ernst Friedrich Ferdinand, 93–94, 102–104
ZFC set theory, 104

Photo courtesy of the author

Jason Socrates Bardi is an award-winning health and science journalist and editorial director in Washington, DC, who has written two previous books about the history of math: *The Calculus Wars* and *The Fifth Postulate*. His original journalism has appeared in the *San Francisco Chronicle*, *Good Morning America*, *U.S. News & World Report*, and *The Lancet*. He has worked as a writer and editor at NASA, Scripps Research Institute, the National Institutes of Health, the University of California, San Francisco, the Council on Foreign Relations, San Francisco General Hospital, the American Institute of Physics, and the magazine *Proto.Life*. In addition to writing and editing, Bardi has published photographs of Nobel laureates in outlets around the world. Contact him at AmericanJournalist@icloud.com.